Climate Change and Adaptation

Climate Change and Adaptation

Edited by
Neil Leary, James Adejuwon, Vicente Barros,
Ian Burton, Jyoti Kulkarni and Rodel Lasco

London • Sterling, VA

First published by Earthscan in the UK and USA in 2008

Climate Change and Adaptation: ISBN 978-1-84407-470-9
Climate Change and Vulnerability: ISBN 978-1-84407-469-3
Two-volume set: ISBN 978-1-84407-480-8

Typeset by FiSH Books, Enfield
Printed and bound in the UK by Antony Rowe, Chippenham
Cover design by Susanne Harris

For a full list of publications please contact:

Earthscan
8–12 Camden High Street
London, NW1 0JH, UK
Tel: +44 (0)20 7387 8558
Fax: +44 (0)20 7387 8998
Email: earthinfo@earthscan.co.uk
Web: **www.earthscan.co.uk**

22883 Quicksilver Drive, Sterling, VA 20166-2012, USA

Earthscan publishes in association with the International Institute for
Environment and Development

A catalogue record for this book is available from the British Library

Library of Congress Cataloging-in-Publication Data

Climate change and adaptation / edited by Neil Leary ... [et al.].
     p. cm.
  Includes bibliographical references.
  ISBN-13: 978-1-84407-470-9 (hardback)
  ISBN-10: 1-84407-470-6 (hardback)
  1. Climatic changes. 2. Climatic changes—Environmental aspects I. Leary,
Neil.
  QC981.8.C5C5113456 2007
  304.2'5—dc22

                                                                    2007023424

The paper used for this book is FSC-certified and
totally chlorine-free. FSC (the Forest Stewardship
Council) is an international network to promote
responsible management of the world's forests.

Contents

List of figures and tables   *viii*

Acknowledgements   *xiii*

Foreword by R. K. Pachauri   *xv*

1   A stitch in time: General lessons from specific cases   1
*Neil Leary, James Adejuwon, Vicente Barros, Punsalmaa Batima,
Bonizella Biagini, Ian Burton, Suppakorn Chinvanno, Rex Cruz,
Daniel Dabi, Alain de Comarmond, Bill Dougherty, Pauline
Dube, Andrew Githeko, Ayman Abou Hadid, Molly Hellmuth,
Richard Kangalawe, Jyoti Kulkarni, Mahendra Kumar, Rodel
Lasco, Melchior Mataki, Mahmoud Medany, Mansour Mohsen,
Gustavo Nagy, Momodou Njie, Jabavu Nkomo, Anthony Nyong,
Balgis Osman-Elasha, El-Amin Sanjak, Roberto Seiler, Michael
Taylor, Maria Travasso, Graham von Maltitz, Shem Wandiga and
Mónica Wehbe*

2   Adapting conservation strategies to climate change
in Southern Africa   28
*Graham von Maltitz, Robert J. Scholes, Barend Erasmus and
Anthony Letsoalo*

3   Benefits and costs of adapting water planning and management to   53
climate change and water demand growth in the Western Cape
of South Africa
*John M. Callaway, Daniël B. Louw, Jabavu C. Nkomo,
Molly E. Hellmuth and Debbie A. Sparks*

4   Indigenous knowledge, institutions and practices for coping   71
with variable climate in the Limpopo basin of Botswana
Opha Pauline Dube and Mogodisheng B. M. Sekhwela

5   Community development and coping with drought in rural Sudan   90
*Balgis Osman-Elasha, Nagmeldin Goutbi, Erika Spanger-Siegfried,
Bill Dougherty, Ahmed Hanafi, Sumaya Zakieldeen,
El-Amin Sanjak, Hassan A. Atti and Hashim M. Elhassan*

6   Climate, malaria and cholera in the Lake Victoria region: 109
     Adapting to changing risks
     Pius Yanda, Shem Wandiga, Richard Kangalawe, Maggie Opondo,
     Dan Olago, Andrew Githeko, Tim Downs, Robert Kabumbuli,
     Alfred Opere, Faith Githui, James Kathuri, Lydia Olaka,
     Eugene Apindi, Michael Marshall, Laban Ogallo, Paul Mugambi,
     Edward Kirumira, Robinah Nanyunja, Timothy Baguma,
     Rehema Sigalla and Pius Achola

7   Making economic sense of adaptation in upland cereal production   131
     systems in The Gambia
     Momodou Njie, Bernard E. Gomez, Molly E. Hellmuth,
     John M. Callaway, Bubu P. Jallow and Peter Droogers

8   Past, present and future adaptation by rural households   147
     of northern Nigeria
     Daniel D. Dabi, Anthony O. Nyong, Adebowale A. Adepetu
     and Vincent I. Ihemegbulem

9   Using seasonal weather forecasts for adapting food production   163
     to climate variability and climate change in Nigeria
     James Oladipo Adejuwon, Theophilus Odeyemi Odekunle and
     Mary Omoluke Omotayo

10   Adapting dryland and irrigated cereal farming to climate change   181
     in Tunisia and Egypt
     Raoudha Mougou, Ayman Abou-Hadid, Ana Iglesias, Mahmoud
     Medany, Amel Nafti, Riadh Chetali, Mohsen Mansour and Helmy
     Eid

11   Adapting to drought, *zud* and climate change in Mongolia's   196
     rangelands
     Punsalmaa Batima, Bat Bold, Tserendash Sainkhuu and
     Myagmarjav Bavuu

12   Evaluation of adaptation options for the Heihe river basin   211
     of China
     Yongyuan Yin, Zhongming Xu and Aihua Long

13   Strategies for managing climate risks in the lower Mekong   228
     river basin: A place-based approach
     Suppakorn Chinvanno, Soulideth Souvannalath, Boontium
     Lersupavithnapa, Vichien Kerdsuk and Nguyen Thuan

14   Spillovers and trade-offs of adaptation in the   247
     Pantabangan–Carranglan watershed of the Philippines
     Rodel D. Lasco, Rex Victor O. Cruz, Juan M. Pulhin and
     Florencia B. Pulhin

15   Top–down, bottom–up: Mainstreaming adaptation in Pacific   264
     island townships
     Melchior Mataki, Kanayathu Koshy and Veena Nair

16   Adapting to dengue risk in the Caribbean   279
     Michael A. Taylor, Anthony Chen, Samuel Rawlins, Charmaine
     Heslop-Thomas, Dharmaratne Amarakoon,
     Wilma Bailey, Dave Chadee, Sherine Huntley, Cassandra Rhoden
     and Roxanne Stennett

17   Adaptation to climate trends: Lessons from the Argentine   296
     experience
     Vicente Barros

18   Local perspectives on adaptation to climate change: 315
     Lessons from Mexico and Argentina
     Mónica Wehbe, Hallie Eakin, Roberto Seiler, Marta Vinocur,
     Cristian Ávila, Cecilia Maurutto and Gerardo Sánchez Torres.

19   Maize and soybean cultivation in southeastern South America: 332
     Adapting to climate change
     Maria I. Travasso, Graciela O. Magrin, Walter E. Baethgen,
     José P. Castaño, Gabriel R. Rodriguez, João L. Pires, Agustin
     Gimenez, Gilberto Cunha and Mauricio Fernandes

20   Fishing strategies for managing climate variability and change   353
     in the estuarine front of the Río de la Plata
     Gustavo J. Nagy, Mario Bidegain, Rubén M. Caffera,
     Walter Norbis, Alvaro Ponce, Valentina Pshennikov and
     Dimitri N. Severov

Index   371

List of Figures and Tables

Figures

2.1 The increase in conservation areas and the number of reserves   33
in seven southern African countries

2.2 A decision tree for selecting adaptation strategies for different   35
surrogate species based on their response to climate change

3.1 The Berg River Spatial Equilibrium Model (BRDSEM)   56

4.1 The Limpopo basin area of Botswana and case study sites: 73
Northeast District, Bobirwa Sub-District and Kgatleng District

4.2 Mean annual rainfall over three stations in the Limpopo basin   74
area of Botswana: Francistown, Northeast District; Bobonong,
Bobirwa Sub-District; and Mochudi, Kgatleng District

5.1 Contour bunds for water harvesting and tree planting in Arbaat   97

6.1 Map showing the Lake Victoria highland malaria region and   112
the studied villages

6.2 Map showing the Lake Victoria cholera region and the studied   113
villages

6.3 Total death toll due to malaria for Ndolage Hospital, Tanzania,   115
in 2001

6.4 Total death toll due to malaria for Rubya Hospital, Tanzania,   116
in 2001

6.5 Sources of information on consequence of cholera   122

7.1 Analytical framework for assessing economic feasibility of   134
climate change adaptation

8.1 Communities surveyed in northern Nigeria   150

8.2 Number of households classified as very vulnerable, vulnerable   153
and less vulnerable in the hamlets of Zangon Buhari, Dabai and
Takwikwi

8.3 Reasons for food storage   156

8.4 Reasons for planting early maturing crop varieties   156

8.5 Reasons for planting high yield crop varieties   156

8.6 Household vulnerability levels for the five livelihood capitals   157

8.7 Farmers' willingness to change practices to reduce vulnerability   160
to drought

10.1 Wheat yield and spring precipitation in Tunisia, 1961–2000   184

10.2 Simulated changes in irrigated wheat yields and   191
evapotranspiration under different climate conditions for (a) water
conservation measures and (b) variations in sowing dates

10.3	Simulated changes in rain-fed wheat yields for (a) different rainfall scenarios and (b) variations in sowing dates	193
12.1	Framework for multi-criteria evaluation of climate change adaptation options	215
12.2	Location of the Heihe river basin	218
13.1	Study sites in Lao PDR, Thailand and Vietnam	229
14.1	Location of the Pantabangan–Carranglan watershed	249
14.2	Land use map of the Pantabangan–Carranglan watershed	249
14.3	Summary of effects of adaptation strategies in one sector on other sectors	262
15.1	Observed rainfall anomalies for Navua, 1960–2003	268
15.2	Maximum daily rainfall during March and April, 1960–2003	269
16.1	Annual variability of the reported cases of dengue and the rate of change (increase or decrease from previous year) for the Caribbean	280
16.2	Monthly variability of the reported dengue cases, rainfall and temperature from 1996 to 2003 in Trinidad and Tobago	282
16.3	Schematic of a possible early warning system	292
17.1	Linear trends of annual precipitation (mm/year), 1959–2003	298
17.2	Isohyets in mm: 1950–1969 (solid line) and 1980–1999 (dashed line)	299
17.3	Percentage change in the rate between standard deviation and mean value in the 1980–1999 period with respect to 1950–1969	300
17.4	Annual precipitation in Chile: La Serena (29.9°S, 71.2°W) (left); Puerto Montt (41.4°S, 73.1°W) (right)	305
17.5	Mean annual streamflow (m^3/s) of a representative river of the Cuyo region, Los Patos river, 1900–2000	306
17.6	Mean annual streamflow (m^3/s) of rivers of the Comahue region, 1900–2000; note that the Negro river starts at the junction of the Limay and Neuquén rivers	307
17.7	Number of events with precipitation greater than 100mm in no more than two days in periods of four years	308
17.8	Annual frequency of cases with precipitation over 150mm in less than two days (left); for the same threshold (150mm), the ratio of the annual frequencies between the 1983–2002 and 1959–1978 periods (right)	309
17.9	The Plata river estuary	310
19.1	Study area and study sites	334
19.2	Changes in monthly precipitation (%) projected by HadCM3 under SRES A2 and B2 for 2020, 2050 and 2080	336
19.3	Changes in irrigated maize and soybean yields (%) under different scenarios and CO_2 concentrations	338
19.4	Changes in rain-fed maize and soybean yields (%) under different scenarios and CO_2 concentrations	339

19.5   Changes in the duration of planting–flowering (P-F) and   340
       flowering–maturity (F-M) periods, expressed as mean values for
       the six sites, for maize and soybean crops under different SRES
       scenarios and time periods
19.6   Maize: Yield changes (%) for different planting dates   342
       (ac = current, –20 and –40 days) in the six sites under different
       scenarios (A2 in grey, B2 in black for 2020, 2050 and 2080)
       and CO_2 concentrations
19.7   Soybean: Yield changes (%) for different planting dates (ac =   343
       current, ± 15, 30 days) in the six sites under different scenarios
       (A2 in grey, B2 in black for 2020, 2050 and 2080) and CO_2
       concentrations
19.8   Adaptation measures for maize: Yield change (%) under   344
       optimal planting dates/nitrogen rates and supplementary
       irrigation for the six sites without considering CO_2 effects
19.9   Adaptation measures for soybean: Yield changes (%) under   344
       optimal planting dates and supplementary irrigation for the
       six sites without considering CO_2 effects
20.1   Vulnerability and adaptation framework   355
20.2   River flow corridors and fronts of the Río de la Plata   357
20.3   Long-term gross income of fishermen (local currency-1999)   358
20.4   Unfavourable days for fishing activity on a monthly basis from   362
       October 2000 to March 2003, based on a lower threshold of
       8m/s wind speed

Tables

2.1   The area as a percentage conserved in southern African   31
       countries in IUCN reserves (IUCN classes I–V), IUCN
       sustainable resource use areas (IUCN class VI), and other
       non-IUCN conservation areas
2.2   The amount of conservation per ecoregion   32
2.3   Extent of conservation versus 'need' for conservation   32
2.4   Relative financial costs compared to the advantages and   36
       disadvantages of differing adaptation options
3.1   Framework for estimating benefits and costs associated with   57
       climate change and climate change adaptation
3.2   Net returns to water and optimal storage capacity of the   61
       Berg River Dam
3.3   Adapting to development pressure and climate change under   64
       the existing water allocation system: net returns to water
       (present value, R billion)
3.4   Adapting to development pressure and climate change by   66
       switching to water markets and adding storage capacity:
       net returns to water (present value, R billion)

4.1   Examples of government policies relevant to vulnerability and   78
      drought impact reduction
5.1   Adaptation measures in Gireighikh Rural Council, Bara Province,   94
      North Kordofan State
5.2   Adaptation measures in Arbaat, Red Sea State   98
5.3   Adaptation measures in El Fashir Rural Council,   102
      North Darfur State
6.1   Percentage responses of how malaria is treated at the   118
      household level by people with different levels of education
6.2   Percentage of reasons/explanations for not treating/boiling   120
      drinking water in Chato village
6.3   Cholera control strategies suggested by stakeholders   123
7.1   Average millet yields (kg/ha) and variability (CV) for current   138
      and future climates with business-as-usual and adaptive
      management strategies
7.2   Climate change damages and costs and benefits of fertilization –   141
      Average annual values in millions of US dollars
7.3   Climate change damages and costs and benefits of irrigation –   142
      Average annual values in millions of US dollars
8.1   Indices and weights for vulnerability assessment in northern   152
      Nigeria
8.2   Respondents' coping strategies   154
9.1   Skill assessment of quint forecast categories   171
9.2   Skill assessment of tercile forecast categories   171
9.3   Organizational skill performance assessment of the June, July,   172
      August and September annual rainfall totals
9.4   Regional disparities in forecasting skill   173
10.1 Average cereal yields during wet and dry years at four sites in   184
      the Kairouan region of Tunisia
10.2 Assumptions for simulations of irrigated wheat production in   189
      the Nile Delta region
10.3 Results of simulations for irrigated wheat in the Nile Delta   190
      region (percentage changes are relative to current climate)
11.1 Evaluation of adaptation options   203
12.1 Indicators used to evaluate adaptation options in the   217
      Heihe river basin
12.2 Water availability, water withdrawals and water withdrawal   219
      ratio in the Heihe river region, 1991–2000
12.3 Water shortage/surplus in Heihe river basin under climate   220
      change to 2040
12.4 Example of AHP comparison table   224
12.5 Overall rank and score of adaptation options for the   224
      Heihe region
13.1 Multiple orders of climate impacts on rain-fed farms in the   230
      lower Mekong region
13.2 Household-level on-farm measures for managing climate risks   232
13.3 Household-level off-farm measures for managing climate risks   235

13.4	Community-level measures for managing climate risks	237
13.5	National-level measures for managing climate risks	239
14.1	Adaptation options for agriculture and forestry by land use category	253
14.2	Options for adapting water resource supply and use in response to climate variations	254
14.3	Adaptation strategies of different institutional organizations	255
14.4	Cross-sectoral impacts of forest/agriculture sector adaptations on other sectors	256
14.5	Cross-sectoral impacts of water sector adaptations on other sectors	258
14.6	Cross-sectoral impacts of adaptations by institutions on other sectors	259
14.7	Adaptation strategies common to multiple sectors	260
15.1	Flood extent, duration and rainfall in five recalled flooding episodes in Navua	267
16.1	Distribution of epidemic peaks among ENSO phases, 1980–2001	283
16.2	Socio-economic characteristics of three communities in Western Jamaica and survey sample size	288
16.3	Adaptation strategies matrix	288
17.1	Cultivated areas	299
17.2	Density of rural roads in six provinces	301
17.3	Major monthly streamflow anomalies (m³/s) at Corrientes	303
17.4	Largest daily discharge anomalies (larger than 3 standard deviations) of the Uruguay river at the Salto gauging station, 1951–2000	303
17.5	Programmes funded by international banks to ameliorate and prevent damages from floods in Argentina	304
18.1	Farmers' socioeconomic characteristics	318
18.2	Synthesis of adaptation options	325
19.1	Projected changes in mean temperature (°C) for the warm semester (October–March) according to HadCM3 under SRES A2 and B2 scenarios for 2020, 2050 and 2080	335
19.2	Length (days) of planting–flowering (P–F) and flowering–maturity (F–M) periods for maize at current planting date and 20 and 40 days earlier under SRES A2 scenario for 2020, 2050 and 2080	341
20.1	Freshwater inflow to the Río de la Plata from the River Uruguay and total	357
20.2	Fishing activity, capture and income: Comparison between a good year (1988–89), long-term average and model results for a low-typical year (1), a bad year (2) and results with change in fishing behaviour (3)	365
20.3	Type II adaptation measures by scale of implementation and objectives	366

Acknowledgements

The two volumes *Climate Change and Vulnerability* and *Climate Change and Adaptation* are products of Assessments of Impacts and Adaptations to Climate Change (AIACC), a project that benefited from the support and participation of numerous persons and organizations. AIACC was funded by generous grants from the Global Environment Facility, the Canadian International Development Agency, the US Agency for International Development, the US Environmental Protection Agency and the Rockefeller Foundation. The initial concept for the project came from authors of the Third Assessment Report of the Intergovernmental Panel on Climate Change (IPCC) and was championed by Robert Watson, Osvaldo Canziani and James McCarthy, the IPCC chair and IPCC Working Group II co-chairs, respectively, during the Third Assessment Report. The productive relationship between AIACC and IPCC was continued and nurtured by Rajendra Pachauri, the current Chair of the IPCC, and Martin Parry, who joined Dr Canziani as co-chair of IPCC Working Group II for the Fourth Assessment Report.

The project could not have succeeded without the very capable and dedicated work of the more than 250 investigators who undertook the AIACC case studies, many of whom are authors of chapters of the two books. The project also benefited from the valuable and enthusiastic assistance of the many committee members, advisers, resource persons and reviewers. These include Neil Adger, Ko Barrett, Bonizella Biagini, Ian Burton, Max Campos, Paul Desanker, Alex De Sherbinin, Tom Downing, Kris Ebi, Roland Fuchs, Habiba Gitay, Hideo Harasawa, Mohamed Hassan, Bruce Hewitson, Mike Hulme, Saleemul Huq, Jill Jaeger, Roger Jones, Richard Klein, Mahendra Kumar, Murari Lal, Liza Leclerc, Bo Lim, Xianfu Lu, Jose Marengo, Linda Mearns, Monirul Mirza, Isabelle Niang-Diop, Carlos Nobre, Jean Palutikof, Annand Patwardhan, Martha Perdomo, Roger Pulwarty, Avis Robinson, Cynthia Rosenzweig, Robert Scholes, Ravi Sharma, Hassan Virji, Penny Whetton, Tom Wilbanks and Gary Yohe. Patricia Presiren of the Academy of Sciences of the Developing World (TWAS) and Sara Beresford, Laisha Said-Moshiro, Jyoti Kulkarni and Kathy Landauer of START gave excellent support for the administration and execution of the project.

Finally, thanks are owed to Alison Kuznets and Hamish Ironside of Earthscan and to Leona Kanaskie for assistance with copy-editing and production of the books.

Foreword

Climate change is increasingly recognized as a critical challenge to ecological health, human well-being and future development, as underscored by the award of the Nobel Peace Prize for 2007 to the Intergovernmental Panel on Climate Change (IPCC). The award recognizes the substantial advances in our shared understanding of climate change, its causes, its consequences and its remedies, which have been achieved by more than 20 years of work by the thousands of contributors to the IPCC science assessments, and which draw from the research and analyses of an even larger number of scientists and experts. This work has culminated in the unprecedented impact of the Panel's most recent report, the Fourth Assessment Report.

The Fourth Assessment Report advances our understanding on various aspects of climate change based on new scientific evidence and research. A major contribution in this regard has come from the work promoted under the project Assessments of Impacts and Adaptation to Climate Change (AIACC). The AIACC project was sponsored by the IPCC to fill a major gap in the available knowledge about climate change risks and response options in developing countries that existed at the completion of the Third Assessment Report in 2001. Twenty-four national and regional assessments were executed under the AIACC project in Africa, Asia, Latin America and small island states in the Caribbean, Indian and Pacific Oceans. The two volumes *Climate Change and Vulnerability* and *Climate Change and Adaptation* present many of the findings from the AIACC assessments.

The findings not only give us a fuller scientific understanding of the specific nature of impacts and viable adaptation strategies in different locations and countries, but have contributed to a much better appreciation of some of the equity dimensions of the problems as well. In simplified terms, the biggest challenge in confronting the negative impacts of climate change lies in the developing world, where people and systems are most vulnerable. Not only are these negative impacts likely to be most serious in the subtropics and tropics, where most developing societies reside, but the capacity to adapt to them is also limited in these regions.

An important element in understanding vulnerabilities to climate change is in linking current and projected exposures to climate stresses with other existing stresses and conditions that are responsible for hardship and low levels of economic welfare. Climate change often adds to these existing stresses, increasing the vulnerability of such communities and ecosystems. Unfortunately, limited research is carried out in developing countries on likely impacts and appropriate responses related to climate change. This is where the knowledge provided by the assessments of the AIACC has been particularly valuable.

There is considerable interest in interpretation of Article 2 of the United

Nations Framework Convention on Climate Change (UNFCCC), the focus of which defines the ultimate objective of the Convention, namely that of preventing a dangerous level of anthropogenic interference with the climate system. Research on impacts of climate change focusing on specific parts of the world that are highly vulnerable enhances our understanding of what may constitute a dangerous level of anthropogenic interference with the world's climate system. In the absence of such knowledge, any value judgement defining a dangerous level would apply to, and be determined by, knowledge only from particular regions of the world, primarily the developed nations. Understanding the critical nature of impacts in some of the most vulnerable parts of the world, which are largely in developing countries, will assist our determination of what might constitute a dangerous level of interference with the earth's climate system. Such knowledge would help appropriately to include and consider those locations which are perhaps much closer to danger than was known earlier.

The record and outputs of the AIACC are impressive. The project, funded by the Global Environment Facility and coordinated by the Global Change System for Analysis, Research and Training (START), the Academy of Sciences for the Developing World (TWAS) and the United Nations Environment Programme (UNEP), engaged investigators from more than 150 institutions and 60 countries to execute the assessments. The quality of the assessments is demonstrated by the more than 100 peer-reviewed publications produced, which benefited substantially the IPCC's Fourth Assessment Report. In view of this success, it is imperative that we build on the experience and achievements of AIACC and develop the next phase of such work to help advance new knowledge for a possible Fifth Assessment Report of the IPCC.

While the material contained in the two volumes from AIACC and the substantial amount of knowledge developed through the case studies presented in the following pages are valuable, the need for further work is enormous. There remain many countries in the developing world where very little is known about the nature and extent of the impacts of climate change, and these gaps would not permit the development of plans and programmes to address climate change risks or to put in place response measures that would help communities and ecosystems to adapt to the impacts of climate change. These clearly would get much more serious with time unless suitable mitigation measures are taken in hand with a sense of urgency. Yet, even with the most ambitious mitigation actions, the inertia of the system will ensure that the impacts of climate change will continue for centuries, if not beyond a millennium. Knowledge of impacts and the manner in which they would grow over time is therefore critical to the development of capacity and measures for adaptation to climate change. The work of the AIACC provides an extremely important platform to take such steps, but there is yet very far to go to meet the challenges ahead. It is hoped that the material contained in this volume is just the start of a process that must expand and continue in the future.

R. K. Pachauri
Director General, The Energy and Resources Institute (TERI)
and Chairman, Intergovernmental Panel on Climate Change (IPCC)

1

A Stitch in Time:
General Lessons from Specific Cases

Neil Leary, James Adejuwon, Vicente Barros, Punsalmaa Batima, Bonizella Biagini, Ian Burton, Suppakorn Chinvanno, Rex Cruz, Daniel Dabi, Alain de Comarmond, Bill Dougherty, Pauline Dube, Andrew Githeko, Ayman Abou Hadid, Molly Hellmuth, Richard Kangalawe, Jyoti Kulkarni, Mahendra Kumar, Rodel Lasco, Melchior Mataki, Mahmoud Medany, Mansour Mohsen, Gustavo Nagy, Momodou Njie, Jabavu Nkomo, Anthony Nyong, Balgis Osman-Elasha, El-Amin Sanjak, Roberto Seiler, Michael Taylor, Maria Travasso, Graham von Maltitz, Shem Wandiga and Mónica Wehbe

Introduction

We can adapt to climate change and limit the harm, or we can fail to adapt and risk much more severe consequences. How we respond to this challenge will shape the future in important ways.

The climate is already hazardous; indeed it always has been so. Variations and extremes of climate disrupt production of food and supplies of water, reduce incomes, damage homes and property, impact health and even take lives. Humans, in an unintended revenge, are getting back at the climate by adding to heat-trapping gases in the Earth's atmosphere that are changing the climate. But the changes are amplifying the hazards. And we cannot in short order stop this. The physical and social processes of climate change have a momentum that will continue for decades and well beyond.

This undeniable momentum does not imply, however, that efforts to mitigate climate change, meaning to reduce or capture the emissions of greenhouse gases that drive climate change, are wasted. Nor is a call for adaptation a fatalistic surrender to this truth. The magnitude and pace of climate change will determine the severity of the stresses to which the world will be exposed. Slowing the pace of human caused climate change, with the aim of ultimately stopping it, will enable current and future generations to better cope with and adapt to the resulting hazards, thereby reducing the damages and danger. Mitigating climate change is necessary. Adapting to climate change is necessary too.

The challenges are substantial, particularly in the developing world. Developing countries have a high dependence on climate-sensitive natural resource sectors for livelihoods and incomes, and the changes in climate that are projected for the tropics and sub-tropics, where most developing countries are found, are generally adverse for agriculture (IPCC, 2001 and 2007a). Furthermore, the means and capacity in developing countries to adapt to changes in climate are scarce due to low levels of human and economic development and high rates of poverty. These conditions combine to create a state of high vulnerability to climate change in much of the developing world.

To better understand who and what are vulnerable to climate change, and to examine adaptation strategies, a group of case studies was undertaken as part of an international project, Assessments of Impacts and Adaptations to Climate Change (AIACC). The studies span Africa, Asia, Central and South America, and islands of the Caribbean, Indian and Pacific Oceans. They include assessments of agriculture, rural livelihoods, food security, water resources, coastal zones, human health and biodiversity conservation. Results from the studies about the nature, causes and distribution of climate change vulnerability are presented in a companion to this volume (Leary et al, 2008). In this volume, we collect together papers from the AIACC studies that explore the challenge of adaptation.

Comparison and synthesis of our individual contributions have yielded nine general lessons about adaptation, as well as many more lessons that are specific to particular places and contexts. The general lessons, formulated as recommendations, are as follows: (1) adapt now, (2) create conditions to enable adaptation, (3) integrate adaptation with development, (4) increase awareness and knowledge, (5) strengthen institutions, (6) protect natural resources, (7) provide financial assistance, (8) involve those at risk, and (9) use place-specific strategies. The lessons are briefly outlined below, followed by a more detailed examination of their nuances and supporting evidence from the case studies.

The Nine Adaptation Lessons

Adapt now!

The time-honoured proverb 'a stitch in time saves nine' means that immediate action to repair damage (to your clothing in the original context) can avoid the necessity to do much more later on. The expression captures one of the main findings of the AIACC's programme of studies. It can simply be stated as the injunction to adapt now.

Climatic variations and extremes cause substantial damage to households, communities, natural resources and economies. In many places the damage is increasing, giving evidence of an adaptation deficit, meaning that practices in use to manage climate hazards are falling short of what can be done (Burton, 2004). We find evidence in all our case study sites of an adaptation deficit that climate change threatens to widen. Acting now to narrow the deficit can yield immediate benefits. It will also serve as a useful, even essential, first step in a longer-term process of adapting to a changing climate. Failure to tackle adaptation vigorously now is likely to mean that many more than nine stitches will be required in the future.

Create conditions to enable adaptation

In contrast to reducing emissions of the greenhouse gases that drive climate change, a policy that, in the parlance of economists, generates benefits that are substantially external, adaptation generates benefits that are largely internal. This means that the individuals, organizations, communities and countries that take action to adapt will capture for themselves most of the benefits of their actions, creating a strong incentive to adapt. This explains why we can see a wide range of practices being used to manage and reduce climate risks. But why then do we nonetheless observe adaptation deficits? Why doesn't self interest motivate people to do more to protect themselves from climate hazards?

Our case studies identify numerous obstacles that impede adaptation. Common obstacles include competing priorities that place demands on scarce resources, poverty that limits capacity to adapt, lack of knowledge, weak institutions, degraded natural resources, inadequate infrastructure, insufficient financial resources, distorted incentives and poor governance. Obstacles such as these severely constrain what people can do. Intervention by public sector entities, at levels from the local community to the provincial, national and international, can create conditions that better enable people to surmount the obstacles and take actions to help themselves. Enabling the process of adaptation is the most important adaptation that the public sector can make. Specific interventions to enable adaptation are addressed by some of the other lessons that follow.

Integrate adaptation with development

The goals and methods of climate change adaptation and development are strongly complementary. The impacts of current climate hazards and projected climate change threaten to undermine development achievements and stall progress towards important goals. Adaptation can reduce these threats. In turn, development, if appropriately planned, can help to enable climate change adaptation. Integrating adaptation with development planning and actions can exploit the complementarities to advance both adaptation and development goals. To be effective, integration needs to engage ministries that are responsible for development, finance, economic sectors, land and water management, and the provision of public health and other services. It is in agencies such as these that key decisions are taken about the allocation of financial and other resources. And it is within these agencies and among their stakeholders that much of the sector-specific expertise that must be engaged resides.

Increase awareness and knowledge

Nearly all the case studies highlighted knowledge as a critical constraint on adaptation and rank efforts to increase and communicate knowledge as a high adaptation priority. Stakeholders in many of the study areas complained of inadequate or lack of access to information about climate history, projections of future climate change and potential impacts, estimates of climate risks, causes of vulnerability, technologies and measures for managing climate risks, and know-how for implementing new technologies. Uncertainty about the future and about the effectiveness and costs of adaptation options are common obstacles to action. Examination of these and other information problems in the case studies demonstrates the need for programmes to help advance, communicate, interpret and apply knowledge for managing climate risks.

Strengthen institutions

Institutions are found to play important roles in enabling adaptation. Local institutions, including community organizations, farmer cooperatives, trade associations, local government agencies, informal associations, kinship networks and traditional institutions, serve functions in communities that help to limit, hedge and spread risks. They do this by sharing knowledge, human and animal labour, equipment and food reserves; mobilizing local resources for community projects and public works; regulating use of land and water; and providing education, marketing, credit, insurance and other services. Provincial, national and international institutions aid by providing extension services, training, improved technologies, public health services, infrastructure to store and distribute water, credit, insurance, financial assistance, disaster relief, scientific information, market forecasts, weather forecasts, and other goods and services.

In many of our case study sites, key functions for managing risks are absent or are inadequate due to weak institutions that are poorly resourced, lacking in human capacity, overloaded with multiple responsibilities and overwhelmed by the demands of the communities that they serve. Strengthening institutions to fill strategic functions in support of adaptation is needed. In some instances, traditional institutions that have been diminished in role by socioeconomic changes and government policies provide a remnant framework that could be revitalized to facilitate adaptation and the management of climate risks.

Protect natural resources

Developing countries typically are dependent on climate sensitive natural resources for a high proportion of their livelihoods, economic activities and national incomes. Too often these resources are in a degraded state from a combination of pressures caused by human use and climatic and environmental variation and change. Their degraded state makes these resources, and the people who are dependent on them, highly vulnerable to damages from climate change. Rehabilitating and protecting natural resources such as farm lands, grazing lands, forests, watersheds, wetlands, fisheries and biodiversity are a central focus of adaptation strategies in places as varied as the African Sahel, southern Africa, central Asia, southeast Asia, and south-eastern South America. Progress in many of these settings will require changes in incentives, reforms of tenure to land, water and natural products, education, training, and more vigorous enforcement of regulations. These, in turn, are dependent on strong institutions and access to financial resources.

Provide financial assistance

Lack of financial resources is commonly cited as a major obstacle to adaptation. The constraint is particularly binding on the poor and the very poor, who typically are among the most vulnerable to climate change. Poor households and small-scale farmers and enterprise owners obtain finance through community and informal networks to recover from losses and make investments that reduce risks. But more adaptation could take place in impoverished localities and regions with greater financial assistance from provincial and national governments and international sources. Innovative ideas are needed to engage the private sector in financing adaptation. Internationally, some financial assistance is being provided and acts as a catalyst for raising awareness, building capacity and advancing understanding of risks and response options. But the magnitude of financial needs for adaptation is much greater than the current level of assistance. Increased financial assistance over and above normal development assistance is needed. Ultimately, however, financing will need to come from multiple sources, including those internal to developing countries.

Involve those at risk

Involving persons at risk in the process of adaptation, the intended beneficiaries, can increase the effectiveness of adaptation to climate change. Many of our case studies involved at-risk groups in assessment activities. The experiences demonstrate the potential of participatory approaches to adaptation for focusing attention on risks that are priorities to the vulnerable, learning from risk management practices currently in use, identifying opportunities and obstacles, applying evaluation criteria that are relevant and credible to at-risk groups, and drawing on local knowledge and expertise for selecting and designing appropriate strategies, garnering support and mobilizing local resources to assist with implementation. A common result of involving those at risk is that it forces climate risks to be examined in context with other problems and gives emphasis to solutions that can be combined to attain multiple objectives.

Use place-specific strategies

Adaptation is place-based and requires place-specific strategies. This fact has long been recognized in the climate impacts research literature. The general lessons outlined above conceal the much richer content of the case studies and risk presenting an oversimplified story. The ninth lesson is that there are many more lessons and that many are specific to particular contexts of particular places.

For example, in the lower Mekong river basin, rice farmers face similar risks from floods but rely on different strategies for managing the risks that reflect differences in the level of economic development of their surrounding community, strength of community institutions, locally available natural resources and seasonal rain patterns (Chinvanno et al, Chapter 13). Pastoralists in Mongolia, Sudan and Botswana share some strategies for coping with drought that have general characteristics in common, but there are significant differences too that derive from different traditions, resources and climates (Batima et al, Chapter 11; Dube et al, Chapter 4; Osman-Elasha et al, Chapter 5). People living in the Caribbean and the highlands surrounding Lake Victoria both face health risks from mosquito-borne diseases that vary with the climate, but differences in public health infrastructure and access to health care contribute to differences in responses to the diseases (Taylor et al, Chapter 16; Yanda et al, Chapter 6). General lessons can be applied in these different settings to help guide adaptive strategies, but details of the local context will determine the specific approaches and measures that will be most effective in each place.

Adaptation Now and in the Future

What is adaptation?

The Intergovernmental Panel on Climate Change (IPCC) defines adaptation as adjustments in ecological, social or economic systems in response to actual or expected climatic stimuli and their effects (Smit et al, 2001). It includes adjustments to moderate harm from, or to benefit from, current climate variability as well as anticipated climate change. Adaptation can be a specific action, such as a farmer switching from one crop variety to another that is better suited to anticipated conditions. It can be a systemic change such as diversifying rural livelihoods as a hedge against risks from variability and extremes. It can be an institutional reform such as revising ownership and user rights for land and water to create incentives for better resource management. Adaptation is also a process. The process of adaptation includes learning about risks, evaluating response options, creating the conditions that enable adaptation, mobilizing resources, implementing adaptations and revising choices with new learning. We mean all these things by adaptation. But the conception of adaptation as a process is often the most important for formulating public interventions that will have lasting benefits.

Is adaptation new?

Adaptation to climate is not new. People, property, economic activities and environmental resources have always been at risk from climate and people have continually sought ways of adapting, sometimes successfully and sometimes not. The long history of adapting to variations and extremes of climate includes construction of water reservoirs, irrigation, crop diversification, disaster management, insurance and even, on a limited basis, recent measures to adapt to climate change (Adger et al, 2007).

The AIACC case studies document a variety of adaptive practices in use that have reduced vulnerability to climate hazards. In most cases these have been adopted in response to multiple sources of risk and only rarely to climate risk alone. One strategy commonly in use is to increase the capacity to bear losses by accumulating food surpluses, livestock, financial assets and other assets. Risks are hedged by diversifying crops, income sources, food sources and locations of production activities. Exposures to hazards have been reduced by relocating, either temporarily or permanently. Variability of production and incomes derived from natural resources have been reduced by restoring degraded lands, using drought-resistant seed varieties, harvesting rainfall, adopting irrigation and using seasonal forecasts to optimize farm management. Prevention of climate impacts through flood control, building standards and early warning systems is practised. Risk spreading is accomplished through kinship networks, pooled community funds, insurance and disaster relief. In many cases the capacity to adapt is increased through public sector assistance such as extension services, education, community development projects and access to subsidized credit.

Is adapting to climate change different?

Is adapting to climate change different? Yes and no. People have always faced an uncertain future when coping with and adapting to climate. Human societies have long coped with floods, droughts and other climate hazards without knowing when the next event would occur, how big it would be or how long it would last. Past experience provided a basis, albeit an imperfect one, for approximating the frequencies of events of different magnitudes and the likely range of conditions that might be encountered in the coming season, year or decade.

But climate change means that past performance of the climate is becoming a less reliable predictor of future performance. The frequency, variability, seasonal patterns and characteristics of climate events and phenomena will change. Phenomena once alien to a region could become regular features of its climate (for example, extra-tropical storm tracks are projected to move poleward, IPCC, 2007b). An important consequence of climate change for adaptation is that the future climate will be less familiar and in key respects more uncertain.

However, some climate parameters will change with predictable trends as a result of human-driven climate change. Globally averaged surface temperatures are projected to rise by 1.1–6.4°C by the end of the 21st century relative to 1980–1999 temperatures (IPCC, 2007b). Annual and monthly average temperatures can be expected, with a very high degree of confidence, to increase virtually everywhere. Changes in average precipitation are also projected but vary from decreases to increases depending on location and season; while confidence in predictions of precipitation trends is less than for temperature trends, some broad patterns do seem to be robust across climate model projections. For example, precipitation is very likely to increase in high-latitudes while decreases are thought likely in most subtropical land areas. Likely trends for extreme weather include more frequent hot days, heat-waves and heavy precipitation events, more intense tropical cyclones with greater peak wind speeds and heavier precipitation, and increased summer drying and drought risk in continental interiors. The projected trends in temperature, precipitation and extremes will push future climate variations and extremes beyond the bounds of what people and places have been exposed to and had to cope with in the past.

The implication is that current practices, processes, systems and infrastructure that are more or less adapted to the present climate will become increasingly inappropriate and maladapted as the climate changes. Fine tuning current strategies to reduce risks from historically observed climate hazards will not be sufficient in this dynamically changing environment. More fundamental adjustments will be needed. This will require recognizing what changes are happening, predicting the range of likely future changes, understanding the vulnerabilities and potential impacts, identifying appropriate adjustments, and mobilizing the resources and will to implement them.

The experience of Argentina in the last decades of the 20th century is instructive of some of the challenges (Barros, Chapter 17). A number of climate trends are documented that began in the 1960s and 1970s. These include large increases in mean annual precipitation in southern South America east of the Andes Cordillera; increased flows and flood frequencies of the major rivers of the region, the Parana, Paraguay and Uruguay rivers; more frequent heavy rainfall events in central and eastern Argentina resulting in localized flooding; more frequent *sudestadas*, which bring winds from the southeast that cause high tides and flooding in Buenos Aires; and, in western Argentina, declining rainfall and stream flows.

The speed and effectiveness of adaptive responses to these trends varied. In each case there was a lag between the onset of the climate trend and recognition by affected persons, government agencies and the public. The lag varied depending on the perception of impacts, their magnitude, natural variability of the climate phenomenon, adequacy of observational data, and the difficulty of detecting trends in low frequency events. The quickest response came in the case of increased rainfall east of the Andes but west of the traditional crop farming areas. Farmers recognized and acted on the new opportunity created by the greater rainfall, as well as by high soybean prices in international markets, to profitably cultivate lands that were previously too dry for crop farming. This resulted in significant westward expansion of crop farming, particularly of soybeans, but with roughly a ten year lag. Less quick to act was the government, which failed to provide road and other infrastructure to support the westward expansion of crop farming.

Usually emphasis is placed on uncertainty of predicted climate change as a barrier to adaptation. Less appreciated is the barrier created by uncertainty in detecting changes that are already underway and likely to continue. The examples from Argentina demonstrate how delays in recognition and limited awareness of climate trends by key stakeholders delayed adaptive responses. They also suggest that those who have a direct self-interest in adapting may be more astute and quicker to respond.

Biodiversity conservation in southern Africa is an example where climate change will require a fundamental change in approach from current risk management (von Maltitz et al, Chapter 2). In 50 years' time, up to half of South Africa will have a climate that is not currently found in that country. With the changes, many species will need to move across the landscape to track climates that are suitable to their requirements. It will no longer be adequate to protect species where they are currently found – conservationists will have to aim for a moving target.

Some species will be able to tolerate the new climate in their current locations (persisters); some will thrive in new climate niches not currently available and expand their ranges (range expanders); some will no longer be viable in part or all of their current range and must disperse to new areas (partial and obligatory dispersers); and some will find no areas with suitable climate and will go extinct from the region (no-hopers). Modelling of climate change impacts on *Proteaceae*, a surrogate for the highly diverse fynbos vegetation of South Africa, yields estimates that in 50 years approximately 60 per cent of species would be persisters, 30 per cent partial or obligatory dispersers, and 10 per cent no-hopers.

The no-hopers can be preserved only by *ex situ* conservation methods. Migration of the obligatory and partial dispersers over a mixed use, fragmented landscape to track a changing climate is not assured. And successful dispersal 50 years from now does not assure long-term survival, as the climate will continue to change. Multiple strategies will be needed to facilitate migration and minimize species loss. Adding to and reconfiguring land reserves is one element that will be needed, but it is a costly approach and the lands needing protection will change through time. New and more aggressive strategies will be needed to make the landscape more permeable and biodiversity friendly, including private and communal lands that are not in formal reserves.

The terminology from the field of biodiversity conservation – obligatory dispersers and no-hopers, for example – is stark. But are there analogous cases in other contexts? Will climate change make inhabitants of some small islands, coastal areas and arid zones partial or obligatory dispersers? Is the hope for survival of some small island nation states and their cultures dependent on *ex situ* conservation? Do some livelihoods have no hope of persistence in a changing, more hazardous climate? The methods of adaptation to climate change will often be similar to, borrow heavily from and build on current adaptation practice. But as these questions suggest, the challenges and stakes are getting higher.

Is current adaptation enough?

Adaptation to climate variation is a regular feature of our lives and, broadly speaking, we are adapted to cope with a wide range of climatic conditions.

Indicators of successful adaptation include the increase in world food production in pace with population growth, increased life expectancy and decreased weather-related deaths in developed countries (Schneider et al, 2007; McMichael et al, 2001). But variations and extremes do regularly exceed coping ranges, too often with devastating effect. Weather-related hazards such as tropical cyclones, floods and droughts have caused more than one million deaths in the past 20 years, the overwhelming majority of which occurred in developing countries (Pelling et al, 2004). Individual events can cause billions of dollars in damages, and economic and insured losses from natural catastrophes increased more than 6-fold and 24-fold respectively since the 1960s (Munich Re, 2005). While climate impacts can never be reduced to zero, the heavy and rising toll of weather-related disasters and the burden of less severe variations indicate that we are not as well adapted as we might or should be.

All of the AIACC case studies give evidence of an adaptation deficit and identify measures that could reduce current losses. For example, greater reforestation efforts and enforcement of forest protection laws would reduce soil erosion and flood risks in the Pantabangan–Carranglan watershed of the Philippines (Lasco et al, Chapter 14). In the Berg river basin of South Africa, allowing greater flexibility for water transfers or water marketing would enable water to be allocated more efficiently during periods of drought (Callaway et al, Chapter 3). A variety of underutilized options for reducing drought and flood risks are available to farmers in Argentina, Botswana, Cambodia, Egypt, Lao PDR, Mexico, Nigeria, Sudan, Thailand and Tunisia (Barros, Chapter 17; Dube et al, Chapter 4; Chinvanno et al, Chapter 13; Mougou et al, Chapter 10; Wehbe et al, Chapter 18; Dabi et al, Chapter 8; Osman-Elasha et al, Chapter 5). In Jamaica, management of dengue fever risks are largely reactive and could be improved by proactive steps for education, elimination of breeding sites and early warnings (Taylor et al, Chapter 16). Building sturdier houses raised above ground level, improved control of river siltation and more regular dredging of rivers would reduce flood losses in coastal towns of Fiji (Mataki et al, Chapter 15).

The current deficit in adaptation makes it imperative to adapt now. Doing so would have immediate benefits in reduced weather-related impacts and increased human welfare. The need to adapt is made more urgent by climate change, which is now upon us and is widening the deficit. Adapting to current climate is an essential step towards adapting to future climates.

What are the obstacles to adaptation?

People may not adapt, or adapt incompletely, for a variety of reasons. Climate may be perceived, rightly or wrongly depending on the context, to pose little risk relative to other hazards and therefore be given low priority. Knowledge of options to reduce climate risks or the means to implement them may be lacking. Or their expected costs may exceed the expected benefits. The means or capacity to adapt may be lacking. Uncertainty about the future may make it difficult to know what to do or when to do it. Irreversible consequences of some actions may delay choices until some of the uncertainty is resolved.

Incentives may be distorted in ways that discourage choices that reduce risks, or even encourage risk-taking behaviour. Sometimes the action of others, or inaction of others, can be an obstacle. Some may believe that reducing their own risk is the responsibility of others. All these are found to impede adaptation in one or more of the case studies.

The AIACC studies are all set in developing countries and most focus on places and households that are poor. Poverty, in human development as well as economic terms, is a major obstacle to adaptation in these study areas. Indicative of the constraint imposed by poverty is the high proportion of households in East Africa that do not use insecticide treated bed nets as a prevention against malaria, despite their effectiveness and seemingly low cost (Yanda et al, Chapter 6).

The case studies of northern Nigeria (Dabi et al, Chapter 8) and the states of North Kordofan, North Darfur and Red Sea in Sudan (Osman-Elasha et al, Chapter 5) are illustrative of the constraints faced by poor rural households. Households in these study areas, located in the dry and drought prone Sudano-Sahel zone, typically have low capacity to adapt because of very limited financial, natural, physical, human and social capital. They have little or no cash income, financial savings or access to credit with which to purchase seed, fertilizer, equipment, livestock or food. The lands from which they derive their livelihoods have poor fertility, are highly erodable and are degraded from heavy use, clearing of vegetation, declines in average precipitation and increasing frequency of drought. Physical infrastructure for transportation, communication, water supply, sanitation, and other services are lacking. People have knowledge of many traditional practices for coping with drought and other stresses, but often have little knowledge of new or alternate methods due to poor access to education, training or extension services. Kinship networks provide a safety net for food and other necessities in times of crisis, but sometimes a crisis such as drought or violence will strike many members of a network simultaneously. Local institutions for providing community services are generally weak, governance at provincial and national levels is ineffective, and violence and conflict have heightened vulnerability – with devastating impact in Darfur.

Lack of awareness, information and knowledge is a constraint on adaptation in all of the case studies. In Argentina, as noted previously, lags in recognition of climate trends that had begun in the 1960s and 1970s resulted in delayed and incomplete adaptive responses (Barros, Chapter 17). Tunisian farmers are reluctant to change from inherited traditional practices because they lack knowledge and education to evaluate and implement new methods (Mougou et al, Chapter 10). Similarly, in Tamaulipas, Mexico, *ejidatarios* and smallholder farmers lack know-how for adopting irrigation (Wehbe et al, Chapter 18). In Mongolia, herders voiced a strong need for education and training in methods for improving the condition and productivity of their rangelands and livestock (Batima et al, Chapter 11). Participants in the artisanal fishery of the La Plata estuary need better information about the effects of variations in climate on movements of fish stocks, forecasts of fishing conditions, and fishing methods and technologies for managing variability in the fishery (Nagy et al, Chapter 20).

Seasonal weather forecasts and early warning systems are frequently suggested as useful for informing the management of climate risks. But, as shown by Adejuwon et al (Chapter 9), they require an effective knowledge network to deliver their promised benefits. Seasonal forecasts are made for West Africa and Nigeria, but few farmers use them. Their reliability is low, the variables forecast are not ones that are most relevant to farmers' decisions, and the spatial resolution of the forecasts is coarse compared to what farmers need. The forecasts are poorly disseminated, are delivered only shortly in advance of the forecast period, do not regularly reach smallholder farmers and are in forms that are not readily understood by farmers. A number of recommendations are made by Adejuwon et al to improve this knowledge network and support an adaptation process that would provide farmers with more useful forecasts and the knowledge and skills to apply them. Success will be dependent on cooperation and coordination across the regional and national meteorological agencies, agricultural extension agency, local government units, and farmers' associations, which may require changes in responsibilities, accountability and incentives.

Scarce and degraded natural resources contribute to vulnerability and detract from the capacity to adapt in many of the case studies. Insufficient water supplies, and poor quality of existing supplies, prevent Tunisian farmers from expanding irrigation (Mougou et al, Chapter 10). In some instances, treatment of a resource as an open access commons has contributed to its degradation and created disincentives for adaptations to protect it. Following the transition to a market economy in Mongolia, livestock ownership was privatized while pastureland remained state owned and access largely unrestricted (Batima et al, Chapter 11). This has contributed to overstocking of animals, diminished seasonal migration of herds, and lack of investment in land improvements. This situation contrasts with earlier periods during which state collectives, and traditional family groups before that, controlled access to communal pastures.

Social capital, an important resource for coping with risk, has been eroded in many places by social and economic changes and by government policies. In the Limpopo Basin of eastern Botswana, the *Kgotla*, or traditional institution for local decision making and administration of justice, played a central role in adapting the local community to climate variability by regulating resource use and maintaining and disseminating traditional knowledge for the use of veld products (Dube et al, Chapter 4). The *mafisa* system of lending cattle to poorer family members, the marriage institution and family-based user rights to land provided social security and income security that limited risks from climate extremes and other crises. These institutions were weakened during the 20th century, with the result that communities were alienated from decision making about local resource use, income poverty and capability poverty were deepened, and dependence on government interventions increased. The loss of social capital has reduced the capacity of communities to adapt and amplified their vulnerability to climate hazards.

Governance can either constrain or enable adaptation. Financial constraints, already mentioned for households, is one factor that prevents governance from playing a more positive role. Government agencies are often poorly resourced relative to the demands placed on them. Other impediments to government support for adaptation include lack of awareness, knowledge and staff with relevant skills, ineffective administration, poor coordination across departments, inadequate accountability and corruption. Also important is the fact that persons who are most vulnerable to climate risks are often socially and politically marginalized and therefore unable to influence governments to act in their interest.

Climate and Development

What are the impacts of climate on development?

Billions of people in more than 100 countries are exposed to natural disaster risk, including weather-related disasters that take lives, damage infrastructure and natural resources, and disrupt economic activities (Pelling et al, 2004). Economic losses from natural catastrophes over the period 1996–2005 are estimated to be US$575 billion, with record losses of US$210 billion reported in 2005 (Munich Re, 2005). In the aftermath of disasters, human development in the impacted communities and wider region is set back and can take years to recover from the loss of housing, businesses, roads, water systems, schools, hospitals, farm fields and livestock. Events such as Hurricanes Mitchell, George and Katrina can cause economic losses that represent a significant percentage of national or regional income, and repairing the damage diverts scarce capital from new development projects. Recurrent climate anomalies that do not rise to the level of natural disasters also adversely affect supplies of food and water, incomes, livelihoods and health, reduce resilience to future shocks by depleting assets for coping, and place a drag on economic development.

The projected changes in climate will have wide-ranging impacts on development. At risk are the productivity of agricultural lands, natural ecosystems and the livelihoods that are dependent on them. Also at risk are water supplies, human health and populations that inhabit low-lying coasts, floodplains, steep slopes and other exposed locations (McCarthy et al, 2001). The AIACC case studies illustrate these and other climate risks at national and local scales in a variety of developing country contexts. Not all impacts will be negative. For example, a number of studies find that climate change and higher concentrations of carbon dioxide in the atmosphere are likely to increase yields of important crops in parts of South America (Travasso et al, Chapter 19) and West Africa (Njie et al, Chapter 7; Adejuwon et al, Chapter 9). But most studies find that impacts will be predominantly negative in developing regions of the world (IPCC, 2001).

Current climate hazards and the impacts of projected climate change threaten human development (African Development Bank et al, 2003). Climate is linked to all the Millennium Development Goals, but is most directly relevant to the goals to eradicate extreme poverty and hunger, reduce child mortality, combat disease and ensure environmental sustainability (Martin-Hurtado et al, 2002). Agriculture, which is highly sensitive to climate and which is projected to be negatively impacted by climate change in much of the tropics and sub-tropics, is the direct or indirect source of livelihood for about two-thirds of the population of developing countries and is a substantial contributor to their national incomes. About 70 per cent of the world's poor live in rural areas. Progress on all the Millennium Development Goals will be dependent on progress in agricultural development and rural development. And management of climate hazards and climate change impacts in the agriculture sector and rural communities will be critical for success.

How does development affect vulnerability to climate?

There is a clear link between development level and vulnerability to climate and other natural hazards. Disaster risk, measured in mortality from natural hazards, is significantly lower in high income countries than in medium and low income countries. Countries classified as having high human development represent 15 per cent of the population that was exposed to natural disasters in 1980–2000 but account for only 1.8 per cent of the deaths (Pelling et al, 2004). In comparison, countries with low human development represent 11 per cent of the exposed population but account for 53 per cent of the recorded deaths.

The association of poverty and low levels of development with high levels of vulnerability are borne out in the AIACC studies. Failures of development to raise people out of poverty causes people to occupy highly marginal lands for farming and grazing, settle in areas susceptible to floods and mudslides, and live with precarious access to water, healthcare and other services. These conditions contribute to the high degree of vulnerability found among the rural poor of Botswana, Nigeria, Sudan, Thailand, Lao PDR, Vietnam, the Philippines, Argentina and Mexico. Squatter communities in Jamaica and the Philippines are more vulnerable than other communities because of lack of infrastructure, access to basic services and social institutions to support collective efforts for reducing risks (Taylor et al, Chapter 16; Lasco et al, Chapter 14).

Although much of the world continues to live in poverty and at high risk from hunger and disease, human development has greatly reduced vulnerability to climate-driven risks by increasing agricultural productivity, food production and trade, water storage and distribution systems, housing quality, transportation and communication networks, healthcare, education and wealth. The Millennium Development Goals have set a challenge to expand the benefits of development to include those who continue to live in deep poverty. Moving forward, development that is focused on the poor can reduce vulnerability to climate and other stresses by improving the conditions and capacities of poor households, communities and countries so that they are more resilient to shocks and more capable of responding and adapting. If based on sound principles of resource management, development can improve resource-based rural livelihoods so that they are less sensitive to climate variations and more sustainable.

Development can, however, exacerbate pressures that add to vulnerability. Past practice has given scant consideration to climate risks in planning development projects, resulting in greater vulnerability than otherwise could have been achieved, and even increasing vulnerability in some instances through maladaptive choices (Burton and van Aalst, 2004). The unevenly distributed benefits of development can also exacerbate vulnerability. Trade liberalization has brought general increases in economic activity, lower prices and greater overall wealth, but all do not share equally in the benefits and some have suffered harm. Smallholder farmers and livestock raisers in Argentina and Mexico have struggled to compete as output prices fell relative to the costs of inputs, making them more vulnerable to climate shocks (Wehbe et al, Chapter 18). Falling rice prices from greater productivity in Asia and liberalized trade caused rice farming to be abandoned in Navua, Fiji (Mataki et al, Chapter 15). The resulting loss of incomes and lack of maintenance of abandoned irrigation channels have raised vulnerability of inhabitants of the township to flood hazards.

Development in the Heihe river basin of China has brought greater livelihood opportunities and incomes. But development has also increased water demand in this arid basin to the point where water withdrawals are 80 to 120 per cent of average annual flows and conflicts have arisen between competing water users (Yin et al, Chapter 12). Social and economic changes have driven rural-to-urban migrations, often concentrating poorer migrants in settlements that are prone to flooding, as is happening on the outskirts of metropolitan Buenos Aires (Barros, Chapter 17). Increasing market orientation, movements of population and government policies have weakened community institutions and diminished the use of collective strategies for managing climate risks in places such as Botswana (Dube et al, Chapter 4), countries of the lower Mekong (Chinvanno et al, Chapter 13), Mongolia (Batima et al, Chapter 11) and Sudan (Osman-Elasha et al, Chapter 5). Development projects intended to benefit one group can have spillover effects that harm others, as is the case with the Pantabangan dam in the Philippines (Lasco et al, Chapter 14) and the Khor Arbaat dam in Sudan (Osman-Elasha et al, Chapter 5).

Integrating adaptation with development

Sometimes climate change adaptation is seen as competing with the human and economic development needs of the world's poor. Development needs are immediate, the consequences of poverty in countries with low development are appalling, progress is less than desired and allocated resources too little. In comparison, climate change can be perceived as a problem distant in time, uncertain in its effects and less consequential than present day poverty. Adaptation may therefore seem less urgent and less compelling than increasing development efforts for the world's poor. But, as argued above, climate hazards are immediate, they are growing, they threaten the quality of life and life itself, and they directly impact on the goals of development.

In balancing needs for climate adaptation with those of development, it is critical to note that there is strong complementarity between their goals and methods. A society that is made more climate-resilient through proactive adaptation to climate variations, extremes and changes is one in which development achievements and prospects are less threatened by climate hazards and therefore more sustainable. Development can repay the compliment by creating conditions that better enable adaptation. This complementarity implies that integration of adaptation efforts with development can yield synergistic efficiencies and benefits that advance the goals of both agendas. This is not to deny that tradeoffs and hard choices may be required. That is the reality of pursuing multiple goals with limited resources. But there are sufficient complementarities to make integration a workable and desirable strategy.

Adaptation activities carried out in isolation from mainstream development, and external to the authorities normally responsible for managing economic sectors and natural resources, may be practical in some contexts. Adaptation carried out in this manner, while not ideal, can help raise awareness, allow experimentation with different methods and demonstrate effective strategies. But adaptation as a stand-alone function that is implemented without the collaboration of agencies responsible for economic and resource policy and management will fail to mobilize the resources and the full range of actors that are necessary for success. To create a climate-resilient society, adaptation as a process needs to be integrated into policy formulation, planning, programme management, project design and project implementation of the agencies that are responsible for human and economic development, finance, agriculture, forestry, land use, land conservation, biodiversity conservation, water, energy, public health, transportation, housing, disaster management, and other sectors and activities.

At the most basic level, integration would avoid maladaptive actions by development and other agencies that fail to account for climate-related risks and thereby unintentionally increase risks or miss easy opportunities to reduce them. This could be achieved by subjecting policies, programmes and projects to initial scrutiny for exposure to climate risks and modifying them accordingly, similarly to assessments that are carried out for environmental impacts, gender equality and poverty reduction. A further step towards integration would be for public sector agencies to promote and support actions and behaviours by individuals, the private sector and civil society that would narrow the current adaptation deficit. Yet more ambitious, but ultimately essential, are development strategies that proactively create conditions to enable adaptation processes by enhancing the capacities of individuals, strengthening community institutions, advancing knowledge and creating knowledge networks, removing obstacles, and providing appropriate incentives.

Many of the AIACC studies demonstrate the need for comprehensive approaches to adaptation that are integrated with broader development strategies and examine how this might be done. They highlight several characteristics of development that would be complementary to the goals of adaptation. These include development that targets highly vulnerable populations, diversifies economic activities, expands opportunities for livelihoods that are less climate sensitive, improves natural resource management, encourages the development and diffusion of technologies that are robust across a wide range of climate variations and extremes, directs development away from highly hazardous locations toward less hazardous ones, and invests in expanding knowledge that is relevant to reducing climate risks.

An examination by Osman-Elasha et al (Chapter 5) of community development efforts in Sudanese villages of Bara Province in North Kordafan, El Fashir in North Darfur and Arbaat in the Red Sea State demonstrates that development and adaptation to climate risks can be strongly complementary. Community development projects implemented in the villages integrated multiple strategies to improve livelihoods, the quality of life and sustainability of resource use within a context of recurrent drought. Using measures of changes in household livelihood assets (human, physical, natural, social and financial capital), the holistic approach to development taken in the study areas is found to have succeeded in increasing the capacity of households to cope with the impacts of drought. Community participation in the projects and reliance on indigenous technologies for improving cultivation, rangeland rehabilitation and water management that are familiar to the communities are found to be important factors for success. The sustainable livelihood approach appears to be a viable model for integrating development and adaptation to climate hazards at the community scale.

Rice farmers in Thailand, Vietnam and Lao PDR rely primarily on their own capacity to implement strategies for coping with floods and mid-season dry spells; this is strongly limited by the social and economic conditions and natural resources in the surrounding community (Chinvanno et al, Chapter 13). Collective strategies to pool resources within communities and provide buffers against food and income losses that were widely prevalent in the past are now much diminished, though still important in Lao PDR. National policies are in general not supportive of reducing the vulnerability of small rice farmers to climate hazards. A national strategy to integrate climate risk management with rural development, poverty reduction and farm policies is recommended for raising the capacity and resilience of farm households and rural communities. Opportunities for effective interventions by national governments include assisting farm households with financial resources, expanding off-farm income opportunities, marketing of farm products, improving access to water, protecting the natural resource base, developing and promoting new technologies to diversify farm incomes, improving seed varieties, and providing information about current and changing climate hazards. Revitalizing community institutions is seen as important for enabling communities to benefit from national interventions.

An approach to integrating adaptation and development that is being embraced by Pacific Island Countries such as Fiji also combines top–down and bottom–up strategies (Mataki et al, Chapter 15). Top–down actions would be taken by the national government to create a climate-proof society by creating incentives, enforcing regulations, assisting with capital financing and implementing large projects that are beyond the means of local authorities. These actions would encourage and enable development and settlement away from hazardous locations, the building of flood-proof homes, purchase of insurance, better land-use practices, regular dredging of rivers, and maintenance of irrigation channels and floodgates. Bottom–up actions would draw on the communal traditions of Pacific Island societies to engage members of the community in pooling financial and human capital and other local resources, and channelling these in efforts to reduce climate related risks. The current political framework in Fiji does not provide an effective means for local communities to make their concerns felt at the national level and there is lack of communication and coordination across government departments. These obstacles will need to be overcome for the combined top–down and bottom–up integration to be effective.

Evaluating Adaptation Options

What to do, how much and when

Adaptation decisions are made in a context of uncertainty and change. While we can be confident that the climate will change in response to greenhouse gas forcing, there is uncertainty about how it will change and how fast, particularly at the spatial scales that are relevant for adaptation. The impacts are also uncertain, partly because the changes in climate are uncertain, partly because the sensitivities of systems to climate stresses are uncertain, and partly because there is uncertainty about future demographic, social, economic, technological and governance conditions that will shape future exposures, sensitivities, capacities and vulnerabilities. There is also uncertainty about the potential performance of different adaptation options, their costs and possible unintended consequences.

Uncertainty makes it difficult to decide what to do, how much of it to do and when to do it. Many of the choices will have irreversible consequences, so choosing wrong can be costly, even deadly. This is just as true for deciding not to adapt, or to delay adapting, as it is for deciding to adapt now. Delaying adaptation will result in irreversible consequences that could be avoided by adapting now. But not all adaptations could or should be implemented now. Which are appropriate for immediate or near-term action and which should be delayed?

A number of factors are relevant to the selection of options for immediate action. These include the timing of benefits, the dependence of benefits on specific climate conditions, irreversible consequences, option values and thresholds for adverse impacts (Leary, 1999). Characteristics of adaptation measures that warrant consideration for early action include expectation of significant near-term benefits (for example, in narrowing existing adaptation deficits), performance that would produce benefits under a wide range of possible future climates, low capital costs and minimal irreversible consequences. Also of interest for early implementation are actions that would preserve or expand options for future adaptation (for example, purchase of development easements, developing knowledge networks and capacity building), or counteract looming thresholds for adverse impacts (for example, facilitated migration of species that are obligatory dispersers). Characteristics that would suggest delay of some actions while uncertainties are resolved include little near-term benefit, future benefits that depend on a narrow range of climate conditions, high capital costs and large irreversible consequences.

Evaluation of options by AIACC studies

Decision-making criteria for evaluating and selecting adaptation options vary from context to context. Criteria can vary depending on who is making the decision, what stakeholders are affected by the decision, what role stakeholders have in the decision process, the objectives of decision makers and stakeholders, and characteristics of the decision such as the time horizon, uncertainty about outcomes, irreversibility of consequences and consequences of decision errors. Criteria applied in the AIACC studies include net economic benefit, timing of benefits, distribution of benefits, consistency with development objectives, consistency with other government policies, cost, environmental impacts, spill-over effects, capacity to implement, and social, economic and technological barriers. In some cases the criteria are chosen by the investigators, in other cases they are chosen by stakeholders or based on stakeholder input. Methods for their application include formal cost–benefit and multi-criteria analysis, expert judgement, and participatory exercises with selected stakeholders.

Callaway et al (Chapter 3) apply formal cost–benefit analysis to decisions about building water storage and switching water allocation regimes for the Berg river basin in South Africa. The net benefits from choices of reservoir capacity are uncertain and vary depending on how the future unfolds with respect to climate, growth in water demand, and reliance on either the current regulatory regime or water markets for allocating water. The climate scenarios analysed include no change in surface water runoff and reductions of either 11 or 22 per cent. Under the current regulatory regime for water allocation and water demand growth of 3 per cent per year, climate change would cause estimated damages with a present discounted value of 13.4 billion to 27.6 billion rand, or roughly 15 to 30 per cent of the total net benefits of water use in the basin. Adapting by correctly anticipating and adjusting reservoir capacity to the optimal size corresponding to the change in climate would reduce the damages and yield net benefits, but the net benefits are modest and less than 2 per cent of the damages. In contrast, a switch from the current regulatory regime to allocation by water markets would yield net benefits of roughly 10 to 20 per cent by allowing efficient reallocation of scarce water.

Njie et al (Chapter 7) also apply cost–benefit analysis to evaluate adaptations to climate change. They investigate increased use of fertilizers and adoption of irrigation for growing cereals in the uplands of The Gambia. Climate change would cause estimated annual damages to cereal production of roughly US$150 million in 2010–2039 and in excess of US$1 billion in 2070–2079. Increased use of fertilizers would yield net benefits that would reduce climate change damages by 10 per cent or more. Irrigation, however, is found to yield negative net benefits in the 2010–2039 time frame and mixed results in the more distant future. For cereal production, the high cost of pump irrigation relative to cereal prices make irrigation an inefficient adaptation for cultivation of cereals, at least in the near to medium term.

Yin et al (Chapter 12) apply an analytic hierarchy process, a form of multi-criteria analysis, to evaluate adaptation options for the water sector in the

Heihe river basin of north-western China. Stakeholder meetings and surveys were used to elicit judgements about the effectiveness of different options with respect to four decision criteria and the relative importance of the criteria. The criteria include water-use efficiency, economic returns on water use, environmental effects and cost. The results rank intuitional options for managing water demand above engineering measures to increase water supply. Preferred options include economic reforms that would constrain sectors that are large water consumers, water user associations to share information and promote water conservation, and transferable water permits for allocating water use.

Lasco et al (Chapter 14) perform a tradeoff analysis of effects of adaptations in one sector that spillover to and impact on other sectors in the Pantabangan–Carranglan watershed of the Philippines. Options are identified and examined for agro-forestry, water resources and local communities. They find that spillovers are common because the shared water resource creates a high degree of interdependence among people, livelihoods and biophysical resources located within the watershed. The spillovers include both positive and negative externalities. For example, many of the options identified for agro-forestry such as improving water use efficiency and controlling runoff and erosion have beneficial effects on the water sector and on local community institutions. But stricter enforcement of forest protection laws and reforestation to protect water resources can negatively affect incomes and livelihoods of some landowners and cause farmers in informal settlements with insecure land tenure to be forced from their farms. They find that these types of tradeoffs are seldom considered in planning new projects or revising policies, risking negative impacts on others, conflicts among stakeholders in the watershed and missed opportunities for mutually beneficial actions.

In Mongolia, evaluation of adaptation options for the livestock sector applied a two-tiered screening process with participation from herders, scientific experts, and authorities from local, provincial and national offices (Batima et al, Chapter 11). In the first tier, options are screened for satisfying broad criteria for promoting both adaptation and development goals, consistency with government policies, and environmental impacts. Options that pass the first screening are then evaluated against a second tier of six additional criteria. These include capacity to implement, importance of climate as a source of risk, near-term benefits, long-term benefits, cost and barriers. Adaptation strategies that emerge as priorities from this process include measures that generate near-term benefits by improving capabilities for reducing the impacts of drought and harsh winters as well as measures that produce long-term benefits by improving and sustaining pasture yields. Some of the specific measures identified as warranting further consideration include improving pastures by reviving the traditional system of seasonal movement of herds; increasing animals' capacity to survive winters by modifying grazing schedules and increasing use of supplemental feeds; enhancing rural livelihoods by strengthening community institutions to regulate use of pasture and provide local services such as education, training, access to credit and insurance; and research and monitoring to develop and improve forecasting and warning systems.

In the study of dengue fever in the Caribbean, the investigators evaluate adaptation options for cost, effectiveness, social acceptability, environmental friendliness, promotion of local cooperation and technical/socioeconomic challenges (Taylor et al, Chapter 16). Three options of multiple measures are recommended based on these criteria. The first option would refocus current education, disease surveillance and vector control efforts to be more proactive and to address deficiencies in community involvement. Emphasis would be placed on education that stresses individual responsibility and community benefits of measures to reduce human–vector contact. The second option would combine the above measures with designing, producing and promoting the use of low-cost covered containers for storing rainwater. Discarded and uncovered oil drums are the most commonly used means of capturing and storing water and are ideal breeding sites for mosquitoes. The third option would include all the above plus development and implementation of an early warning system. Early warnings to give advance knowledge of the expected severity of possible disease outbreaks would enable responses to be calibrated to the anticipated threat level. Responses to an alert would include more frequent and extensive vector surveillance and control, stepped up education efforts tailored to the threat level, and more diligent efforts to eliminate breeding sites for mosquitoes.

Creating an Enabling Environment

Many studies, including our own, identify numerous options for adapting to existing and changing climate hazards. Some are novel and untested, but many are based on current practices that have been amply demonstrated to reduce risks. As we noted earlier, individuals, communities and nations all have a strong self-interest in adapting. Yet many options go unused, or are used much less extensively or intensively than their benefits would seem to warrant.

It is not for lack of options that adaptation lags. It is lack of determination, lack of cooperation and lack of means that impede it. Deliberate and sustained efforts are needed to create an enabling environment for overcoming these obstacles and facilitating the process of adaptation. The efforts need to engage the general public, as well as stakeholders and authorities, from the many different economic sectors and spheres of activity that are affected by climate and should link across local, provincial, national and international jurisdictions.

Creating the determination to adapt

A primary obstacle is a lack of will, or determination, to adapt. This can happen at the individual level (for example people failing to take simple actions to limit their own exposure to malaria and dengue), the community level (local authorities allowing new development in hazardous locations), the national level (ministries failing to consider climate risks in new programmes and not being held accountable), and the international level (adaptation continuing to receive strong rhetorical support from international environmental and development communities but few resources).

The reasons for lack of will are varied. One is a problem of awareness and understanding. People lack knowledge about, or are uncertain or sceptical about, current climate risks, climate change, options for adaptation and the effectiveness, and the feasibility and cost of adaptation. Another important reason is that people have other objectives that compete with adaptation for attention, priority and resources. In essence, determination to adapt will not gain acceptance unless people find the evidence that climate risks represent a substantial problem compelling, that addressing the risks warrants priority on a par with other objectives, that there are effective, feasible and affordable options, and that we know enough to make wise choices.

Greater awareness and knowledge can help to create the determination to adapt. But it is not enough to simply create more knowledge. It needs to get into the hands, or the heads, of people facing decisions about how to allocate scarce resources to achieve their objectives, objectives that include, but are not limited to, reducing risks from climate and other sources. The knowledge needs to be relevant to the decisions being made and understandable to stakeholders and decision makers, who might be residents of hazardous places, resource users and owners, farmers, business operators, community leaders or government officials. The knowledge also has to be seen as credible and untainted by bias or intent to manipulate.

The different types of knowledge, intended users and applications are too varied for the functions of knowledge creation, collection, communication, integration and interpretation to be done well by a single entity. Networks of knowledge institutions are needed that link the scientists, practitioners and public, the various economic sectors, and local, national and international actors. In each of the AIACC study areas, knowledge networks are very incomplete and not well coordinated, resulting in substantial gaps in the awareness and understanding of climate hazards, climate change and adaptation among many key stakeholders.

This situation can be improved by strengthening knowledge networks. Investments are needed in scientific research, assessment and capacity in areas that are relevant to understanding climate risks and response options. Expanded efforts are needed to collect knowledge from the experiences and practices of at-risk groups, including traditional knowledge. Mechanisms are needed to integrate, interpret and communicate the created and collected knowledge and to assist stakeholders in applying the knowledge in decision making. Avenues are needed for stakeholders to give feedback about the information received and the information required, as well as to share their knowledge with other at-risk groups.

Participatory processes that engage stakeholders and attempt to link the different functions and components of knowledge networks can be effective in generating and communicating knowledge that is relevant, understandable and credible. The AIACC project is one example of such a process and similar projects have been initiated and are underway. Ultimately, though, the generation and communication of knowledge for supporting adaptation needs to be connected with and embedded in ongoing processes of human development, economic planning, poverty reduction and resource management.

Creating cooperation to adapt

What any one person or organization can do to adapt is very much constrained by what others do or do not do. Cooperation among members of a community can mobilize resources to reduce, hedge and spread risks beyond what individuals acting independently might achieve. Cooperation between local and national authorities can rationalize policies and plans so that they work toward common adaptation goals and not at cross purposes. Cooperation among stakeholders and authorities from different economic sectors can increase positive spillovers and avoid negative spillovers of their sector-based strategies. International cooperation can help to ensure that actions are based on the best available science, that information about best practices is shared, that financial resources can be pooled and directed toward common goals, and that efforts under different international agreements contribute to adaptation objectives where possible.

Fostering cooperation on adaptation requires leadership within national governments. An environment or science ministry might play a useful role in raising awareness, sharing information about risks and adaptation options, supporting knowledge networks, assessing the implications of new legislation and policies for narrowing or widening the adaptation deficit, and monitoring overall progress on managing climate risks. But environment and science ministries typically lack the standing to marshal resources at the required scale or to compel other ministries to cooperate. The determination to adapt will need to permeate beyond environment and science ministries and be accepted by other ministries as important to their missions and objectives if there is going to be effective cooperation.

The purpose of integrating or mainstreaming adaptation with development is to enlist the cooperation of these other ministries and associated stakeholders in making adaptation commonplace in economic and sector-, resource- and livelihood-based planning and programmes at national to local scales. Cooperation is not forthcoming when actors and stakeholders in these different spheres of activity view climate change as immaterial to their main objectives and adaptation as a potential new mandate that will divert resources from their priorities. The experience of the AIACC case studies is that stakeholders from varied perspectives often are aware of climate threats to their interests and that, when put in a broad context of managing current climate hazards and not limited to only climate change, are willing to engage with others to assess threat levels and possible responses. Through their participation in an assessment process, many accept, or at least are willing to consider seriously, the need to adapt to narrow the existing adaptation deficit, to limit vulnerability to climate change in the near- to medium-term future and to cooperate with others to move toward a climate-proof society.

Creating the means to adapt

Determination to adapt and cooperation are not sufficient by themselves, however. The means to adapt must also be available. Much of what needs to be done to adapt is at the level of the household and community. But for the most vulnerable households and communities, the means to adapt are in short supply. Often they do not have sufficient resources, knowledge and skills to implement measures that would reduce the risks that they face.

Targeting development to highly vulnerable populations to provide expanded and diversified livelihood opportunities and access to services such as clean water, health care, education and credit can increase the assets of households and bolster their capacity to cope with and adapt to hazards of all types, including climatic hazards. Capacities that are specific to climate adaptation can be increased by providing information, training, technical advice and resources for adopting technologies and practices that can reduce climate-driven damages and variability of production and income. Strengthening and supporting community institutions can increase the capacity for collective action to reduce, hedge and spread risks.

Financing adaptation

Financial resources are also an important part of the means to adapt. At the local level, many communities have been resourceful in operating village funds and other mechanisms to provide access to credit for small-scale farmers, enterprise owners and others that have proven useful for helping to finance risk-reducing investments or recovery from losses. Private sector finance markets play an important role in financing investments by larger enterprises, for example, for large-holder farmers to diversify farm operations, adopt new seed varieties, implement irrigation and provide insurance against losses. Insurance needs particular attention as it is far less prevalent in developing countries than in developed; premiums, already more than can be afforded by poor and vulnerable communities, are rising; and insurers are withdrawing from some markets where climate risks are high. Private sector innovations in micro-credit and micro-insurance can help to increase the access of the poor to financial resources. National governments also assist with direct financial payments and with subsidized credit and insurance, although in many places financial assistance from national governments to both rural and urban poor is diminishing.

At the international level, financial assistance is being provided for adaptation through the Global Environment Facility under the United Nations Framework Convention on Climate Change (UNFCCC) as well as through development assistance from bilateral and multilateral aid agencies. This international funding is acting as a catalyst for raising awareness, building capacity, advancing understanding of risks and response options, and engaging developing country governments in prioritizing and assessing options. Recently, funding is also being made available for experimenting with and implementing selected measures for adapting to climate change.

But the magnitude of the adaptation problem and the likely financial needs in developing countries are far greater than current funding can cover. Compelling arguments have been made that developed countries have a liability to help fund adaptation in developing countries that also exceed current contributions (see, for example, Baer, 2006). International financial assistance for adaptation does appear to be increasing. But it is not clear to what extent these are new resources or reallocations of limited development assistance funds, which is a source of tension for integrating adaptation and development. While the logic for integration is inescapable, there is legitimate concern that this will divert some funds away from critically important development objectives. Ultimately though, financing for adaptation will need to come from multiple sources, including developing country governments and their private sectors, as well as from foreign direct investment, international development assistance, and specialized funds under the UNFCCC and other multilateral sources.

A Final Word

Climate hazards exact a heavy toll, impacting most strongly on the poor and acting as a drag on development. The toll is rising as climate change widens the gap between our exposures to risks and our efforts to mange them. National governments are increasingly aware of the growing risks and are cooperating in the UNFCCC and other processes to cautiously consider how to respond. But there is not yet widespread determination to adapt.

The determination to adapt can be assisted by increasing recognition that closing the current adaptation deficit provides immediate benefits and is a first step toward adapting to climate change, that feasible, effective and affordable options are available, and that these options do not require certainty about how the climate will change to be effective. But beyond determination, the means to adapt need to be enhanced. Knowledge of climate risks and adaptation response strategies need to be increased. Capacities of at-risk households and community institutions need to be raised and access provided to improved technologies. Climate-sensitive natural resources need to be protected and rehabilitated. Financial resources are needed. Most of all, adaptation needs to be integrated with development so that it becomes commonplace in each sector of human activity. The time to act, to make a stitch in time, is now.

References

Adger, W. N., S. Agrawala, M. Mirza, C. Conde, K. O'Brien, J. Pulhin, R. Pulwarty, B. Smit and K. Takahashi (2007) 'Assessment of adaptation practices, options, constraints and capacity', in M. Parry, O. Canziani, J. Palutikof and P. J. van der Linden (eds) *Climate Change 2007: Impacts, Adaptation and Vulnerability*, contribution of Working Group II to the Fourth Assessment Report of the Intergovernmental Panel on Climate Change (forthcoming)

African Development Bank; Asian Development Bank; Department for International Development, UK; Directorate-General for Development, European Commission; Federal Ministry for Economic Cooperation and Development, Germany; Ministry of Foreign Affairs – Development Cooperation, The Netherlands; Organization for Economic Cooperation and Development; United Nations Development Programme; United Nations Environment Programme; and The World Bank (2003) *Poverty and Climate Change, Reducing the Vulnerability of the Poor through Adaptation*, The World Bank, Washington, D.C., US

Baer, P. (2006) 'Adaptation: Who pays whom?', in W. N. Adger, J. Paavola, S. Huq and M. J. Mace (eds) *Fairness in Adaptation to Climate Change*, MIT Press, Cambridge, MA, US

Burton, I. (2004) 'Climate change and the adaptation deficit', in Adam Fenech (ed) *Climate Change: Building the Adaptive Capacity*, papers from International Conference on Adaptation Science, Management, and Policy Options, Lijiang, Yunnan, China, 17–19 May 2004, Meteorological Service of Canada, Environment Canada, Toronto

Burton, I., and M. van Aalst (2004) 'Look before you leap: A risk management approach for incorporating climate change adaptation in World Bank operations', Working Paper No 100, Environment Department, World Bank, Washington, D.C., US

IPCC (2001) 'Summary for policymakers', in J. McCarthy, O. Canziani, N. Leary, D. Dokken and K. White (eds) *Climate Change 2001: Impacts, Adaptation and Vulnerability*, contribution of Working Group II to the Third Assessment Report of the Intergovernmental Panel on Climate Change, Cambridge University Press, Cambridge, UK, and New York

IPCC (2007a) 'Summary for policymakers', in M. Parry, O. Canziani, J. Palutikof and P. van der Linden (eds) *Climate Change 2007: Impacts, Adaptation and Vulnerability*, contribution of Working Group II to the Fourth Assessment Report of the Intergovernmental Panel on Climate Change, Cambridge University Press, Cambridge, UK, and New York

IPCC (2007b) 'Summary for policymakers', in S. Solomon, D. Qin, M. Manning, Z. Chen, M.C. Marquis, K. Averyt, M. Tignor and H. L. Miller (eds) *Climate Change 2007: The Physical Science Basis*, contribution of Working Group I to the Fourth Assessment Report of the Intergovernmental Panel on Climate Change, Cambridge University Press, Cambridge, UK, and New York

Leary, N. (1999) 'A framework for benefit–cost analysis of adaptation to climate change and climate variability', *Mitigation and Adaptation Strategies for Global Change*, vol 4, pp307–318

Leary, N., C. Conde, J. Kulkarni, A. Nyong and J. Pulhin (eds) (2008) *Climate Change and Vulnerability*, Earthscan, London

Martin-Hurtado, R., K. Bolt and K. Hamilton (2002) *The Environment and the Millennium Development Goals*, The World Bank, Washington, DC

McCarthy, J., O. Canziani, N. Leary, D. Dokken and K. White (eds) (2001) *Climate Change 2001: Impacts, Adaptation and Vulnerability*, Contribution of Working Group II to the Third Assessment Report of the Intergovernmental Panel on Climate Change, Cambridge University Press, Cambridge, UK, and New York

McMichael, A., A. Githeko, R. Akhtar, R. Carcavallo, D. Gubler, A. Haines, R. S. Kovats, P. Martens and J. Patz (2001) 'Human health', in J. McCarthy, O. Canziani, N. Leary, D. Dokken and K. White (eds) *Climate Change 2001: Impacts, Adaptation and Vulnerability*, contribution of Working Group II to the Third Assessment Report of the Intergovernmental Panel on Climate Change, Cambridge University Press, Cambridge, UK, and New York

Munich Re (2005) *Topics GEO – Review on Natural Catastrophes 2005*, Munich Re, Munich, Germany

Pelling, M., A. Maskrey, P. Ruiz and L. Hall (2004) *Reducing Disaster Risk: A Challenge for Development*, United Nations Development Bank, Bureau for Crisis Prevention and Recovery, New York

Schneider, S. H., S. Semenov, A. Patwardhan, I. Burton, C. Magadza, M. Oppenheimer, A. B. Pittock, A. Rahman, J. B. Smith, A. Suarez and F. Yamin (2007) 'Assessing key vulnerabilities and the risk from climate change', in M. Parry, O. Canziani, J. Palutikof and P. J. van der Linden (eds) *Climate Change 2007: Impacts, Adaptation and Vulnerability*, contribution of Working Group II to the Fourth Assessment Report of the Intergovernmental Panel on Climate Change (forthcoming)

Smit, B., O. Pilifosova, I. Burton, B. Challenger, S. Huq, R. Klein and G. Yohe (2001) 'Adaptation to climate change in the context of sustainable development and equity', in J. McCarthy, O. Canziani, N. Leary, D. Dokken and K. White (eds) *Climate Change 2001: Impacts, Adaptation and Vulnerability*, contribution of Working Group II to the Third Assessment Report of the Intergovernmental Panel on Climate Change, Cambridge University Press, Cambridge, UK, and New York

2

Adapting Conservation Strategies to Climate Change in Southern Africa

Graham von Maltitz, Robert J. Scholes, Barend Erasmus and Anthony Letsoalo

Introduction

Global climate change is predicted to have substantial impacts on southern Africa's biodiversity, including wide-scale extinctions over the next 50 years (Rutherford et al, 1999; Hannah et al, 2002a and b; Gitay et al, 2001 and 2002; Midgley et al, 2002a and b; MA, 2005). At a global scale, Thomas et al (2004) have predicted that 15–37 per cent of species in their sample (which covered 20 per cent of the Earth's surface) may be at risk of premature extinction due to anthropogenically caused global change by 2050. The Millennium Ecosystem Assessment, using different models and assumptions based largely on habitat loss, reached similar conclusions (MA, 2005). Within South Africa, a reduction in size and an eastward shift for current biomes is predicted and up to half of the country will likely have a climatic regime that is not currently found in the country (Rutherford et al, 1999). The succulent karoo biome, (a succulent-dominated semi-desert located on the southwestern coast of southern Africa) is projected to be the most severely impacted, with the grassland and *fynbos* (a Mediterranean-climate sclerophyllous thicket that approximates to the Cape Floristic region) biomes also likely to suffer from high climate change impacts (Rutherford et al, 1999; Midgley et al, 2002a and b). *Fynbos* and succulent karoo are biodiversity hotspots of international importance (Myers et al, 2000), with the latter being one of only two globally important arid-climate biodiversity hotspots.

Two climate parameters critical for animal and plant species distributions are temperature and water balance (a combination of precipitation and evaporation, which, in turn, is directly influenced by temperature) (Cubasch et al, 2001). The dynamics of plant and animal populations change at the edge of individual species' distribution, as net mortality becomes larger than net fecundity, with a spatial gradient of declining population numbers as a result. In a scenario of climate change, the direct influence of temperature and water balance in combination with the indirect influence of interspecies competition, fire frequency, pollinator distribution, herbivory and predation, food availability, soil type, topography and so forth could lead to the progressive extinction of nonvagile species in their natural range, beginning with population dieback in the so-called 'trailing edge' of the historical distribution range (Davis and Shaw, 2001; Gaston, 2003).[1]

In southern Africa, global circulation models project the greatest increases in temperature (2–4°C this century) for the inland areas, while the coastal areas are predicted to experience somewhat lesser increases (1–3°C), due to the thermal buffering effect of the oceans (Cubasch et al, 2001; Scholes and Biggs, 2004). Changes to precipitation are more difficult to predict, and there is less agreement between models. For southern Africa the majority of models predict about a 10 per cent reduction in annual precipitation during the 21st century in the western two-thirds of the continent south of 15°S , while the eastern one-third may see an increase of the same order (Scholes and Biggs, 2004; Hewitson and Crane, 2006). A combination of increased temperature (and thus increased evaporative demand) with decreased rainfall will increase the aridity of affected environments, notwithstanding the slight offsetting beneficial effect of elevated CO_2 on plant water use efficiency (Scholes and Biggs, 2004). A combined increase in rainfall and temperature will increase primary plant production, but will still be detrimental to specific species (Gitay et al, 2001; Gelbard, 2003).

Excluding evolutionary adaptations, species can be classified into four functional groups based on their response to climate change as follows:

1 *Persisters*: These species have tolerance for the new climate of their current location.
2 *Obligatory dispersers*: These species will have to physically move with the changing climate to track areas with suitable climates (autonomous dispersers), or alternatively will have to be moved artificially to new areas with suitable climates if they are unable to move on their own (facilitated dispersers).
3 *Range expanders*: These species may expand into new climatic envelopes that are not currently available, but to which the species are already well adapted.
4 *No-hopers*: These species cannot do any of the above and will become prematurely extinct, although they may persist under unsuitable climates for some time.

Some species, referred to as partial dispersers, will experience range shifts causing them to persist in parts of their previous range while dispersing into new areas. The time span involved and the intensity of the climate change experienced (or modelled) will determine the extent to which species may persist or are obliged to disperse.

Detailed modelling on the impacts of climatic change on individual species has been conducted in the *fynbos* and succulent karoo regions. The Proteaceae was studied as a surrogate for the *fynbos* vegetation to understand individual species response to changing climate over the next 50 years and thus to evaluate future conservation strategies. The model predicted that 57 per cent were persisters, 26 per cent partial dispersers, 6 per cent obligatory dispersers and 11 per cent no-hopers (Williams et al, 2005). In the karoo region, it was found that the riverine rabbit (*Bunolagus monticularis*) is likely to become extinct because of its specialized food and habitat requirements, while the tortoise (*Homopus signatus),* which is less selective, is likely to survive in the 50-year study timeframe (G. O. Hughes, personal communication, 2005).

The combined impacts of climatic change and CO_2 effects have been modelled for the lowveld savanna regions of South Africa (R. J. Scholes, personal communication, 2005).[2] Preliminary results suggest that the decrease in soil moisture and the increase in temperature overwhelm the small elevated CO_2 advantage that trees have, given that C3 and C4 plants respond differently to these factors.[3] The model predicts that the structural and functional habitat suitability for browsers and grazers will likely remain relatively constant in the 50-year timeframe, provided that fire and elephant management are appropriate. Overall, the carrying capacity for large herbivores is projected to decrease by about 10 per cent. Although this study does not consider individual species, it suggests that the functional integrity of the savanna habitat can be maintained near to current conditions through appropriate management.

A Brief History of Conservation in Southern Africa

Extensive tracts of land are managed as conservation areas in southern Africa (see Table 2.1). About half the countries in the region exceed the International Union for the Conservation of Nature (IUCN) guidelines of 10 per cent of land area under formal conservation. Over the entire region, approximately 10 per cent of land is conserved in IUCN categories I–V reserves (reserve categories set up strictly for conservation) and another 8 per cent conserved in areas managed for sustainable use, i.e. IUCN category VI areas. Some countries fall far short of the IUCN guidelines; for example, in Lesotho only 0.2 per cent of the surface area is conserved (Scholes and Biggs, 2004; WDPA, 2005).

Even where countries have a relatively high level of land conserved, the fraction of *biodiversity* conserved may be substantially less (Rodrigues et al, 2004; Orme et al, 2005). This is because, historically, conservation has been based on the availability of land and in many instances the presence of big game species rather than strategic conservation objectives (Pressey et al, 1993; Heywood and Iriondo, 2003). As a result, of the 52 unique ecoregions identified in southern Africa (Olson et al, 2001), 23 per cent (15 per cent of land area) have less than 3 per cent conservation (see Table 22.). Forty per cent of ecoregions, representing 35 per cent of the land area, have less than 5 per cent formally conserved in IUCN reserves. Southern Africa has an exceptionally high biodiversity, including a number of centres of endemism and three biodiversity hotspots (Myers et al, 2003). The Madagascar hotspot has only 2.9 per cent of the area conserved in IUCN reserves with a further 1 per cent conserved outside of IUCN reserves. The succulent karoo hotspot has only 1 per cent conserved, although there are proposals to conserve an additional 19 per cent. The Cape floristic region is well conserved in the mountainous areas

Table 2.1 *The area as a percentage conserved in southern African countries in IUCN reserves (IUCN classes I–V), IUCN sustainable resource use areas (IUNC class VI), and other non-IUCN conservation areas*

Country	IUCN VI	IUCN I–V	Total IUCN	non-IUCN	Total
Angola	0.0	6.7	6.7	5.5	12.2
Botswana	0.0	18.0	18.0	12.7	30.7
Burundi	0.0	3.7	3.7	0.0	3.7
Congo	0.5	9.3	9.8	8.5	16.6
Congo (DRC)	3.6	4.7	8.2	3.1	10.6
Equatorial Guinea	0.0	17.2	17.2	0.0	17.2
Gabon	0.0	2.5	2.5	14.5	16.4
Kenya	1.6	5.6	7.1	2.5	9.6
Lesotho	0.0	0.3	0.3	20.8	21.0
Madagascar	0.6	2.4	2.9	1.0	4.0
Malawi	0.0	8.5	8.5	0.0	8.5
Mozambique	1.4	4.0	5.4	5.9	11.1
Namibia	0.7	13.2	13.8	3.6	16.7
Rwanda	0.0	11.1	11.1	0.0	11.1
Seychelles	0.0	59.2	59.2	0.0	59.2
South Africa	0.0	5.5	5.5	0.8	6.2
Swaziland	1.0	2.1	3.0	0.0	3.0
Tanzania	0.1	14.8	14.9	16.0	27.8
Uganda	12.6	7.4	20.0	6.1	23.8
Zambia	18.8	8.1	26.8	9.5	35.4
Zimbabwe	4.8	7.9	12.7	15.3	27.9
Total	2.6	7.6	10.2	6.0	15.6

Note: Non-IUCN conservation areas are mostly forest reserves. All data presented in this table are based on WDPA (2005).

but poorly conserved on the flats (see Table 2.3). By comparison, the *mopane* savanna regions (not a biodiversity hotspot) are well preserved, largely due to their low economic value for agriculture (see Table 2.3).

Formal conservation began in the late 19th century. From about 1910 to 1970, there was a steady expansion of protected areas (see Figure 2.1), which fell into two categories: the forest reserves, managed for sustainable wood extraction and/or catchment protection, and the game and nature reserves for hunting, which originally tended to be centred in areas with high wildlife populations. Game and nature reserves are currently managed for biodiversity conservation and ecotourism (von Maltitz and Shackleton, 2004). During this period, reserves enjoyed strong state support and were well maintained. The postcolonial period saw a shift in government focus to social development

Table 2.2 *The amount of conservation per ecoregion*

Percentage conserved per ecoregion	Conservation in IUCN category I–VI reserves			Total conservation including IUCN and non-IUCN reserve areas (some of which are only in the planning stage)		
	Total number of eco-regions	Cumulative percentage of eco-regions	Cumulative percentage of total land area	Total number of eco-regions	Cumulative percentage of eco-regions	Cumulative percentage of total land area
<3%	12	23.1	15.1	8	15.4	10
3–5%	9	40.4	35.1	5	25	19.5
5–10%	10	59.6	53.2	4	32.7	27.1
10–15%	10	78.8	83	12	55.8	60.4
15–20%	3	84.6	86.1	8	71.2	68.7
>20%	8	100	100	15	100	100

Note: All data presented are based on ecoregions studied in Olson et al (2001) and the WPDA (2005) database of protected areas. This is for the same set of southern and east African countries, including Madagascar, as listed in Table 2.1. Note that non-IUCN areas include some planned areas that have as yet not been proclaimed. Most of the non-IUCN areas are forest reserves.

issues and budgets for protected areas diminished. Due to population growth, there was increasing pressure on reserve borders and increasing conflict over resources. In some cases, local communities invaded the reserves and settled there (Fabricius et al, 2004; von Maltitz and Shackleton, 2004; Child, 2004).

In response to the budget constraints and growing negative perception of conservation areas, government policies regarding resource conservation were altered. The trend from the 1980s was towards delegating ownership of wildlife and forestry resources from the state to those owning or resident on the land. This would make it possible for communities on communal land to enter into community-based natural resource management (CBNRM) programmes (Fabricius et al, 2004; Child 2004; Hutton et al, 2005), and for the establishment of private wildlife ranches on commercial land (ABSA, 2003). This approach

Table 2.3 *Extent of conservation versus 'need' for conservation*

Vegetation type	Centre of endemism	Area in km² thousands	Percentage transformed	Percentage conserved
Mopane Shrubveld	no	26	0	99.8
Mopane Bushveld	no	209	8	38
West Coast Renoster veld	yes	61	97	1.7
Mountain Fynbos	yes	247	11	26.2

Note: Two extremes shown are based on south African statistics; all data presented are based on Low and Rebelo (1996).

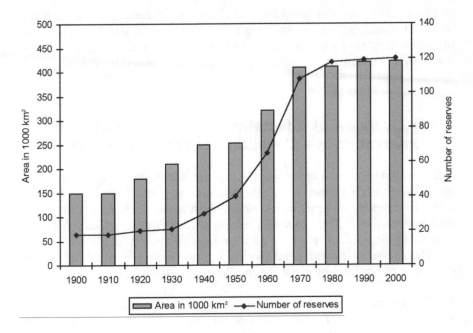

Figure 2.1 *The increase in conservation areas and the number of reserves in seven southern African countries*

Note: Only national parks and large reserves in South Africa have been included.
Source: Based on Cumming (2004).

was expected to promote biodiversity conservation in the communal and private areas by creating an economic incentive for conservation (Fabricius, 2004; Child, 2004; ABSA, 2003). However, the programme has had mixed success, largely due to a lack of appropriate capacity, both in government departments and in communities (Hutton et al, 2005), resulting in increased criticism of the co-management and resource sharing strategy (Wilshusen et al, 2002; Hutton et al, 2005; Büscher, 2005).

The most recent trend is towards international assistance for conservation in Africa, and millions of dollars have been contributed for this purpose by the Global Environmental Facility (GEF) and by first world countries. For the first time in decades, new areas are being proposed for conservation, and existing conservation is being strengthened. Strategic conservation planning tools such as Worldmap (www.nhm.ac.uk/science/projects/worldmap/index.html) and C-plan are making it possible to plan the location of reserves in a scientific and defensible manner to achieve agreed conservation targets (Pressey et al, 1993; Margules and Pressey, 2000; Pressey and Cowling, 2001), for example, in the *fynbos*, thicket and succulent karoo regions of South Africa (Cowling and Pressey, 2003). A number of transnational megaparks (sometimes referred to as 'peace parks') are also being developed such as the Limpopo, Kalagadi and Maluti-Drakensberg Transfrontier Parks (van der Linde et al, 2001). The possible consequences of climate change to biodiversity are also beginning to be considered (Hannah et al, 2002a and b; Midgely et al, 2003; Williams et al, 2005).

An Overview of Adaptation Options for Biodiversity Conservation in a Climatically Changing Environment

Conservation becomes a moving target in a climatically changing environment, and although current reserve systems are a starting point, there is no clear end point. Biodiversity patterns in 50 years' time represent only one period in an environment that is likely to see increasing temperature for at least 200 years because of the residual effect of CO_2 increases (Cubasch et al, 2001).

The following potential adaptation options were identified to prevent extinction of biodiversity given the predicted climate change:

- Do nothing (i.e. maintain the current conservation strategy).
- Reconfiguration of reserve system to strategically conserve areas that accommodate climate change.
- Matrix management, i.e. managing the biodiversity in areas outside of reserves.
- Translocation of species into new habitats.
- *Ex-situ* conservation, for example, gene banking, cryopreservation, zoos and botanical gardens.

Current understanding of ecosystem response to climate change, based both on historical data and modelled predictions, suggests that individual species will respond at different rates. As a consequence, entire ecosystems will not move in unison, but species will move independently, leading to altered community composition (Huntley, 1991; Graham, 1992; Gitay et al, 2001; Williams, 2005; Thuiller et al, 2006; Bush, 2002). It is therefore important that, in attempting to minimize losses, conservation strategies must also account for individual species in addition to the need to maintain entire habitats (ecosystems), which would be likely to have a different composition in the future, although in some instances the functional attributes may be similar (see description of lowveld savanna modelling study above).

On the basis of individual species responses to climate change, a set of adaptation options are identified in Figure 2.2 and their relative constraints and benefits are compared in Table 2.4.

Conservation of species that persist or expand their range

Where a species persists in large populations in an already-conserved area under future climates, there is no strong basis for concern. However, if the species becomes invasive and its range expands then it may become a threat to

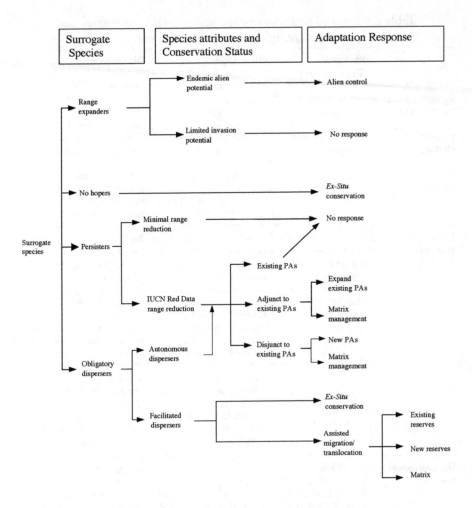

Figure 2.2 *A decision tree for selecting adaptation strategies for different surro-gate species based on their response to climate change*

other species and may need control. If the species is already threatened under current conditions, even if it persists, it might warrant extra conservation atten-tion, especially if it is not currently found in existing conservation areas.

Conservation of obligatory dispersers

For the autonomous obligatory dispersers, a climatically and environmentally suitable migratory pathway must exist to allow the species to move through the landscape to track the changing climate. The extent of land transformation in dispersal corridors is a major concern (Hannah et al, 2002a). There are two

Table 2.4 *Relative financial costs compared to the advantages and disadvantages of differing adaptation options*

	Relative financial cost	Advantages	Disadvantages
Do nothing, i.e., maintain the current conservation strategy	Zero additional cost but there is an existing high current cost of conservation management	The current reserve system is in place and funded. No new land needed. Easier to justify than new land acquisition. Will preserve a large percentage of current biodiversity. Maintains intact habitats and ecological interactions.	Not optimized for climate change. No provision is made for protection in a changing climate, so extinction of some species is inevitable. In most areas, the current reserves do not optimize biodiversity conservation, even for a static climate.
Reconfigure reserves	Very high additional cost if multiple small reserves are added, more cost-effective if existing reserves are expanded or realigned.	Ensures high conservation levels for a changing climate. Allows full state control and management of the land. If adequately funded reserves remain the most secure mechanism for ensuring biodiversity conservation. Maintains intact habitats, ecological process, and a large proportion of biodiversity. Most affordable when linked to existing reserves and for large areas. Best suited to land with high agricultural or development potential.	The high cost. The political aspects relating to acquiring land from private individuals or communities. Very difficult to acquire new land once the land is settled (as it is in many priority areas). Poor predictive capacity currently on how species will respond to climate change; therefore, it is difficult to know which land to include. Requires strategic planning to identify priority areas. Unlikely to ever conserve more than a small percentage of the total biodiversity.
Use contractual reserves	Less expensive per hectare than state-run reserves, especially if small areas involved.	No capital cost for land acquisition. A more cost-effective strategy to deal with small parcels of land than formal reserves. May be less detrimental to other land-based economic activities (e.g. it may be possible to mix agriculture with strategically configured migratory corridors. Similar benefits to reserve expansion, though slightly less secure. Does not require relocation of current land owners and therefore politically sounder option. Cheaper than reserve expansion, especially on agriculturally marginal land.	Less state control over the land. May require expensive administration and other infrastructure to administer. A recurring state budgetary item that may be cut in the future. May be difficult to secure long-term (indefinite) funding. This is still potentially an expensive option, particularly on land where high-value alternative land use options are available. Requires strategic planning to identify priority areas. Easier to implement on private land than on communal land. May be less effective at conserving some ecosystem processes than conventional reserves.

Table 2.4 (*continued*)

	Relative financial cost	Advantages	Disadvantages
Matrix Management Conservation outside of reserves	Some options are very inexpensive. All options are less costly than formal reserves. Because of the land area involved (potentially 5 to 10 times greater than conservation areas), the overall cost may be high.	Ensures migratory pathways even if limited information is available on priority areas. Potentially conserves the greatest amount of biodiversity. May be relatively inexpensive.	State has limited control. Land conversion will continue to threaten some species. Some species cannot be accommodated in populated areas due to human–animal conflicts.
Translocation	Relatively cheap compared to the above options, but actual costs will depend on the number of samples translocated and the species involved.	The only option for facilitated dispersers, i.e., where habitat cannot be reached by natural distribution mechanisms. Far cheaper than ensuring migratory corridors. Will still require a conservation network into which the species can be reintroduced.	Only conserves a fraction of the genetic diversity within a species. Competitive interactions with other species will be an unknown element. Does not conserve ecosystem processes, but only species. Will need a sound understanding of individual species habitats. Will require extensive research and monitoring to know which species to move, where to move them to, and what species need to be moved jointly (e.g., pollinators or seed dispersers). Potential negative impacts of translocated species on the existing species in the new habitat.
***Ex situ* conservation**	Relatively cheap once the infrastructure is in place, but varies between different types of species.	An 'insurance policy' when there is uncertainty as to how species will respond in the natural environment. The only option for 'no-hoper' species. The only option where there is total habitat loss. Relatively cheap (but the cost cannot be compared directly with *in situ* conservation as different objectives are achieved).	Conserves only a tiny fraction of genetic diversity. Conserves no ecosystem processes.

options for protecting migratory pathways: expand the existing reserve network or ensure that the matrix (in other words those areas outside of formal reserves) is sufficiently protected by measures that do not require state ownership and exclusive use of the landscape for conservation objectives. The time-slice methodology of Williams et al (2005) provides a way of identifying key areas that need conservation to ensure the movement of these species and also for identifying those species that will require facilitated dispersal.

For facilitated obligatory dispersers, the only option for maintaining wild populations is to physically move the species to the new suitable habitat (Hossell et al, 2003). While this has been undertaken to reintroduce large mammal and bird species to locations of their historic occurrence or to increase genetic exchange, the introduction of plant and invertebrate species to places where they probably did not exist within the recorded past is a new concept. Facilitated dispersal will have ethical and practical considerations such as follows:

• What is the number of individual organisms per species that need to be moved to establish a new viable population, and how should individuals for translocation be selected (Heywood and Iriondo, 2003)?
• Under what circumstances should a species be moved to an area where it did not historically exist, and what impact will this have on the species currently occurring in that area (or which will occur there naturally as a consequence of climate change) (Sakai et al, 2001; Hossell et al, 2003; Radosevich et al, 2003)?
• Which species need to be moved together, in order to preserve the community structure?
• How is the pattern of genetic variability within the population to be maintained?

Conservation of no-hopers

For the no-hopers, the only nonfatalistic option is to maintain the biodiversity in artificial situations such as zoos, botanical gardens, seed banks and through cryopreservation, in the hope of perhaps introducing them to the wild at some distant future time. Such *ex situ* conservation practices are also a wise 'insurance policy' for species with some hope of surviving in the wild.

The threat of invasive species

Some persisters, autonomous dispersers and facilitated dispersers are likely to become 'weeds', in other words overabundant in their new habitats, to the detriment of other species (McDonald, 1994). We will need to reconsider the concept of invader species given climatic change. The most likely candidates to invade are primary succession species that are well adapted to dispersal into new habitats. Weed outbreaks will be further encouraged by the disruption of communities in the receiving environment, directly or indirectly due to climate change, and by the possibility that the invasive species will travel faster than their natural competitors and controlling agents (Malcolm and Markham,

2000). Range expansion is a potential threat to existing species in the new areas, and may indirectly prevent their survival in that habitat (even if they can persist from a climatic perspective). Additionally, climate change may well favour introduced exotic species, increasing their chance of becoming invasive. A more aggressive control of invasive species may therefore be needed.

Interventions to facilitate biotic adaptation

From Figure 2.2, it is clear that no-hopers and facilitated obligatory dispersers require direct human intervention to prevent extinction. For the survival of the remaining species, conserving key areas of distribution both now and in the future and maintaining permeability of migratory pathways between protected areas would be necessary. For the autonomous obligatory dispersers the strategic combination of both ensuring conservation outside of protected areas (matrix management) and reconfiguring or expanding the conservation area would be the most effective.

Economic Considerations Relating to Adaptation Options

The costs and benefits of the various adaptation options discussed above were investigated for the *fynbos* biome, and particularly for the conservation of members of the Proteaceae. A modelling process was used to identify areas critical for conserving migratory pathways, and to identify disjunct habitats and no-hoper species (Williams et al, 2005).

Reserve expansion was found to be a very expensive option if it is used as the only mechanism of protection. Reserve costs comprise the costs of land acquisition and the annual cost of land management. Both operational costs and land management costs per unit area decrease substantially as reserve size increases.[4] Therefore, from a cost-efficiency perspective, a few large reserves are better than many small reserves (Frazee et al, 2003; Balmford et al, 2003).

In most cases contractual reserves (on private land) are more cost-effective than forming state reserves (Pence et al, 2003; A. Letsoalo, personal communication, 2005). The opportunity cost of managing the land for conservation is typically low where land is presently used for extensive rangelands or for dryland grain production but high where land is used for high value crops.[5] In the latter case a formal reserve would be more cost effective (A. Letsoalo, personal communication, 2005). In many instances, rangeland management is already biodiversity-friendly to many species, and conservation may be achieved with little or no increased cost to the rancher. Where dryland cropping is involved, a spatially explicit strategic approach would be needed to ensure that viable biodiversity corridors are achieved.

Where facilitated translocation becomes necessary, the cost is dependent on the number of organisms translocated and the establishment costs involved. Simultaneous translocation of communities of mutually interdependent organisms may have to be considered, including pollinators and seed dispersers in the case of plants.

Gene-banking and other *ex situ* conservation will not achieve the same level of biodiversity conservation as *in situ* conservation, but will remain a fall-back position when other opportunities are not available, as well as an insurance measure when they are. Where *in situ* conservation typically targets the conservation of at least 10 per cent of the historical population, *ex situ* conservation only conserves a small number of organisms for each species. Gene fingerprinting to ensure that the collection represents the broader population is therefore a significant cost consideration. Table 2.4 compares the relative economic advantages and disadvantages of the different conservation strategies.

Planning and Design Considerations in Implementing Adaptation Options for Biodiversity Conservation

Considerations for migratory corridors

In general, the movement of species will be poleward or to higher altitudes in response to global warming, but it will also be affected at the local level by changes in precipitation and microclimatic influences (Gitay et al, 2001). Species are expected to respond individually, and gradually, per generation. Therefore, resources in any designated migratory corridor must be sufficient to sustain a lifecycle, not just an individual passing through (Simberloff et al, 1992). Both Halpin (1997) and Noss (2001) have emphasized the need for firm ecological evidence on which to base corridor and buffer zone design.

Convincing ecological evidence can only be obtained by collating explicit studies on habitat use and habitat preference for a large number of species in any particular ecosystem. A key development in this field is the comprehension of the spatially explicit nature of habitat use. However, for effective corridor design, we need to understand fluxes of organisms and matter in the landscape in a spatially explicit manner. The intuitive ecological advantages of wildlife corridors suffer from a lack of empirical supporting evidence (Saunders et al, 1991; Simberloff et al, 1992).[6]

Connectivity and corridor design in a landscape with varying habitat suitability depends on the definition of habitat for a particular species. Any analysis must account for a large number of species, or groups of species, and the variables that influence habitat selection for each of them. An alternative approach is to use *processes* in landscapes as spatial planning units, and design reserves and corridors to maintain local and regional processes.[7] The assumption is that if the processes thought to be responsible for the observed heterogeneity are preserved, then heterogeneity will be maintained in the face of climate change. However, apart from knowledge of previous disturbance events, the measure of the heterogeneity to be maintained is unknown. This level of heterogeneity has been termed functional heterogeneity in the context of savanna herbivore assemblages (Owen-Smith, 2004).

Reconfiguring the reserve network

The benefits of existing formal conservation areas can be enhanced by ensuring that they are well configured to best conserve biodiversity given the impacts of climate change, for example, by the conservation of potential refugia, environmental gradients and likely migratory corridors as adaptations to the current reserve network. Systematic conservation planning can now provide land-parsimonious algorithms to prioritize new areas quantitatively for addition to the existing reserve network (Pressey and Taffs, 2001; Pressey et al, 2000 and 2001; Reyers, 2004; Rodriguez et al, 2004). The inclusion of a climate change component is, however, still in its infancy (Cowling et al, 2003; Hannah et al, 2002a and b; Williams et al, 2005). In many situations, current reserve networks are poorly planned to conserve current biodiversity patterns, let alone the additional requirements stemming from climate change.

In a first for southern Africa, Williams et al (2005) developed a method based on time-slice analysis of potential climate change-induced species migrations to understand how best to locate conservation areas in the *fynbos* biome. For the Proteacea species in this study area and a 50-year time frame, they recommend an approximate doubling of the current reserve network to achieve the required level of conservation in a changing climate, although some of this also reflects the inadequacy of existing reserve networks to conserve current biodiversity. Despite many limitations and assumptions, this study provides a powerful tool for objectively considering climate change impacts in reserve planning.

Managing areas outside of reserves (the matrix)

Matrix management should be a complementary activity to formal conservation, and one way to do this is by the creation of contractual reserves outside of formal reserves (Pence et al, 2003). Changes to legislation in South Africa now make it possible for the state to enter into a contractual arrangement with landowners to ensure conservation (Pence et al, 2003). This is potentially cheaper than outright purchase of land and, for many landowners, non-agricultural activities such as ecotourism and wildlife ranching are economically attractive because they provide better returns, especially in drier areas.

Areas outside formal reserves generally contain a significant portion of their biodiversity, often indeed more than in the reserves (Rodrigues et al, 1999). For instance, R. Biggs, B. Reyers, and R. J. Scholes (2006) estimated that 80 per cent of South Africa's biodiversity is outside of formally protected areas, despite the high levels of degradation and land transformation. It has been shown that protecting 10 per cent of the land area in the savanna landscape, as per IUCN (1993) guidelines, may only represent 60 per cent of species in an area and exclude up to 65 per cent of rare and endangered species (Reyers et al, 2002). In fact, up to 50 per cent of the land area may be needed to preserve a representative portion of species (Soule and Sanjayan, 1998).

Although South Africa has only 5.4 per cent of its land area under state conservation, it is estimated that an additional 13 per cent is currently managed as private wildlife ranches (Bond et al, 2004, updated from Cumming, 1999). Not all game-ranching practices automatically result in improved biodiversity conservation, but it is argued that, on balance, greater biodiversity benefits are achieved through this land use versus alternative agricultural practices (Taylor, 1974; Child, 1988; Bond et al, 2004). However, biodiversity is most threatened in the higher-rainfall areas where crop agriculture or forestry are more economically attractive options. In these areas, greater direct intervention may be needed to maintain biodiversity and migratory corridors.

The landscape comprising the boundaries and edges between conserved areas and the matrix is considered dynamic and permeable to water, matter, species and energy fluxes (Saunders et al, 1991).[8] These spatial linkages of energy, matter and species fluxes across edges provide additional support for biodiversity-friendly, matrix management as part of formal reserve management.

When considering the likely impacts of climate change on biodiversity,[9] matrix management practices need to anticipate an increased movement of species through the landscape, and therefore connectivity between suitable habitat patches is important. This connectivity may translate into buffer zones around existing suitable patches or linear corridor features that link suitable patches (in this chapter fragmented landscapes have been taken as a given and important component for consideration in conservation planning). For implementing matrix management for species movement an integrated procedure for determining land use, based on robust ecological evidence, is needed. Buy-in from local stakeholders is also critical since the decision to use or not use any piece of land will affect individuals.

To achieve an effective climate adaptation strategy for biodiversity using matrix management, both of the following options are considered important (adapted from Frazee et al, 2003):

1   *Strategic conservation of critically important areas of the matrix*: These are areas identified as critically important for conservation, but cannot be included into the formal conservation network for financial or other reasons. In these circumstances, the state can enter into a contractual agreement with the landowner that the land be managed for conservation purposes and the farmer could be compensated based on the opportunity cost of not undertaking the next best agricultural practice.
2   *General enhancements to biodiversity conservation on all non-reserve land*: In this instance, less costly incentives could be used to promote more biodiversity-friendly farming practices. This could include incentives as discussed below for commercial land or the establishment of CBNRM in the communal areas.

Policy mechanisms for facilitating biodiversity conservation within the matrix

Matrix management for biodiversity conservation while sustaining economic benefits may involve a variety of strategies such as setting aside riparian strips or woodland corridors, reducing the use of pesticides and fertilizers, reducing animal stocking rates, or reintroducing necessary disturbances such as fire. The wrong mix of land uses in the matrix can be inimical to conservation, for instance by increasing alien plant invasion, or by causing a retreating forest edge (Gascon et al, 1999). Due to poorly developed markets for ecological services, there is minimal incentive for landowners to promote biodiversity or maintain migratory corridors. Perverse policies for protecting threatened species that disadvantage landowners may even result in them deliberately reducing biodiversity on their land. Land tenure is important for developing matrix management interventions and a different set of incentives and approaches would be applicable for private as opposed to communal lands.

Incentives for matrix management on privately held land

Shogren et al (2003) and Doremus (2003) suggest the following policy and economic incentive systems for promoting biodiversity on private land:

- *Education*: Many landowners have a conservation ethic, and if educated about the pertinent issues, may change land management practices to meet biodiversity conservation needs, provided costs are low.
- *Direct incentives*: Positive incentives include cash payments, zero rating land tax on key conservation areas (Pence et al, 2003) or debt forgiveness. An example of negative incentives is taxes for poor land use.
- *Approval and recognition*: Competitive awards for conservation activities can be an incentive. An example, here is the landowner-targeted programme to promote raptor breeding in Kimberley, South Africa.
- *Market creation or improvement*: The state can create markets for environmental services, for example carbon credits, promotion of ecotourism, provision of information on markets, and the introduction of certification schemes (such as 'badger friendly' honey).
- *Tradable development rights*: Landholders are granted a fixed amount of tradable development rights. This creates a market value for resources.
- *Regulatory control*: The enactment and enforcement of laws, including the types of social prohibitions that served this function historically.

Inappropriate agricultural subsidies need to be removed. For example, large direct and indirect subsidies previously made cattle ranching economically viable (Child 1988; Bond et al, 2004), but now wildlife management is a better option for the arid and semi-arid areas in South Africa, Namibia and Zimbabwe.

Matrix management on communal land

The management of shared resources is referred to as 'common property resource management'. A common property resource is defined as any resource that is subject to individual or group use but not to individual ownership and is used under some arrangement of community or group management (Mol and Wiersum, 1993). Despite concerns about overexploitation of such communal resources (Hardin, 1968), there is evidence that degradation is not an inevitable outcome of group management (Bromley and Cernea, 1989; Lawry, 1990; Ostrom 1992). A number of criteria have been identified under which group management is most likely to be successful (see, for example, Baland and Platteau, 1996; IFAD, 1995; Ostrom, 1992; Wade, 1987; Lawry, 1990; Cousins, 1996; Shackleton et al, 2002).

Changes in human population density and resource use patterns have resulted in the evolution of the communal property resource management theory into a new paradigm of community-based natural resource management (CBNRM). Most southern African states now have some form of CBNRM programme (Murphy, 1997; Fabricius et al, 2004), partly due to support from official development aid agencies.[10] Early CBNRM programmes emphasized the need to devolve ownership and management to the lowest possible level but it was later recognized that though important, this devolution of power on its own was not sufficient to initiate successful CBNRM. Even though the CBNRM programmes have not always been successful, Fabricius et al (2004) believe they can potentially achieve both community development and increased sustainability of natural resources. They identify seven principles paramount to a sustainable CBNRM:

1   a diverse and flexible range of livelihood options is maintained;
2   the production potential of the resource base is maintained or improved;
3   institutions for local governance and resource management are in place and effective;
4   economic and other benefits to provide an incentive for wise use of resources exist;
5   there are effective policies and laws, these are implemented, and the authority is handed down to the lowest level where there is the capability to apply it;
6   there is sensible and responsible outside facilitation; and
7   local-level power relations are favourable to CBNRM and are understood.

In southern Africa, the principles of CBNRM are presently being implemented in all of the transfrontier parks; the Wild Coast Initiative in South Africa; Administrative Management Design for Game Management Areas (Zambia); Communal Areas Management Program for Indigenous Resources (CAMP-FIRE) (Zimbabwe); Community-Based Natural Resource Management Program in Conservancies (Namibia); and Community-Based Natural Resource Management Program in Controlled Hunting Areas (Botswana).

Conclusions

The conservation network in southern Africa, although extensive, is poorly configured to adequately conserve biodiversity, even less so under a climatically changing environment. The largest proportion of biodiversity is still found outside of the reserve areas, despite the impacts of land transformation and degradation. With the anticipated impacts of climate change over the next 50 to 200 years, many species will have to move from their current locations to track areas with suitable climates. To facilitate this process and minimize species loss, a multitude of strategies will be necessary, such as realignment of reserves; ensuring that land use outside of reserves is biodiversity friendly; facilitated translocations, where species are unable to move on their own; and *ex situ* conservation as a precautionary measure and for species with no future habitats.

In a climatically changing environment, strategic conservation becomes a shifting target, and it is therefore important to protect the migratory corridors and not simply a single end point. It is also important to realize that the entire habitat will not move, but rather individual species will move at different rates, creating new habitat structures. Cost considerations and difficulties in acquiring new areas for reserves may, however, pose barriers in reconfiguring the reserve network. The most cost-effective mechanism to both conserve biodiversity and allow species movement to new habitats is to ensure the permeability of areas between reserves to species migration. Economically, formal reserve expansion is viable where opportunity costs of alternative land use options are high, where large areas are involved, where there are clearly defined gradients needing protection, and where high levels of biodiversity loss can be prevented through reserve realignment. Contractual reserves should be considered for more marginal areas.

Within southern Africa, the mechanisms to ensure biodiversity-friendly management of the matrix are likely to differ significantly between privately owned areas and communal lands. Direct incentives such as tax rebates, assistance with vegetation management and education may be sufficient to change behaviour on private land. Allowing private ownership of wildlife has greatly increased the extent of private game ranches. Contractual reserves, where the state compensates private land owners to manage portions of their farms as areas for biodiversity conservation, are also an option. On communal land, practices based on CBNRM principles are the likely solutions.

Biodiversity-friendly practices may also be promoted by the removal of distorted market forces that encourage inappropriate agricultural practices. Use of management tools such as fire and grazing intensity (including grazing by mega-herbivores, such as elephants), can help maintain habitat functionality in a state similar to the present. It is the landscapes profoundly transformed for crop production that pose the greatest challenges for biodiversity conservation.

Where the movement of species to new suitable habitats must be facilitated, of greater concern will be the movement of smaller animals, insects and plants. For species with no suitable future habitats, *ex situ* conservation is the only option to prevent extinction, although it can only conserve small populations of individual species and does not conserve ecological function.

A radical change in current thinking about conservation planning will therefore be necessary since simply maintaining the status quo might result in species extinction under a climatically changed future. Ongoing monitoring, research and model improvement will be necessary to better understand the response of biodiversity to a changing climate and address the present uncertainties. Fortunately, there are many areas in which our current understanding is sufficient for us to begin planning for biodiversity conservation in a climatically changing environment.

Notes

1   Species typically tend to occupy their 'realized niche', which is a subset of their 'fundamental niche' (the range determined by their physiological tolerance limits) resulting from the outcome of interactions with other species. The degree to which species distribution can be predicted based on their climatically defined habitat niche differs between species (see, for example, Thuiller et al, 2006).
2   Studies show that the fertilization effect of increasing atmospheric concentration of CO_2 (primarily responsible for the current anthropogenic climate change) starts to saturate in natural ecosystems at around 500 ppm (Scholes et al, 1999).
3   The C3 plants produce a three-carbon compound in the photosynthetic process and include most trees and common crops like rice, wheat, barley, soybeans, potatoes and vegetables. The C4 category of plants produce a four carbon compound in the photosynthetic process and include grasses and crops like maize, sugar cane, sorghum and millet. Under increased atmospheric concentrations of CO_2, C3 plants have been shown to be more responsive than C4 plants (see IPCC, 2001).
4   On the basis of South African National Park data, a 1km² park has a US$104,793 annual operational cost, while a 100,000 km² park only costs US$66/km² (Martin, 2003). The land management cost per hectare decreases nonlinearly as the reserve size increases.
5   The cost of managing the land for conservation is the opportunity cost of lost income to the farmer for not using the land for the most profitable alternative land use activity. This cost will be very low for extensive rangelands, low for dryland grain production, but high for irrigated crops and speciality crops such as horticulture.
6   An often-stated example of the usefulness of corridors is riparian vegetation. However, Simberloff et al (1992) state that riparian vegetation does not constitute a typical corridor from a management point of view, as it is a unique habitat in itself that happens to be linear, and it does not connect discrete patches of like habitat.
7   An excellent example of using such processes in conservation planning is found in Rouget et al (2003a and b).
8   The process of forming such a landscape has been termed habitat variegation (McIntyre and Barrett, 1992) and Murphy and Lovett-Doust (2004) have expressed a similar viewpoint that a binary approach of suitable habitat versus the matrix is not a true reflection of landscape dynamics.
9   Biodiversity responses to climate change may take a variety of forms, and our current ability to predict this is limited due to uncertainties in both the climate scenarios and in how species will react to the change (reviewed by Walther et al, 2002; McCarty, 2001; Hughes, 2000; Parmesan and Yohe, 2003; Root et al, 2003).
10 The Communal Areas Management Programme for Indigenous Resources (CAMPFIRE) was initiated in Zimbabwe in the early 1980s as one of the first experiments in this regard.

References

ABSA (2003) *Game Ranch Profitability in Southern Africa*, ABSA Group Economic Research, South Africa Financial Sector Forum, Rivonia, South Africa, available at www.finforum.co.za/absa/investment.html#game%20ranch%20profitability

Baland, J. and J. Platteau (1996) *Halting Degradation of Natural Resources. Is there a Role for Rural Communities?*, Food and Agricultural Organization and Clarendon Press, Oxford, UK

Balmford, A., L. Moore, T. Brooks, N. Burgess, L. A. Hansen, P. H. Williams and C.

Rahbek (2001) 'Conservation conflicts across Africa', *Science*, vol 291, pp2616–2619

Balmford, A., K. J. Gaston, S. Blyth, A. James and V. Kapos (2003) 'Global variation in terrestrial conservation costs, conservation benefits, and unmet conservation needs', *Proceedings of the National Academy of Sciences*, vol 100, pp1046–1050

Biggs, R., B. Reyers and R. J. Schole (2006) 'A biodiversity intactness score for South Africa', *South African Journal of Science*, vol 102, pp277–283, July/August

Bond, I., B. Child, D. de la Harpe, B. Jones, J. Barnes and H. Anderson (2004) 'Private land contribution to conservation in South Africa', in B. Child (ed) *Parks in Transition: Biodiversity, Rural Development and the Bottom Line*, IUCN, SASUSG and Earthscan, London and Sterling, VA

Bromley, D. W. and M. M. Cernea (1989) 'The management of common property natural resources: Some conceptual and operational fallacies', World Bank discussion papers, no 57, World Bank, Washington, DC

Büscher, B. (2005) 'Conjunctions of governance: The state and the conservation-development nexus in southern Africa', *Journal of Transdisciplinary Environmental Studies*, vol 4, pp1–15

Bush, M. B. (2002) 'Distributional change and conservation on the Andean flank: A palaeoecological perspective', *Global Ecology and Biogeography*, vol 11, pp463–473

Child, B. (1988) 'The role of wildlife utilization in the sustainable development of semi-arid rangelands in Zimbabwe', D.Phil. thesis, Oxford University, Oxford, UK

Child, B. (ed) (2004) *Parks in Transition: Biodiversity, Rural Development and the Bottom Line*, Earthscan, London

Cousins, B. (1996) 'Livestock production and common property struggles in South Africa's agrarian reform', *The Journal of Peasant Studies*, vol 23, pp166–208

Cowling, R. M., R. L. Presseyb, M. Rougetc and A. T. Lombarda (2003) 'A conservation plan for a global biodiversity hotspot—The Cape Floristic Region, South Africa', *Biological Conservation*, vol 112, pp191–216

Cowling, R. M. and R. L. Pressey (2003) 'Introduction to systematic conservation planning in the Cape Floristic Region, *Biological Conservation*, vol 112, pp1–13

Cubasch, U., G. A. Meehl, G. J. Boer, R. J. Stouffer, M. Dix, A. Noda, C. A. Senior, S. Raper and K. S. Yap (2001) 'Projections of future climate change', in J. T. Houghton, Y. Ding, D. J. Griggs, M. Noguer, P. van der Linden, X. Dai, K. Maskell and C. I. Johnson (eds) *Climate Change 2001: The Scientific Basis*, Contribution of Working Group I to the Third Assessment Report of the Intergovernmental Panel on Climate Change, Cambridge University Press, New York, pp525–582

Cumming, D. (1999) 'Study on the development of transboundry natural resource management areas in Southern Africa, environmental context: Natural resources, land use, and conservation', Biodiversity Support Program, Washington, D.C.

Davis, M. B. and R. G. Shaw (2001) 'Range shifts and adaptive responses to 36 Quaternary climate change', *Science*, vol 292, pp673–679

Doremus, H. (2003) 'A policy portfolio approach to biodiversity protection on private lands', *Environmental Science and Policy*, vol 6, pp217–232

Dudley, N. and S. Stolton (2003) 'Ecological and socioeconomic benefits of protected areas in dealing with climate change', in L. J. Hansen, J. L. Biringer and J. R. Hoffmans (eds) *Buying time: A User's Manual for Building Resistance and Resilience to Climate Change in Natural Systems*, World Wildlife Fund, Gland, Switzerland

Fabricius, C., E. Koch, H. Maome and S. Turner (eds) (2004) *Rights Resources and Rural Development: Community-based Natural Resource Management in Southern Africa*, Earthscan, London and Sterling, VA

Frazee, S., R. M. Cowling, R. L. Pressey, J. K. Turpie and N. Lindenberg (2003) 'Estimating the costs of conserving a biodiversity hotspot: A case study of the Cape Floristic region, South Africa', *Biological Conservation*, vol 112, pp275–290

Gascon, C., T. Lovejoy, R. O. Bierregaard, J. R. Malcolm, P. C. Stouffer, H. L.

Vasconcelos, W. F. Laurance, B. Zimmerman, M. Tocher and S. Borges (1999) 'Matrix habitat and species richness in tropical forest remnants', *Biological Conservation*, vol 91, pp223–229

Gaston, K. J. (2003) *The Structure and Dynamics of Geographic Ranges*, Oxford Series in Ecology and Evolution, Oxford University Press, New York

Gelbard, J. L. (2003) 'Grasslands at a crossroads: Protecting and enhancing resilience to climate change', in L. J. Hansen, J. L. Biringer and J. R. Hoffmans (eds) *Buying Time: A User's Manual for Building Resistance and Resilience to Climate Change in Natural Systems*, World Wildlife Fund, Gland, Switzerland

Gitay, H., S. Brown, W. Easterling and B. Jallow (2001) 'Ecosystems and their goods and services', in J. J. McCarthy, O. F. Canzini, N.A. Leary, D.J. Dokken and K.S. White (eds) *Climate Change: Impacts, Adaptations, and Vulnerability*, Cambridge University Press, Cambridge, pp235–342

Gitay, H., A. Suarez, R. Watson and D. J. Dokken (eds) (2002) *Climate Change and Biodiversity*, IPCC Technical Paper V, Intergovernmental Panel on Climate Change, Geneva, Switzerland

Graham, R. W. (1992) 'Late Pleistocene faunal changes as a guide to understanding effects of greenhouse warming on mammalian fauna of north America', in R. L. Peters and T. Lovejoy (eds) *Global Warming and Biological Diversity*, Yale University, New Haven, CT, pp76–87

Graham, R. W. and E. C. Grimm (1990) 'Effects of global climate change on the patterns of terrestrial biological communities', *Trends in Ecology and Evolution*, vol 5, pp289–292

Halpin, P. N. (1997) 'Global climate change and natural-area protection: Management responses and research directions', *Ecological Applications*, vol 7, pp828–843

Hannah, L., G. F. Midgley, T. Lovejoy, W. J. Bond, M. Bush, J. C. Lovett, D. Scott and F. I. Woodward (2002a) 'Conservation of biodiversity in a changing climate', *Conservation Biology*, vol 16, pp264–268

Hannah, L., G. F. Midgley and D. Millar (2002b) 'Climate change-integrated conservation strategies', *Global Ecology and Biogeography*, vol 11, pp485–495

Hardin, G. (1968) 'The tragedy of the commons', *Science*, vol 162, pp1243–1248

Hewitson, B. C. and R. G. Crane (2006) 'Consensus between GCM climate change projections with empirical downscaling', *International Journal of Climatology*, in press

Heywood, V. H. and J. M. Iriondo (2003) 'Plant conservation: Old problems, new perspectives', *Biological Conservation*, vol 113, pp321–335

Hossell, J.E., B. Briggs and I. R. Hepburn (2003) *Climate Change and UK Nature Conservation: A Review of the Impact of Climate Change on UK Species and Habitat Conservation Policy*, Department of Transport, Environment and the Regions Wildlife and Countryside Directorate, Bristol, UK

Hughes, L. (2000) 'Biological consequences of global warming: Is the signal already apparent?', *Trends in Ecology and Evolution*, vol 15, pp56–61

Huntley, B. (1991) 'How plants respond to climate change: Migration rates, individualism and the consequences for plant communities', *Annals Botany.*, vol 67 (Supplement 1), pp15–22

Hutton, J., W. M. Adams and J. C. Murombedzi (2005) 'Back to the barriers? Changing narratives in biodiversity conservation', *Forum for Development Studies*, vol 32, pp341–370, Norsk Utenrikspolitisk Institutt, www.nupi.no

IFAD (International Fund for Agricultural Development) (1995) *Common Property Resources and the Rural Poor in Sub-Saharan Africa*, Special Programme for Sub-Saharan African Countries Affected by Drought and Desertification, International Fund for Agricultural Development., IFAD, Rome

IPCC (Intergovernmental Panel on Climate Change) (2001) 'Glossary of terms', in J. McCarthy, O. Canziani, N. Leary, D. Dokken and K. White (eds) *Climate Change*

2001: Impacts, Adaptation, and Vulnerability, Contribution of Working Group II to the Third Assessment Report of the Intergovernmental Panel on Climate Change, Cambridge University Press, Cambridge, UK and New York, US

IUCN (1993) *Parks for Life*, report of the Fourth World Congress on National Parks and Protected Areas, IUCN, Gland, Switzerland

Lawry, S. W. (1990) 'Tenure policy towards common property natural resources in sub-Saharan Africa', *Natural Resources Journal*, vol 30, pp403–422

Lindenmayer, D. B. and H. Nix (1993) 'Ecological principles for the design of wildlife corridors', *Conservation Biology*, vol 7, pp627–630

Low, B. and A. G. Rebelo (eds) (1996) *Vegetation of South Africa, Lesotho and Swaziland*, Department of Environmental Affairs and Tourism, Pretoria, South Africa

Malcolm, J. R. and A. Markham (2000) *Global Warming and Terrestrial Biodiversity Decline*, World Wildlife Fund, Washington, D.C.

MA (2005) 'Millennium ecosystem assessment', www.millenniumassessment.org

Margules, C. R. and R. L. Pressey (2000) 'Systematic conservation planning', *Nature*, vol 405, pp243–253

Martin, R. (2003) 'Annual recurrent expenditure for conservation and management: A preliminary investigation of the use of a spreadsheet model to generate staff establishment and operating budget for individual parks', South African National Parks, Pretoria, South Africa

McCarty, J. P. (2001) 'Ecological consequences of recent climate change', *Conservation Biology*, vol 15, pp320–331

McDonald, I. A. W. (1994) 'Global change and alien invasion, implications for biodiversity and protected area management', in O. T. Solbrig, P. G. van Emden and W. J. van Oordt (eds) *Biodiversity and Global Change*, CAB International, Wallingford, Oxon, UK

McIntyre, S. and G. W. Barrett (1992) 'Habitat variegation: An alternative to fragmentation', *Conservation Biology*, vol 6, pp146–147

Midgley, G. F., L. Hannah, D. Millar, W. Thuiller and A. Boot (2002a) 'Developing regional species-level assessments of climate change impacts on biodiversity in the Cape Floristic region', *Biological Conservation*, vol 112, pp87–97

Midgley, G. F., L. Hannah, D. Millar, M. C. Rutherford and L. W. Powrie (2002b) 'Assessing the vulnerability of species richness to anthropogenic climate change in a biodiversity hotspot', *Global Ecology and Biogeography*, vol 11, pp445–451

Midgley, G. F., L. Hannah, D. Millar, W. Thuiller and A. Boot (2003) 'Developing regional and species-level assessments of climate change impacts on biodiversity: A preliminary study in the Cape Floristic region', *Biological Conservation*, vol 112, pp87–97

Mol, P. W. and K. F. Wiersum (1993) 'Common forest resource management: Asia', in D. A. Messerschmidt (ed) *Common Forest Resource Management: Annotated Bibliography of Asia, Africa and Latin America*, Community Forestry Note No 11, Food and Agricultural Oganization of the United Nations, Rome

Murphy, H. T. and J. Lovett-Doust (2004) 'Context and connectivity in plant metapopulations and landscape mosaics: Does the matrix matter?' *OIKOS*, vol 105, pp3–14

Murphy, M. W. (1997) 'Congruent objectives, comparing interests and strategic compromise: Concepts and processes in the evolution of Zimbabwe's CAMPFIRE programme', Community Conservation in Africa Working Paper No 2, Institute for Development Policy and Management (IDPM),University of Manchester, Manchester, UK

Myers, N., R. A. Mittermeier, C. G. Mittermeier, G. A. B. da Fonseca and J. Kent (2000) 'Biodiversity hotspots for conservation priorities', *Nature*, vol 403, pp853–858

Naidoo, R. and W. L. Adamowicz (2006) 'Modeling opportunity costs of conservation in transitional landscapes', *Conservation Biology*, vol 20, pp490–500

Noss, R. F. (2001) 'Beyond Kyoto: Forest management in a time of rapid forest change', *Conservation Biology*, vol 15, pp578–590

Olson, D. M., E. Dinerstein, E. D. Wikramanayake, N. D. Burgess, G. V. N. Powell, E. C. Underwood, J. A. D'amico, I. Itoua, H. E. Strand, J. C. Morrison, C. J. Loucks, T. F. Allnutt, T. H. Ricketts, Y. Kura, J. F. Lamoreux, W. W. Wettengel, P. Hedao and K. R. Kassem (2001) 'Terrestrial ecoregions of the world: A new map of life on Earth', *BioScience*, vol 51, pp933–938

Orme, C. D. L., R. G. Davies, M. Burgess, F. Eigenbrod, N. Pickup, V. A. Olson, A. J. Webster, T. S. Ding, P. C. Rasmussen, R. S. Ridgely, A. J. Stattersfield, P. M. Bennett, T. M. Blackburn, K. J. Gaston and I. P. F. Owens (2005) 'Global hotspots of species richness are not congruent with endemism or threat', *Nature*, vol 436, pp1016–1019

Ostrom, E. (1992) 'The rudiments of a theory of the origins, survival, and performance of common property institutions', in D. W. Bromley (ed) *Making the Commons Work: Theory, Practice, Policy*, Institute for Contemporary Studies, San Francisco, CA

Owen-Smith, N. (2004) 'Functional heterogeneity in resources within landscapes and herbivore population dynamics', *Landscape Ecology*, vol 19, pp761–771

Parmesan, C. and G. Yohe (2003) 'A globally coherent fingerprint of climate change impacts across natural systems', *Nature*, vol 421, pp37–42

Pence, G. Q. K., M. A. Botha and J. K. Turpie (2003) 'Evaluating combinations of on- and off-reserve conservation strategies for the Agulhas Plain, South Africa: A financial perspective, *Biological Conservation*, vol 112, pp253–273

Pressey, R. L., C. J. Humphries, C. R. Margules, R. I. Vane-Wright and P. H. Williams (1993) 'Beyond opportunism: Key principles for systematic reserve selection, *Trends in Ecology and Evolution*, vol 8, pp124–128

Pressey, R. L. and K. H. Taffs (2001) 'Scheduling conservation action in production landscapes: priority areas in western New South Wales', *Biological Conservation*, vol 100, pp355–376

Pressey, R. L., R. M. Cowling and M. Rouget (2001) 'Formulating conservation targets for biodiversity pattern and process in the Cape Floristic region, South Africa', *Biological Conservation*, vol 112, pp99–127

Pressey, R. L, T. C. Hager, K. M. Ryan, J. Scharz, S. Wall, S. Ferrier and P. M. Creaser (2000) 'Using abiotic data for conservation assessments over extended regions: Quantitative methods applied across New South Wales, Australia', *Biological Conservation*, vol 96, pp55–82

Pressey, R. L. and R. M. Cowling (2001) 'Reserve selection algorithms and the real world', *Conservation Biology*, vol 15, pp275–277

Radosevich, S. R., M. M. Stubbs and C. M. Ghersa (2003) 'Plant invasions – Processes and patterns' *Weed Science*, vol 51, pp254–259

Reyers, B. (2004) 'Incorporating anthropogenic threats into evaluations of regional biodiversity and prioritisation of conservation areas in the Limpopo Province, South Africa', *Biological Conservation*, vol 118, pp521–531

Rodrigues, A. S. L., R. Tratt, B. D. Wheeler and K. J. Gaston (1999) 'The performance of existing networks of conservation areas in representing biodiversity', *Proceedings of the Royal Society, London*, vol B 266, pp1453–1460.

Rodrigues, A. S. L., S. J. Andelman, M. I. Bakarr, L. Boitani, T. M. Brooks, R. M. Cowling, L. D. C. Fishpool, G. A. B. Da Fonseca, K. J. Gaston, M. Hoffmann, J. S. Long, P. A. Marquet, J. D. Pilgrim, R. L. Pressey, J. Schipper, W. Sechrest, S. N. Stuart, L. G. Underhill, R. W. Waller, M. E. J. Watts and X. Yan (2004) 'Effectiveness of the global protected area network in representing species diversity', *Nature*, vol 428, pp640–643

Root, T. L., J. T. Price, K. R. Hall, S. H. Schneider, C. Rosenzweig and A. J. Pounds (2003) 'Fingerprints of global warming on wild animals and plants', *Nature*, vol 421, pp57–60

Rouget, M., D. M. Richardson, R. M. Cowling, J. W. Lloyd and A. T. Lombard (2003a) 'Current patterns of habitat transformation and future threats to biodiversity in terrestrial ecosystems of the Cape Floristic region, South Africa', *Biological Conservation*, vol 112, pp63–85

Rouget, M., D. M. Richardson and R. M. Cowling (2003b) 'The current configuration of protected areas in the Cape Floristic region, South Africa – Reservation bias and representation of biodiversity patterns and processes', *Biological Conservation*, vol 112, pp129–145

Rutherford. M. C., G. F. Midgeley, W. J. Bond, L. W. Powrie, R. Roberts and L. Allsopp (1999) *Plant Biodiversity: Vulnerability and Adaptation Assessment. South African Country Study on Climate Change*, National Botanical Institute, Cape Town, South Africa

Reyers, B., D. H. K. Fairbanks, K. J. Wessels and A. S. Van Jaarsveld (2002) 'A multi-criteria approach to reserve selection: Addressing long-term biodiversity maintenance', *Biodiversity and Conservation*, vol 11, pp769–793

Sakai, A. K., F. W. Allendorf, J. S. Holt, D. M. Lodge and K. A. J. Molofsky (2001) 'The population biology of invasive species', *Annual Review of Ecology and Systematics*, vol 32, pp305–332

Saunders, D. A., R. J. Hobbs and C. R. Margules (1991) 'Biological consequences of ecosystem fragmentation: A review', *Conservation Biology*, vol 5, pp18–32

Scholes R. J., E. D. Schulze, L. F. Pitelka and D. O. Hall (1999) 'The biogeochemistry of terrestrial ecosystems', in B. H. Walker, W. L. Steffen, J. Canadell and J. S. I Ingram (eds) *The Terrestrial Biosphere and Global Change*, Cambridge University Press, New York, pp88–105

Scholes, R. J. and R. Biggs (eds) (2004) *The Regional Scale Component of the Southern African Millennium Ecosystem Assessment*, CSIR, Pretoria, South Africa

Shackleton, S., G. P. von Maltitz and J. P. Evans (2002) 'Factors, conditions and criteria for the successful management of natural resources held under a common property regime: A South African perspective', occasional report no 8, Plaas School of Government, University of the Western Cape, South Africa

Shogren J. F., G. M. Parkhurst and C. Settle (2003) 'Integrating economics and ecology to protect nature on private lands: Models, methods, and mindsets', *Environmental Science and Policy*, vol 6, pp233–242

Simberloff, D., J. A. Farr, J. Cox and D. W. Mehlman (1992) 'Movement corridors: Conservation bargains or poor investments?' *Conservation Biology*, vol 6, pp6493–6505

Soule, M. E. and M. A. Sanjayan (1998) 'Conservation targets: Do they help?' *Science*, vol 279, pp2060–2061

Taylor, R. D. (1974) 'A comparative study of land use on the cattle and game ranch in the Rhodesian lowveld', MSC thesis, University of Rhodesia, Harare, Zimbabwe

Thomas, C. D., A. Cameron, R. E. Green, M. Bakkenes, L. J. Beaumont, Y. C. Collingham, B. F. N. Erasmus, M. F. de Siqueira, A. Grainger, L. Hannah, L. Hughes, B. Huntley, A. S. van Jaarsveld, G. F. Midgley, L. Miles, M. A. Ortega-Huerta, A. Townsend Peterson, O. L. Phillips and S. E. Williams (2004) 'Extinction risk from climate change', *Nature*, vol 427, pp145–148

Thuiller, W., O. Broennimann, G. Hughes, J. R. W. Alkemade, G. F. Midgley and F. Corsi (2006) 'Vulnerability of African mammals to anthropogenic climate change under conservative land transformation assumptions', *Global Change Biology*, vol 12, pp12,424–12,444

van der Linde, H., J. Oglethorpe, T. Sandwith, D. Snelson and Y. Tessema (with contributions from A. Tiéga and T. Price) (2001) *Beyond Boundaries: Transboundary Natural Resource Management in Sub-Saharan Africa*, Biodiversity Support Program, Washington, DC

von Maltitz, G. P. and S. E. Shackleton (2004) 'Use and management of forests and woodlands in South Africa: Stakeholders, institutions and processes from past to present', in M. J. Lawes, H. A. C. Ealey, C. M. Shackleton and B. G. S. Geach (eds)

Indigenous Forests and Woodlands in South Africa, Policy, People and Practice, University of KwaZulu-Natal Press, Scottsville, South Africa

Wade, R. (1987) 'The management of common property resources: Collective action as an alternative to privatisation or state regulation', *Cambridge Journal of Economics*, vol 11, pp95–106

Walther, G. R., E. Post, P. Convey, A. Menzel, C. Parmesan, T. J. C. Beebee, J. M. Fromentin, O. Hoegh-Guldberg and F. Bairlein (2002) 'Ecological responses to recent climate change', *Nature*, vol 416, pp389–395

WDPA (2005) *2005 World Database on Protected Areas*, Centre for Applied Biodiversity Science, Conservation International, Washington, DC

Williams, P., L. Hannah, S. Andelman, G. Midgley, M. Araujo, G. Hughes, L. Manne, E. Marinez-Meyer and R. Pearson (2005) 'Planning for climate change: Identifying minimum-dispersal corridors from the Cape Proteaceae', *Conservation Biology*, vol 19, pp1063–1074

Wilshusen, P. R., S. R. Brechin, C. L. Fortwangler and P. C. West (2002) 'Reinventing a square wheel: Critique of a resurgent "protection paradigm" in international biodiversity conservation', *Society and Natural Resources*, vol 15, pp17–40

3

Benefits and Costs of Adapting Water Planning and Management to Climate Change and Water Demand Growth in the Western Cape of South Africa

John M. Callaway, Daniël B. Louw, Jabavu C. Nkomo,
Molly E. Hellmuth and Debbie A. Sparks

Introduction

The Berg river basin, located in the Western Cape Region of South Africa, provides the bulk of the water for household, commercial and industrial use in the Cape Town metropolitan region as well as irrigation water to the lower part of the basin to cultivate roughly 15,000 hectares of high-value crops. Since the early 1970s, water consumption in municipal Cape Town has grown at an average annual rate of about 3 per cent. As the population of the metropolitan Cape Town region grows, the competition for water in the basin has become intense, and farmers have responded by dramatically improving their irrigation efficiencies and shifting even more land into the production of high-value export crops. Meanwhile, over the past two decades, a number of national and regional commissions have been set up to investigate options for coping with the long-term water supply problems in the basin.

One outcome of these efforts was the authorization of the Berg River Dam. In June 2004, after almost 20 years of debate about its economic feasibility and environmental impacts, final agreement was reached on construction of the dam. It will consist of a 130.1 million cubic metre storage reservoir and a pumping site to pump water from below the dam back to it. The dam is expected to be operational sometime during the period 2008–2010.

The region has also recently experienced a number of unusually dry years, the most recent in the summer of 1994–1995, when peak storage in the upper basin was only about one-third of average. At the same time, concerns about the effects of global warming on basin runoff have been growing, along with suggestions that recent climatic anomalies may be associated with regional climate change. Not surprisingly, planning for the Berg River Dam and other water supply and demand options in the basin has, up until this point, failed to take into account the possibility that the build-up of greenhouse gases in the global atmosphere is already affecting and will continue to affect the regional climate by reducing runoff in the basin.

The context for our analysis of water planning and management in the Berg river basin consists of three main elements. The first is the increasing competition for water between urban and agricultural water users because of growing urban water demands; second is the threat of unusual climate variability and/or climate change to exacerbate that competition; and third is the planning and policy responses to these issues. To address these three elements, we developed a policy-planning model that can be used to evaluate a wide range of structural, non-structural and technological measures for coping with basin water shortages due to demand growth, climate variability and climate change. The model we have developed to do this is called the Berg River Dynamic Spatial Equilibrium Model (BRDSEM) and is described in detail in Callaway et al (2006). In this chapter we describe very briefly the model; illustrate some of the ways the model can be used to assess the benefits, costs and risks of avoiding climate change damages by increasing the maximum storage capacity of the reservoir and/or implementing a system of efficient water markets; and present the results and major conclusions of our analysis for three deterministic climate scenarios. We also describe the limitations of the current version of the model and analysis methods and outline future plans for improving the model and analytical methods.

The Model

BRDSEM is a dynamic, multiregional, nonlinear programming model patterned after the hydro-economic surface water allocation models developed by Vaux and Howitt for California (1984), Booker (1990) and Booker and Young (1991 and 1994) for the Colorado River Basin, and Hurd et al (1999 and 2004) for the Missouri, Delaware and Apalachicola-Flint-Chattahoochee river basins in the United States. This type of model is a more specific application of spatial and temporal price and allocation models that originated from the work of Samuelson (1951) and Takayama and Judge (1971), which have been widely applied in many natural resource sectors (McCarl and Spreen, 1980). Hydro-economic models have been used by Hurd et al (1999 and 2004) to estimate the economic impacts of climate change in the four large US river basins. In these two studies, the authors provided estimates for the economic losses due to climate change that took into account only short-run adjustments to climate variability. They did not specifically assess the benefits of long-run measures for avoiding climate change damages, such as investments in storage capacity and changes in water allocation institutions and laws.

Building on this past work, BRDSEM was designed specifically to estimate the economic value of the damages due to climate change with and without long-run adaptation measures and thereby to isolate the benefits and costs of avoiding climate change damages. This chapter represents the first attempt, to the best of our knowledge, to quantify the benefits of avoiding climate change damages in economic terms in a water basin context using such a model.

BRDSEM is also an extension of a static spatial equilibrium model developed by Louw (2002) for the Berg river basin to examine the potential of water markets in the region (Louw and Van Schalkwyk, 2001). Much of the data from that model is used in BRDSEM, but is being updated on a continuous basis. Significant modifications were made to Louw's model to add the spatial relationships between runoff, water storage, water conveyance, transfers, return flows, and water use in the natural and man-made hydrological systems and to account for the intertemporal aspects of reservoir operation in the upper and lower parts of the basin.

One of the important features added to BRDSEM is that it can determine, endogenously, the optimal (in other words economically efficient) capacity of planned reservoirs and other structural works. The model does this by satisfying an economic efficiency (Kuhn-Tucker) condition, namely that the marginal capital cost associated with the optimal storage capacity is equal to the present value of the shadow price of storage in all periods where storage levels are at this maximum. The maximum capacity is determined in year one for all future periods (i.e. perfect foresight) and remains fixed thereafter.

The structure of the BRDSEM model is shown in Figure 3.1. The core of the model consists of four main elements linked together in a nonlinear mathematical programming framework. These include a nonlinear (quadratic) objective function that characterizes the normative objectives of the agents in the model; an intertemporal, spatial equilibrium module that characterizes the spatially distributed flow of water and water storage in the basin; an urban water demand and supply module; and a regional farm irrigation demand module. External information inputs to BRDSEM include downscaled climate scenarios from a regional climate model; monthly runoff, surface water evaporation coefficients and crop water use adjustment factors from a regional hydrologic model (WATBAL, Yates, 1996); and information about policies, plans and technologies used to alter various parameters in the programming model to reflect alternative demand- and supply-side choices and constraints. Outputs of the model include measures of the economic value of water, optimal storage capacity of the reservoir, shadow prices for water transfers, monthly reservoir storage, releases and transfers, and monthly water diversions and consumptive use by urban and irrigation users by region.

Model Application: Methods and Scenarios

Economic framework and methods

The economic framework used in this chapter for evaluating the costs and benefits of measures to avoid climate change damages was first presented in Callaway et al (1998). It was extended to link adaptation to climate variability and climate change and to situations in which 'regrets' occur when the climate that is realized, ex post, is not the same as the climate planners and policy

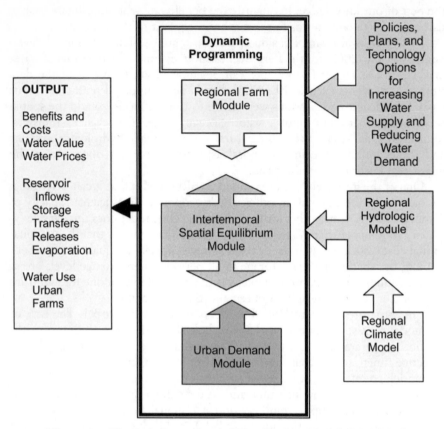

Figure 3.1 *The Berg River Spatial Equilibrium Model (BRDSEM)*

makers anticipated in formulating and implementing their plans and policies, ex ante (Callaway, 2003 and 2004a and b).

Table 3.1 illustrates the basic framework for estimating various benefits and costs associated with climate change and adaptation for a simple case of two climate states, the existing climate (C_0) and climate change (C_1), and a single long-run adaptation option, investments in reservoir storage capacity, K(C), which is climate sensitive. The framework can be extended to multiple climate states and measures, including those that are not sensitive to climate in both deterministic and stochastic settings (Callaway, 2004b).

The value of the net returns to water in each of the four cells, represented as W[C, K(C)], depends on the climate state and on the reservoir storage capacity, which is, in turn, determined, in part, by the climate state. From an ex ante planning perspective, the upper left cell characterizes welfare for the existing climate, with current water storage capacity adapted to the existing climate. The cell in the upper right corresponds to the case in which the climate changes but planners fail to plan for it. Economic agents make ex ante short-run partial adjustments to the climate change, but there is no new investment in reservoir storage (or institutional change). Note that water users and managers can adapt to many forms of extreme climate variability without knowing whether the climate is changing or not, by using short-run measures. For example, even if reservoir storage capacity is fixed at K_0, operating policies can be changed, cities can institute water restrictions and farmers can change their management. But unless people can detect climate change or are confident that it will occur, or estimate that the costs of planning for climate change are small if it does not occur, generally they will not undertake long-run measures. The cell in the lower right depicts the ex ante long-run welfare consequences that take place when economic agents fully adjust using long-run measures to adapt to the expected climate. The cell in the lower left, depicts an ex ante situation where water storage capacity is increased in anticipation of climate change that does not occur.

Table 3.1 *Framework for estimating benefits and costs associated with climate change and climate change adaptation*

Ex Ante Climate Planning Options	Ex Post Climate States(C)	
	Existing Climate (C_0)	Climate Change (C_1)
Plan for current climate, reservoir capacity = $K(C_0)$	Existing climate with current reservoir capacity. Net returns to water = $W[C_0, K(C_0)]$	Do not plan for climate change that occurs. Net returns to water = $W[C_1\ K(C_0)]$
Plan for climate change, reservoir capacity = $K(C_1)$	Plan for climate change that does not occur. Net returns to water = $W[C_0, K(C_1)]$	Plan for climate change that occurs. Net returns to water = $W[C_1, K(C_1)]$

Table 3.1 can also be used to measure the economic costs, or regrets, of implementing planning decisions based on ex ante climate expectations that are wrong in terms of ex post climate outcomes. There are two kinds of regrets: those associated with caution, which involve not planning for climate change that is occurring, and those associated with precaution, which involve planning for climate change that is not occurring. In both cases, the net returns to water will be lower than if the ex post climate realization matched the ex ante climate expectation. In other words, regrets lead to costs. Estimates of the costs of regrets can be made without assigning any explicit probabilities to different climate states but can also be cast in a Bayesian framework when better information is available (Lindgren, 1968). From Table 3.1, we construct the following definitions of climate change damages, net benefits of adaptation, and ex post costs of errors of caution and precaution.

Climate change damages (CCD) are the ex ante welfare losses caused by climate change when economic agents only make short-run adjustments to climate variability or climate change (in other words they do not adjust reservoir capacity) to cope with climate change compared to the reference case and are calculated as:

$$CCD = W[C_1, K(C_0)] - W[C_0, K(C_0)].$$

Net benefits of adaptation (NBA) are the ex ante welfare gains associated with reducing climate change damages by implementing long-run measures (in other words adjusting the reservoir capacity) compared to the partial adjustment case:

$$NBA = W[C_1, K(C_1)] - W[C_1, K(C_0)].$$

Imposed climate change damages (ICCD) are the ex ante welfare losses relative to the reference case that cannot be avoided by implementing long-run measures. They correspond to the residual climate change damages that are not avoided by full adaptation and are calculated as:

$$ICCD = W[C_1, K(C_1)] - W[C_0, K(C_0)].$$

Cost of precaution (CP) is the ex post welfare loss that would occur as a result of planning for climate change that does not occur:

$$CP = W[C_0, K(C_1)] - W[C_0, K(C_0)].$$

Cost of caution (CC) is the ex post welfare loss that would occur as a result of not planning for climate change that does occur. It is equivalent to the net benefits of adaptation, with the sign reversed:

$$CC = W[C_1, K(C_0)] - W[C_1 K(C_1)].$$

Policy, climate and water demand scenarios

Two examples of adaptation policy scenarios or measures that are relevant and important to the basin are selected for analysis: building and optimally sizing the Berg River Dam and replacing the existing regulatory framework for allocating water in the basin with a system of efficient water markets. As already noted, the dam is under construction and expected to be in operation soon. Implementing water markets is a policy that is currently being investigated in the context of the new national water law. Existing water use regulation is represented in the BRDSEM model by placing upper bounds on summer and winter diversions by the seven regional farms as in Louw (2001 and 2002) and on water diversions from storage sites at Theewaterskloof and Wemmershoek, consistent with current water allocation policy. Efficient water markets are introduced to the model by removing both the agricultural and urban allocation constraints to simulate the economically efficient allocation of water to both urban and agricultural users.

We simulated the welfare consequences of the different policy options under both full (optimal) and partial adjustment to three deterministic, transient climate scenarios. A detailed explanation of the climate scenarios, the downscaling from the global circulation models to the basin weather stations, and the generation of the monthly hydrologic data used by BRDSEM is discussed in detail in Hellmuth and Sparks (2005) and Nkomo et al (2006). The hydrology-climate scenarios are constructed with information provided by WATBAL and derived from downscaled outputs of the CSIRO general circulation model experiment for the SRES B2 scenario of greenhouse gas emissions. Three different time-slices of the CSIRO experiment are used to construct our scenarios: 1961–1990 for the reference case (REF), 2010–2039 for the near future case (NF) and 2070–2099 for the distant future case (DF). Relative to the reference case, average annual runoff in the basin is reduced by 10.7 and 22.0 per cent in the NF and DF scenarios respectively.

While the climate scenarios are time dependent and correspond to specific years, we apply all of the scenarios to the same period, 2010–2039, to avoid having to develop even more uncertain future scenarios for the 2070–2099 period. Thus REF is a counterfactual reference case, assuming the same underlying runoff as in the period 2010–2039 as 1961–1990, while DF, instead of being a longer-term continuation of CSIRO B2, can be viewed as a more adverse climate scenario, producing lower runoff and higher evaporation, compared to NF for the same time period.

CSIRO climate experiments for A2 and B2 emission scenarios project reductions in runoff for the region relative to the reference period for both the NF and DF time periods, with the B2 scenarios being slightly more severe. In comparison, projections from the Hadley Centre general circulation model are mixed: the A2 scenarios show virtually no changes in runoff for the region in the NF and DF time periods, whereas the B2 scenarios show reductions in runoff in the NF time period, but increases in the DF time period. Our initial economic analysis uses only the CSIRO projections for the SRES B2 emissions scenario. There are two reasons for this. First, these scenarios showed the most severe effects on basin runoff, yet appear to be more consistent with recent trends. Evaluating severe, yet plausible, scenarios is useful for benchmarking the upper range of potential impacts. Second, the long solution times of the model in conjunction with the large number of runs forced us to limit the analysis for the time being.

Agricultural area in the basin has been relatively stable for the last half-decade and is not expected to grow much more due to limited land availability (Louw 2001 and 2002). Most of the irrigated land in the basin is already under drip irrigation and the potential for efficiency gains is quite small. Dryland agriculture is basically limited to areas without access to irrigation water and should conditions become drier and hotter, this lack of access is likely to drive this land out of production. For these reasons, we did not make any exogenous changes to the irrigated land base in the model. The only changes in water demand for irrigation are those that are introduced by WATBAL for changes in crop water use factors in response to changes in potential evapotranspiration and changes in irrigation efficiencies in response to shadow prices that are determined endogenously by BRDSEM.

However, urban water consumption in Cape Town has been growing rapidly at an average annual rate of about 3 per cent since the mid-1970s (Louw 2001 and 2002). In the real world and in our model, increases in water consumption over time will increase net returns to water as the sectoral water demand curves shift to the right. In other words, development increases welfare. But if we are to correctly estimate the costs and benefits associated with climate change and adaptation, we need to be able to separate not only the effects of climate from the effects of development, but also the adjustments to development from the adjustments to climate as a result of the implementation of various adaptation options.

To illustrate how one can control for development and climate separately, we developed two urban water demand scenarios by adjusting the slopes of the monthly urban water demand functions using consumption data provided by the Cape Metropolitan Council (CMC). The first scenario is a no-demand growth case in which water demand curves are held fixed. The second scenario shifts water demand curves such that, at a constant water price, the quantity of water demanded by urban users grows at a rate of 3 per cent per year.

Results

Table 3.2 shows the simulated present values for the total net returns to water for four policy options under three climate change and two urban water demand scenarios. The policy options are (A) water is allocated by existing regulations and no dam is constructed, (B) water is allocated by efficient water markets and no dam is constructed, (C) water is allocated by existing regulations and the dam is constructed and optimally sized for the ex post climate and water allocation regime, and (D) water is allocated by efficient water markets and the dam is constructed and optimally sized for the ex post climate and water allocation regime. The table also presents the optimal storage capacity of the Berg River Dam for options C and D. All of the net returns to water depicted in this table represent the optimum values that can be achieved by a policy option for each climate and urban demand scenario.

Economic values in this chapter are calculated as constant South African rand, based on the year 2000 (Louw and Van Schalkwyk, 2001; Louw, 2002). This assumes that all input and output prices in the model are inflating at the same, constant rate. A constant, real discount rate of 6 per cent is used to convert future value flows into constant present values. In sensitivity trials, reducing (increasing) the discount rate had predictable effects on water use, increasing (reducing) future consumption and, thus, increasing (reducing) the endogenously determined maximum optimal storage capacity of the Berg River Dam.

In all four policy cases, the net returns to water decrease as the annual average annual runoff in the climate scenarios decreases, holding urban water demand growth constant, and increase as urban water demand growth increases, holding climate constant. Moreover, the percentage reductions in the

Table 3.2 *Net returns to water and optimal storage capacity of the Berg River Dam*

	Climate Scenarios No Urban Demand Growth			Climate Scenarios 3% Urban Demand Growth Per Year		
	REF[1]	NF[1]	DF[1]	REF[1]	NF[1]	DF[1]
Ave. Annual Runoff (m³ – 10⁶)	75.5	67.4	58.9	75.5	67.4	58.9
(% Reduction from REF)		(–10.7)	(–22.0)		(–10.7)	(–22.0)
Allocation/Reservoir Scenarios	*Present Value of Net Returns to Water[2] (R 10⁹)*					
A. Existing water use regulation No Berg Dam	58.7	55.8	52.8		58.0	44.9[3]
(% Reduction from REF)		(–5.1)	(–10.1)	74.9	(–22.6)	(–40.1)
B. Efficient water markets No Berg Dam	59.8	56.3	53.1	92.0	83.9	76.5
(% Reduction from REF)		(–5.9)	(–11.2)		(–8.7)	(–16.8)
C. Existing water use regulation Optimal storage for Berg Dam	58.7	55.8	53.6	90.1	76.9	62.7
(% Reduction from REF)		(–5.0)	(–8.9)		(–14.7)	(–30.5)
D. Efficient water markets Optimal storage for Berg Dam	59.9	56.8	54.4	96.3	89.7	83.4
(% Reduction from REF)		(–5.1)	(–9.2)		(–6.9)	(–13.3)
Options	*Optimal Berg River Dam Capacity m³ × 10⁶*					
C	0	15	69	151	272	240
D	6	84	109	138	128	178

[1]REF: reference climate scenario, CSIRO model simulation for 1961–1990; NF: near future scenario, CSIRO model simulation for period 2010–2039 for SRES B2 emission case; DF: distant future scenario, CSIRO model simulation for period 2070–2099 for SRES B2 emission case. All climate scenarios are assumed to apply to 2010–2039 period.
[2]All monetary estimates are expressed as present values in billions of constant rand (R) for the year 2000, discounting over 30 years at a real discount rate of 6 per cent.
[3]This scenario was unfeasible because the urban regulation constraints were tight; consequently their right-hand side values were increased by 5 per cent in every year.

net returns to water become much sharper as urban demand growth increases. This directly raises the important question of how we can disentangle the effects of urban water demand growth from climate change in assessing these two options.

The results in Table 3.2 also show a consistent relationship between the net returns to water and the type of allocation method. Allocation by water markets in cases B and D produces higher net returns than allocation by regulations in cases A and C. This should not be surprising, as BRDSEM is an optimization model and the only differences between these two pairs of scenarios is that A and C are more highly constrained than B and D.

More importantly, as water demand increases, switching to efficient water markets reduces the economic impacts of climate change, compared to the existing regulated system. In the no urban demand growth cases, the reductions in the net returns to water from the base case are around 5 per cent for the DF climate and 10 per cent for the RF climate for both allocation systems. However, when urban demand growth increases substantially, allocation by markets serves to moderate the decreases in the net returns to water as climate change becomes more severe. Thus, simulating the substitution of efficient water markets in case B for the existing allocation system in case A reduces the percentage of welfare losses in basin-wide welfare from –22.6 per cent (NF) and –40.1 per cent (DF) in case A to –8.7 per cent (NF) and –16.8 per cent (DF) in case B. For cases C and D, the corresponding reductions in the percentage changes in basin welfare due to market substitution are –14.7 per cent (NF) and –30.5 per cent (DF) for case C (existing allocation system) and –6.9 per cent (NF) and –13.3 per cent (DF) for case D (water markets).

The last conclusions demonstrate forcefully that simulated water markets have a substantial moderating effect on the impact of climate change on basin-wide welfare as urban demand growth increases. But this is also true of dams, since the absolute values for the net returns to water are higher, and the percentage reductions due to climate change are smaller when storage capacity is added behind the Berg River Dam compared to the no dam cases, when urban demand growth is 3 per cent per year. The fact that both dams and water markets influence the effects of climate change on basin-wide welfare again raises the issue of how we separate out both the effects of climate change and economic development on basin-wide welfare and the effects of these measures on climate change and development.

However, before we do this, we want to take one last look at the optimal reservoir capacity simulated for cases C and D due to climate change and development (see Table 3.3). One thing is clear: as urban development increases, the need for additional storage capacity not only increases, but is also economically feasible. Nevertheless, two partly counter-intuitive results need further explanation. First, under the no urban demand growth scenario, optimal reservoir capacity is higher for water markets (case D) than for the existing allocation system (case C). This is because, at low urban water demand levels in the upper basin, reducing consumption is expensive in terms of consumer welfare losses, and the implementation of markets increases the marginal present value of storage capacity (or the opportunity cost of the lack of storage) more than the capital cost in this capacity range, leading to higher capacity levels in the market situation. As urban demand growth increases, markets tend to become better substitutes for storage capacity, because the reductions in consumption (from very high prices) cost less, at the margin, than the additions to storage capacity at these very high storage capacity levels. Second, the pattern of storage capacity additions under high urban water demands is nonlinear with respect to climate change and the nature of this nonlinearity differs depending on the water allocation method.

For option C, with higher water demand growth, the optimal storage capacity of the Berg River Dam peaks under the NF climate scenario and then falls from 272,000m^3 to 240,000m^3. This is because the marginal yield of the reservoir system due to increases in capacity falls sharply as runoff is reduced and development increases, leading to problems in filling the reservoir under the DF scenario and reducing the benefits of additional capacity. For option D at the same level of water demand, the optimal capacity decreases under the NF scenario and then rises. This is because lower basin runoff in the NF scenario varies in such a way relative to the upper basin that water can be moved around by transfers much more easily to satisfy demand than by adding storage capacity, but this is not the case for the existing water allocation system.

Table 3.3 presents the welfare results for incrementally adapting storage capacity of the dam to development pressure and climate change under the existing regulatory system for allocating water. Adaptation is presented in two steps, first to development pressure and then to climate change by adding the economically efficient amount of water storage capacity in both steps. The upper part of Table 3.3 shows results for partially and fully adjusting reservoir capacity to development pressure if there is no change in climate. Development, represented by 3 per cent annual growth in urban water demand, with no addition to reservoir capacity would raise net returns to water from R58.7 billion to R74.9 billion. The net increase in welfare, R16.196 billion, represents the ex ante net benefits of adapting to development pressure, holding climate constant at REF, and only allowing short-run adjustments to take place in response to urban water demand growth. If basin planners correctly anticipated and fully adjusted to this development pressure by adding 151 million m^3 of storage capacity, net returns to water would increase to R90.1 billion. This increase of R15.2 billion represents the ex ante net benefits of optimally (instead of partially) adjusting (or adapting) to development pressure. This same amount, but in minus also represents the ex ante, ex post cost of not planning for development that does occur, or the cost of caution associated with development.

If, on the other hand, the Berg River Dam was built at this level of capacity, but no increase in water demand growth occurred, ex post basin-wide welfare would drop to R58.3 billion, which is R400 million less than the base case. This net welfare loss is the ex ante, ex post cost of planning for development pressure that does not occur, or the cost of precaution associated with development pressure.

The second part of Table 3.3 shows results for partially and fully adapting to climate change for two climate scenarios, NF (by adding 272 million m^3 of storage capacity) and DF (by adding 240 million m^3 of storage capacity), under the existing water allocation system. The reference case used for measuring the effects of adapting to climate change is the post-development adjustment in reservoir capacity to 151 million m^3 and the corresponding welfare level of

Table 3.3 *Adapting to development pressure and climate change under the existing water allocation system: net returns to water (present value, R billion)*

I. Adjustment to Development: Change Storage Capacity		
	Ex Post Scenario: No Development and 3% Development	
Ex Ante Action: Change in Berg Reservoir Capacity Only	**No Development, REF Climate 3%**	**Development, REF Climate**
Partial adjustment to development Capacity = 0	58.7	74.9
Full adjustment to development Capacity = 151 × 10⁶m³	58.3	90.1

II. Adjustment to Climate Change: Change Storage Capacity		
	Ex Post Scenario: REF Climate and NF Climate	
Ex Ante Action: Change in Berg Reservoir Storage Capacity Only	**3% Development, REF Climate**	**3% Development, NF Climate**
Partial adjustment to NF climate change Capacity = 151 × 10⁶m³	90.1	76.7
Full adjustment to NF climate change Capacity = 272 × 10⁶m³	89.9	76.9
	Ex Post Scenario: REF Climate and DF Climate	
Ex Ante Action: Change in Berg River Storage Capacity Only	**3% Development, REF Climate**	**3% Development, DF Climate**
Partial adjustment to DF climate change Capacity = 151 × 10⁶m³	90.1	62.5
Full adjustment to DF climate change Capacity = 240 × 10⁶m³	90.0	62.7

R90.1 billion. If the climate changes and the rate of urban demand growth is 3 per cent and reservoir capacity is not enlarged from 151 million m³, simulated net returns to water fall to R76.7 billion for the NF climate scenario (10.7 per cent reduction in average annual runoff) and R62.5 billion for the DF climate scenario (22.0 per cent reduction in average annual runoff). The resulting welfare losses of R13.4 billion (NF) and R27.6 billion (DF) represent the ex ante values of climate change damages. However, if basin storage capacity is increased from 151 million m³ to 272 million m³ (optimal for NF) or 240 million m³ (optimal for DF), ex ante net returns to water would increase to R76.9 billion and R62.7 billion respectively. The increases in net welfare of R0.209 billion (NF) or R0.198 billion (DF) represent the ex ante values for the net benefits of adaptation under the two different climate scenarios, with urban demand growth at 3 per cent. But these adaptation benefits are quite small relative to the value of climate change damages, so the simulated, ex ante imposed climate change damages are quite large, namely R13.2 billion (NF) and R27.4 billion (DF).

Following on from this analysis, we can also estimate the ex ante, ex post cost measures associated with climate regrets. As already stated, the cost of not planning for climate change that does occur (caution) are the same as the net benefits of adaptation but in minus, –R0.209 billion (NF) or –R0.198 billion (DF). If the Berg River Dam is built but the climate does not change to NF or DF, then the ex post value of the net returns to water would be R89.9 billion (NF) or R90.0 billion (DF), which are R0.204 billion (NF) and R0.142 billion (DF) less than the reference case of R90.1 billion. These are the estimated costs of precaution associated with the two climate change scenarios.

In Table 3.4 we show the results for adapting to development pressure and climate change by substituting efficient water markets for the existing allocation system and increasing the storage capacity. Adaptation is again represented in sequential steps, first for adjusting to development pressure alone and then for adjusting to the change in climate to either NF or DF scenarios. This time, the adjustment to development is decomposed into two separate steps: first by switching to efficient water markets and then by adjusting the storage capacity to be optimal for efficient water markets.

Part I of Table 3.4 shows that, if urban water demand growth increases at 3 per cent per year, the ex ante net present value of welfare in the basin will increase from R58.7 billion to R74.9 billion, without any long-run adjustments by basin planners. Thus the ex ante net benefits of increased water demand growth are R16.2 billion. If basin planners and managers adjust to this development pressure by instituting a system of efficient water markets, but do not build a storage reservoir, ex ante basin welfare will further increase to R92.0 billion. If a storage reservoir is also built, and its capacity is optimal for the market allocation system and 3 per cent urban water demand growth, then ex ante basin welfare rises even further to R96.3 billion. The net benefits of adjusting to the development pressure by switching to a system of efficient water markets to cope with the development pressure are R17.1 billion in step A. Adjusting the reservoir capacity to 138 million m^3 in step B so that it is optimal for the higher level of water demand growth and the new allocation system produces another R4.3 billion in ex ante basin welfare, or R21.4 billion compared to the reference case.

Since the cost of caution is the cost of avoiding an error of caution (in other words not anticipating the higher level of development pressure), one just needs to reverse the signs on the estimates for the net benefits of development. If efficient water markets are implemented, ex ante, but the ex post level of demand growth is zero, net welfare in the basin would be R59.8 billion and, if 138 million m^3 of storage capacity were developed, ex ante, but the ex post level of demand growth remains at zero, basin welfare would increase just slightly from that to R59.6 billion. These welfare levels are actually higher than the reference case. Thus, the resulting costs of precaution would be ex post

Table 3.4 *Adapting to development pressure and climate change by switching to water markets and adding storage capacity: net returns to water (present value, R billion)*

I. Adjustment to Development: Efficient Water Markets		
	Ex Post Scenario: No Development and 3% Development	
Ex Ante Action: Change Allocation System Only Change Storage Capacity	No Development, REF Climate	3% Development, REF Climate
Partial adjustment to development: Existing Regulations Capacity = $0 \times 10^6 m^3$	58.7	74.9
A. Full adjustment to development: Efficient Water Markets Capacity = $0 \times 10^6 m^3$	59.8	92.0
B. Full adjustment to development: Efficient Water Markets Capacity = $138 \times 10^6 m^3$	59.6	96.3

II. Adjustment to Climate Change: Efficient Water Market + Change in Storage Capacity		
	Ex Post Scenario: REF Climate and NF Climate	
Ex Ante Action: Change in Berg Reservoir Storage Capacity with Efficient Water Markets in Place	3% Development, REF Climate	3% Development, NF Climate
Partial adjustment to NF climate change Capacity = $138 \times 10^6 m^3$	96.3	89.7
Full adjustment to NF climate change Capacity = $128 \times 10^6 m^3$	96.3	89.7
	Ex Post Scenario: REF Climate and DF Climate	
Ex Ante Action: Change in Berg Reservoir Storage Capacity with Efficient Water Markets in Place	3% Development, REF Climate	3% Development, DF Climate
Partial adjustment to DF climate change Capacity = $138 \times 10^6 m^3$	96.3	83.4
Full adjustment to DF climate change Capacity = $178 \times 10^6 m^3$	96.3	83.5

benefits (not costs) of R1.1 billion and R0.9 billion, illustrating nicely the 'no regrets' character of these adjustments to development pressure.

The second part of Table 3.4 presents results for partially and fully adapting to climate change for two climate scenarios, NF and DF, under the revised water allocation system via markets. The adjustments can be interpreted in the same way as part II of Table 3.3, and so we will highlight only the major differences in the two strategies as shown in the results from Tables 3.3 and 3.4.

There are three main conclusions that can be drawn from these tables. First, the market-oriented strategy (Table 3.4) increases basin welfare by a larger amount over the whole range of development and climate impacts and adjustments than the strategy relying on the existing allocation system (Table 3.3). The welfare difference is roughly R12.8 billion (NF) to R20.8 billion (DF), in favour of the market-oriented strategy, depending on the climate scenario. The net benefits of adjusting to development pressure using the market-oriented strategy are R1.8 billion to R6.2 billion higher than those of the other strategy, depending on whether one just switches to efficient water markets or does this and also adjusts the reservoir capacity for the change in development pressure and water allocation system.

Second, the development adjustment step in the market-oriented strategy also reduces the level of climate change damages by more than 50 per cent compared to the other strategy for both the RF and DF climates. After the adjustment to development pressure, but before the adjustment to climate change, the level of climate change damages is R6.8 billion (NF) to R14.8 billion (DF) lower than for the other strategy, depending on the climate change scenario. As a result, switching to water markets acts like a 'no regrets' insurance policy against having to adjust further to climate change.

Third, once water markets are instituted, the net benefits of adapting to climate change (in the market-oriented strategy) are quite small both in absolute terms and compared to the other strategy. This is because of the large reduction in climate change damages that occurs as a result of adjusting to development pressure. For this reason, and because the costs of both caution and precaution associated with climate change adaptation for the market-adjustment strategy are very low in both absolute terms and compared to the other strategy, the least-risky and most economically efficient option would involve adjusting to development pressure and climate change by implementing a system of efficient water markets (step A) and increasing the reservoir capacity to 128 million m^3 (step B), without developing any additional storage capacity until more information is available about climate change. Adding additional storage capacity above this level produces only small net benefits, but also has low costs of caution and precaution.

Limitations in the Analysis and Future Research

BRDSEM, like many policy models, is a work in progress. Typically, models like this are never final and undergo numerous revisions as new data become available and new questions are asked. BRDSEM has reached the point in its development where it can be used to illustrate some of its policy uses, as in this chapter. Nevertheless, we would be remiss not to identify its current limitations and how we plan to remedy them.

The current version of BRDSEM and its application in this chapter has the following limitations. First, the parameters of the urban water demand functions are not estimated from empirical time series and/or panel data, but are fit algebraically using assumed elasticities and estimates of base-level consumption. There is no upward sloping waterworks supply function to capture resource costs in purifying and distributing water to urban consumers, and we lack reliable data regarding the water supply elasticity to even 'fit' these functions algebraically. The model lacks a full set of demand- and supply-side options for coping with demand growth, climate variability and climate change. The transactions costs associated with markets are ignored, and the analysis is deterministic with regard to climate scenarios.

With future support from all of the elements of the regional and South African water and agricultural resources community, we plan to address the first four of these issues as follows. Parameters of the urban demand and water works supply functions will be estimated with empirical observations, using suitable econometric estimators. We plan to extend the model to the larger Boland Region to include representations of all current and alternative water supply sources, reservoirs and conveyance structures, and users in the region, as well as a larger set of demand- and supply-side options. In the current analysis, we assumed that the transition from the current regulatory system to a system of efficient short- and long-run water markets would be frictionless and without cost. In the future, we plan to include a more realistic treatment of water transfers, including differences between long- and short-term water transfers and transaction costs.

The final limitation, the deterministic nature of the illustrative analysis in this paper, requires further consideration and investigation. We used deterministic climate change scenarios because downscaled stochastic climate scenarios do not currently exist for the region. The use of deterministic climate scenarios is adequate for demonstrating the model and the economic framework for estimating the benefits and costs associated with adjusting to climate change. But the results are unreliable because they do not reflect the underlying means, variances and other moments of the partial and joint distributions of relevant meteorological variables at the appropriate spatial scales, nor does the runoff, also generated using the deterministic climate inputs. When such information becomes available, it will be possible to propagate the runoff, evaporation and crop water use distributions through BRDSEM by maximizing the expected value of net returns to water for a single or for mixed climate distributions using the methods illustrated in Callaway (2004b). This will also allow us to explore the economic and physical consequences of runoff sequences that depart from mean values, that are drier and wetter than average periods than reflected in mean runoff. Finally, it will allow us to explore more thoroughly the stochastic nature of ex ante, ex post regrets and the possibility of minimizing these regrets by policies and plans that are flexible over a wide range of mixed runoff distributions.

References

Booker, J. F. (1990) 'Economic allocation of Colorado River water: Integrating quantity, quality, and instream use values', PhD thesis, Department of Agricultural and Resource Economics, Colorado State University, Fort Collins, CO

Booker, J. F. and R. A. Young (1994) 'Modeling intrastate and interstate markets for Colorado River water resources', *Journal of Environmental Economics and Management*, vol 26, pp66–87

Booker, J. F. and R. A. Young (1991) *Economic Impacts of Alternative Water Allocations in the Colorado River Basin*, Report 161, Colorado Water Resources Institute, Fort Collins, CO

Callaway, J. M., L. Ringius and L. Ness (1998) 'Adaptation costs: A framework and methods', in J. Christensen and J. Sathaye (eds) *Mitigation and Adaptation Cost Assessment Concepts, Methods and Appropriate Use*, UNEP Collaborating Centre on Energy and Environment, Risø National Laboratory, Roskilde, Denmark

Callaway, J. M. (2003) *Adaptation Benefits and Costs – Measurements and Policy Issues*, Report NV/EPOC/GSP (2003) 10/FINAL, Environment Directorate, Environment Policy Committee, OECD, Organisation for Economic Co-operation and Development, Paris

Callaway, J. M. (2004a) 'Adaptation benefits and costs: Are they important in the global policy picture and how can we estimate them', *Global Environmental Change: Human and Policy Dimensions*, vol 14, pp273–282

Callaway, J. M. (2004b) 'The benefits and costs of adapting to climate variability and change', in J. C. Morlot and S. Agrawala (eds) *The Benefits and Costs of Climate Change Policies: Analytical and Framework Issues*, Organisation for Economic Co-operation and Development, Paris

Callaway, J. M., D. B. Louw, J. C. Nkomo, M. E. Hellmuth and D. A. Sparks (2006) 'The Berg River Dynamic Spatial Equilibrium Model: A new tool for assessing the benefits and costs of alternatives for coping with water demand growth, climate variability and climate change in the Western Cape', AIACC Working Paper No 31, International START Secretariat, Washington, DC

Hellmuth, M. E. and D. Sparks (2005) 'Modeling the Berg river basin: An explorative study of impacts of climate change on runoff', AIACC Project Completion Report, Project No 47, UNEP Collaborating Centre on Energy and Environment, Risø National Laboratory, Roskilde, Denmark

Hurd, B. J., J. M. Callaway, P. Kirshin and J. Smith (1999) 'Economic effects of climate change on US water resources', in R. Mendelsohn and J. Neumann (eds) *The Impacts of Climate Change on the US Economy*, Cambridge University Press, London

Hurd, B. J., J. M. Callaway, P. Kirshin and J. Smith (2004) 'Climatic change and US water resources: From modeled watershed impacts to national estimates', *Journal of the American Water Resources Association*, vol 22, pp130–148

Lindgren, B. W. (1968) *Statistical Theory*, Macmillan, New York

Louw, D. B. (2001) 'Modelling the potential impact of a water market in the Berg river basin', PhD thesis, University of the Orange Free State, Bloemfontein, South Africa

Louw, D. B. (2002) 'The development of a methodology to determine the true value of water and the impact of a potential water market on the efficient utilisation of water in the Berg River Basin', Water Research Commission Report (WRC) No 943/1/02, Pretoria, South Africa

Louw, D. B. and H. D. Van Schalkwyk (2001) 'Water markets: An alternative for central water allocation', *Agrekon*, vol 39, pp484–494

McCarl, B. A. and T. H. Spreen (1980) 'Price endogenous mathematical programming as a tool for sector analysis', *American Journal of Agricultural Economics*, vol 62, pp88–102

Nkomo, J. C., J. M. Callaway, D. B. Louw, M. E. Hellmuth and D. A. Sparks (2006) *Adaptation to Climate Change: The Berg River Basin Case Study*, Energy Research Center, University of Cape Town, Cape Town, South Africa

Samuelson, P. A. (1952) 'Spatial price equilibrium and linear programming', *American Economic Review*, vol 42, pp283–303

Smith, J. B. and S. S. Lenhart (1996) 'Climate change adaptation policy options', *Climate Research*, vol 6, pp193–201

Takayama, T. and G. G. Judge (1971) *Spatial and Temporal Price and Allocation Models*, North Holland Press, London

Vaux, H. J. and R. E. Howitt (1984) 'Managing water scarcity: An evaluation of inter-regional transfers', *Water Resources Research*, vol 20, pp785–792

Yates, D. N. (1996) 'WATBAL: An integrated water balance model for climate impact assessment of river basin runoff', *International Journal of Water Resources Development*, vol 12, no 2, pp121–139

4

Indigenous Knowledge, Institutions and Practices for Coping with Variable Climate in the Limpopo Basin of Botswana

Opha Pauline Dube and Mogodisheng B. M. Sekhwela

Introduction

Widespread poverty, high reliance on natural resources and low adaptive capacity contribute to conditions of high vulnerability to climate variability and climate change in developing countries such as Botswana (Kates, 2000; Desanker and Magadza, 2001; Mirza, 2003; Dube and Moswete, 2003). Lack of choice to meet basic needs such as nutrition, shelter and clothing at the household and individual levels in Botswana is attributed to income poverty and capability poverty (Jefferies, 1997). Income poverty is the inability to command the level of income or tangible resources needed to meet basic needs, while capability poverty involves the lack of human capabilities or intangible resources such as education and good health that enables one to escape poverty (Ministry of Finance and Development Planning, 1998). Both income poverty and capability poverty undermine the capacity to adapt to environmental and non-environmental stresses in Botswana.

Despite the conditions of income and capability poverty that have been noted in recent years, rural communities of semiarid areas of Botswana evolved local institutions for managing natural resources and adaptable lifestyles marked by multiple livelihood strategies that enabled the communities to cope with variable climate and variable supplies of natural resources. This has been noted for similar natural resource-dependent communities (Burton, 2004; Hulme, 2004; Thomas and Twyman, 2005). The institutions and livelihood practices, supported by an indigenous knowledge system that evolved through accumulated experiences of changing environmental conditions, imparted a capacity or resilience to withstand drought and other climate variations over a relatively wide coping range.

The indigenously developed institutions, livelihoods and knowledge base served well until recent years, when droughts exposed an increasing vulnerability among rural communities (Hulme, 2004; UNEP, 2001). The increasing vulnerability may indicate a limit to coping strategies for natural resource-dependent systems that are highly exposed to environmental stresses. However, other evidence suggests that the decline in resilience among natural resource-dependent communities signals major structural changes in their livelihood systems resulting from the way these systems interface with Western-oriented democratic systems. External interventions have been made in response to the increasing vulnerability, but their effectiveness has been minimal and they may even have inadvertently further weakened the adaptation capacity of communities in general (Sporton and Thomas, 2002; Warren, 2005).

This chapter examines the effectiveness of past strategies used by rural communities in the Limpopo basin part of Botswana to reduce the negative impacts of climate variability. An assessment is made of the interplay between past strategies and the new socio-political and economic frameworks introduced since the beginning of the 20th century and the implications for evolving a capacity to adapt to climate change. In particular, attention is given to the role of the government's rural development and disaster relief programmes in transforming coping capacity at the community level. The potential is explored for using remnant community coping strategies and institutions as bases for enhancing adaptation to climate change and stresses among rural communities. A synthesis of multiple factors that need to be considered when building adaptation capacity at the community level in Botswana, as well as recommendations for future actions, are presented in the closing section of the chapter.

The Study Area

The study area, shown in Figure 4.1, covers the Botswana part of the Limpopo river basin. The basin as a whole extends over 3,720,000 square kilometers across semiarid lands in Botswana, Zimbabwe and South Africa, as well as humid and sub-humid areas in Mozambique (Sharma et al, 1996). In Botswana, the Limpopo river basin forms 20 per cent of the eastern section of the country known as the hardveld, an area of active erosion formed by igneous and metamorphic rocks overlaid by loamy soils. The hardveld has the highest density of surface drainage in the country, but tributaries have an average flow period of 10–70 days a year (B. P. Parida, 2005, personal communication). The Limpopo catchment yields, on average, about 10mm of surface water annually (Parida et al, 2005).

The hardveld is subject to frequent climate extremes such as drought. Some of the worst droughts documented occurred in 1935, 1965, 1984 and 1991 and resulted in significant losses of livestock (Bhalotra, 1989). Average annual rainfall in the hardveld ranges from 400 to 450mm with 35 per cent variability (Figure 4.2). Temperatures can be as high as 38°C during the wet season, resulting in high potential evapotranspiration, with an estimated average of 1400mm per year, which reduces rainfall effectiveness (Parida et al, 2005).

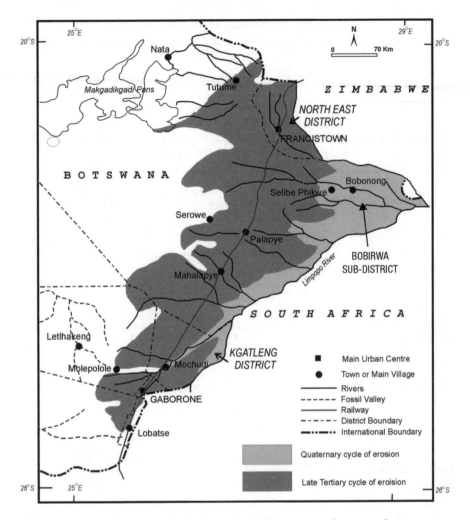

Figure 4.1 *The Limpopo basin area of Botswana and case study sites:*
Northeast District, Bobirwa Sub-District and Kgatleng District

Most of the land in the hardveld is communally owned, but there are freehold farms along the main Limpopo river channel and in the Northeast District. The basin has a longer history of livestock rearing and dryland farming than the remaining 80 per cent of the western part of the country covered by the Kalahari aeolian sands. The Tswana pastoralists living in villages separated from cattle posts and fields originally occupied the drier southern part of the basin during the 17th century. In the wetter central to northern parts, the Ikalanga people lived in scattered settlements and practised mainly arable agriculture (Dube, 1984). The crops grown included millet, sorghum, beans, melons and maize. Mining of first gold, then coal and later copper and nickel have had a major impact on economic activities in the basin.

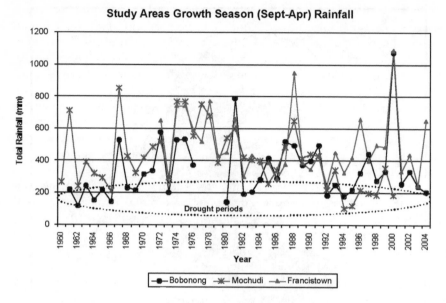

Figure 4.2 *Mean annual rainfall over three stations in the Limpopo basin area of Botswana: Francistown, Northeast District; Bobonong, Bobirwa Sub-District; and Mochudi, Kgatleng District*

Source: Data from Department of Meteorological Services, Botswana.

Projections of future climate change indicate a warmer and drier climate in the basin that may be accompanied by greater variability and more frequent extremes (Scholes and Biggs, 2004). Temperatures over southern Africa are expected to rise by 2 to 5°C by 2050, affecting most of the central land mass of the region occupied by Botswana (Desanker and Magadza, 2001). Precipitation patterns may shift and reduce growing season rainfall by up to 15 per cent (Hulme et al, 2001). The noted climate changes are likely to affect water availability with implications for plant, livestock and wildlife productivity and ultimately, human livelihoods (Desanker and Magadza, 2001). Not all of these changes will be new to the basin, but the magnitude may be different. Some of the relevant past experiences in coping with such variability will be crucial to efforts geared towards enhancing adaptation to these elevated impacts.

Traditional Institutions and Coping with Variability

Past socioeconomic frameworks in the hardveld were in many ways born out of years of exposure to unpredictable semiarid climate and the resulting uncertainty in the supply of natural resources. Community coping measures were mostly integrated in the everyday social interactions and economic structures, allowing communities to be ready for difficulties rather than treating them with emergency measures introduced in times of difficulties and later forgotten until another disaster strikes. Consequently, institutions of local administration, marriage and extended family were adapted to be resilient to climate shocks. These practices have passed from one generation to another as part of a strategy to minimize the impacts of climate variability in the Limpopo river basin, as was the case for the rest of the country. However, different factors operating over the 20th century have seen most of these strategies fade away, and those that remain are distorted and becoming less effective. This section outlines examples of some of the past coping strategies, reasons for their success and factors that contributed to their weakening.

Coping mechanisms were built into the traditional political and administrative framework of each settlement. Among the pastoral Tswana group, local decision making and the administration of justice were implemented through a centralized political institution formed by the chief (*Kgosi*), members of the royal family (*Dikgosana*) and other grades of council, such as adult men of the village (Schapera, 1951). Major decisions were made at the *Kgotla*, or assembly point, regarding user rights to communal lands, livestock and other resources, mobilization of labour, seasonal migration to cattle posts, villages and cultivation areas, settlement of disputes and security. This political and administrative system provided for local management of resources that helped to buffer drought impacts by securing access to land for women and poor households, producing and storing community food reserves and reducing grazing pressure.

The institutions of the family and marriage shielded members from a variety of environmental stressors and also reduced the potential to degrade environmental resources (Khama, 1971). For example, the bride price entitled the bride to user rights to land and other household property such as cattle for subsistence purposes at her family of origin if her marriage failed to materialize. This reduced the possibility for the development of poor female-headed households, which is common nowadays as more and more women lose user rights for key resources such as land. The extended family structure still prevails, albeit subject to continuous weakening.

At a broader community level, cattle were loaned by livestock owners to poorer households and relatives under a system called *mafisa*. The beneficiary family would take care of the cattle and in exchange would have use of the animals' labor and milk, plus a payment at the end of the lease that would depend on the condition of the animals and whether the herd had increased in number. The *mafisa* system represented a win–win coping strategy that provided income and food security, as well as an opportunity to own cattle, to families that borrowed cattle, spatially diversified the risk to livestock owners from variable water supplies, pasture productivity and disease, and reduced grazing pressures (Tlou, 1990).

The traditional institutions of the *Kgotla* and the extended family were weakened and the *mafisa* system was discontinued in the Limpopo basin as a result of political, social and economic changes in the 20th century. These changes have contributed to widespread poverty and increased vulnerability of communities to stresses.

Following a policy common to southern Africa, the colonial administration in Botswana reduced the powers of traditional authorities such as the *Kgosi* and the *Kgotla*, a process that was intensified after independence (Sithole, 1993). The independence and community influence of these institutions, and their role in decision making about natural resource use and management, have been lost. As a result, communities have been alienated from decisions concerning their resources in their immediate environment and, consequently, unregulated use of their communal land, depletion of resources and loss of self-reliance. The role of the extended family and practices such as bride price and polygamous marriage have been lessened by growing influences of Christianity, the school system, Western social values, the market system, wage labour and urbanization. Results of these changes are increases in the number of female-headed households, loss of user rights and unrelenting poverty.

Reasons for discontinuing the *mafisa* system include access to ground water as cattle spread to the western Kalahari sandveld, commercialization of cattle that diminished incentives to lend cattle to poorer farmers; expansion of wage labour that provided other means of acquiring cattle, adoption of the ox-drawn plough and subsequent use of tractors, which enabled the production of surplus grain in good years that could be exchanged for livestock, and veterinary services that reduced cattle mortality (Campbell, 1986; Dube, 1995; Dube and Kwerepe, 2000; Tlou, 1990). Termination of the *mafisa* system meant reduced milk and meat intake leading to protein-energy malnutrition in poorer families, particularly in drought periods.

Veld Products and Coping with Climate Stresses

People of the hardveld harvested and used a variety of natural veld products that reduced their vulnerability to drought by diversifying their sources of food and income. Veld products such as meat from wild animals, edible insects, honey, roots, melons, seeds and wild fruits were an important part of the diet of all of the communities in the Limpopo river basin, as was the case for the rest of African societies (Campbell, 1986). Increased access to other food sources, as well as depletion of veld resources from increased population and use of new technologies, have diminished the contribution of these products to the local diet. But despite these changes, communities still resort to the harvesting of veld products for consumption and income generation during, for example, drought periods. Poorer households, the majority of which are female-headed households, currently dominate the harvesting of veld products.

Fuel wood trade in the Limpopo river basin is common and is dominated by crop farmers who resort to this activity in response to failed crop yields and eventually adopt the trade as part of their strategy to diversify risks (Kgathi, 1984 and 1989a and b). Traders in the northern parts of the Limpopo river basin earn an average of US$29 and US$38 gross per week for rural and urban sellers respectively, with some wholesale distribution networks developing (White, 1999). However, the increase in fuel wood trade due to persistent drought and unemployment has raised environmental concerns. Normally, wood is sourced from old and dead trees. However, in areas of high demand, living trees are now being cut down for fuel wood, thus promoting deforestation. This has resulted in the government imposing bans on the sale of natural products without permit (Botswana Government, 2004).

Phane caterpillar, the larvae of the *Imbrasia belina* moth, is a cherished food product of the veld that is harvested by local harvesters and traded commercially between Botswana and South Africa. On average, 150–250kg is produced by each harvester in one harvest period, 80 per cent of which typically is sold for cash to traders, at a price of roughly US$45 per 50kg bag, who resell the dried Phane for several times that price. Around 4,000 tons of air-dried Phane caterpillar worth around US$9 million was exported to South Africa between 1991 and 1994 (Moruakgomo, 1996).

Medicinal plants represent another set of natural resources closely linked to indigenous knowledge. A number of medicinal plants are traded locally by herbalists and traditional doctors in both rural and urban areas (Botswana Government, 2003). A few medicinal plants, such as the Grapple plant (*Harpagophytum procumbens*), also known as the Kalahari Devil's Claw, have reached international markets. Extracts from the Grapple plant are used in the industrial production of drugs for the relief of arthritis-related pains. Further research is being conducted on the sustainability of the plant, as well as the domestication of the plant for agricultural production (Kgathi, 1989b).

Government Interventions

Increasing vulnerability in rural areas in the past decades necessitated the intervention of the government to alleviate the impact of the stresses such as drought. One of the first well-documented droughts is that of 1964–1965, which resulted in large numbers of deaths of both humans and livestock and was associated with increased incidents of foot and mouth disease outbreaks (Campbell, 1978 and 1986; Dube, 1995). Numerous government interventions were made to alleviate the impact of the drought, including a borehole-drilling scheme to move more cattle to the Kalahari Desert, intensified veterinary services and establishment of a vaccine production center, and food relief through organizations from the UK and the US.

Since then, the government has implemented a number of programmes in order to increase the resilience of rural areas to climate and other stresses by supporting and diversifying economic development, improving resource management and providing social security (see Table 4.1). The Drought Relief Programme (DRP) was implemented from 1979 to provide temporary supplements to rural incomes in times of drought. The programme accounted for 14 to 18 per cent of total government development expenditure between 1984 and 1987 (MFDP, 1998). In addition, vulnerable group feeding programmes targeting preschool and school-going children, as well as pregnant and lactating mothers, have existed since the 1960s (Ohiokpehai et al, 1998). A Destitute Policy was formulated in 1980 to reduce death due to hunger among those who had completely lost the means of sustenance. Despite all of these efforts,

Table 4.1 *Examples of government policies relevant to vulnerability and drought impact reduction*

Category	Policy	Year	Objectives	Weaknesses
Agriculture Production	Tribal Grazing Land Policy (TGLP)	1975	Increase livestock productivity; prevent overgrazing	Lack skill development in range management, had a weak link to climatic factors, and did not reduce over-grazing
	Arable Lands Development programmes (ALDEP)	1977–2000	Reduce food grain deficits and achieve food self-sufficiency in rural areas	Top–down approach, weak link to climatic factors, and less skills development in conservation tillage; no additional manpower to implement and monitor
	Accelerated Rain-fed Agriculture Programme (ARAP)	1985–1990	Reverse impacts of long period of drought on farmers	Destumping without condition for cultivating ultimately contributed to land degradation
	National Master Plan for Agricultural Development (NAMPAAD)	2002	Food security from a diversified sustainable production base	The manpower, expertise and infrastructure required to implement is limited, and climatic issues are not adequately factored in
	Game Ranching Policy	2002	Economic use of wildlife through game ranching industry	Out of rural community reach due to high capitalization costs
Resource Management	Wild Life Conservation Policy	1986	Wildlife conservation through sustainable utilization, including game ranching	Exclusion of communities in management of resources in their areas
	Community-Based Natural Resource Management (CBNRM)	2000 (draft)	Decentralized control of natural resources, greater local benefits from resources and improved management	Needs well-defined property rights for full re-empowerment and management skills
Income and Employment Generation Support Policies	Tourism Policy	1990	Promote tourism	Currently dominated by foreign operated companies. Limited skill among locals.
	Financial Assistance Policy (FAP)	1992–2001	Citizen-owned businesses, employment creation and economic diversification	Lacked strong skills development in rural areas where business culture was nonexistent

Table 4.1 *(continued)*

Category	Policy	Year	Objectives	Weaknesses
	Small, Medium and Microenterprises (SMME)	2001	Support creation of own income-generating projects at micro level	Strengths in target small-scale income generation
	Citizen Entrepreneurial Development Agency (CEDA)	2002	Focused development of citizen-owned businesses	Still new to evaluate
Rural Development	National Policy for Rural Development	2002	Enhance quality of life for all living in rural areas	Mainly strengths of recognizing the diversity of potential economic pursuits in rural areas
Social Protection	Drought Relief Programme	1980	Temporary supplement to rural incomes during drought	Welfare relief rather than develop adaptation capacity
	Destitute Persons Policy	1980	Address plight of extreme poverty	Does not address capacity poverty
	Old-Age Pension Scheme		Financial assistance to elderly	A good start but more resources needed to improve the situation
	Policy on Disaster Management	1996	Preparedness for timely and adequate response to emergencies	Lack skilled manpower, monitoring and ready resources to tackle quick-spreading disasters

Source: Botswana government (2002a–c), Ministry of Finance and Development Planning (1998 and 2002) and Gichangi and Toteng (2004).

however, Buchanan-Smith (1998) concluded that the Drought Relief Programme had not succeeded in preventing malnutrition in all parts of the country.

Other programmes focused on agriculture. For example, the Tribal Grazing Land Policy (TGLP), introduced in 1975, allowed the demarcation of ranches for improved livestock production. However, the policy did not substantially improve the productivity of livestock or rangelands because, among other factors, it failed to account for climate variability (White, 1993). To compensate for this, ranch owners, who are mostly large cattle owners, graze their livestock in both communal areas and in their private ranches when convenient. It is ultimately the small farmers, confined only to communal areas, who are most disadvantaged (Dube and Pickup, 2001).

The Arable Land Development Programme (ALDEP) was implemented to reduce food grain deficits and achieve food self-sufficiency in rural areas. Limitations of the programme include weak links to climatic factors, top–down approach, lack of skills development and competing agricultural programmes that resulted in manpower shortages for implementation and monitoring. The Accelerated Rain-fed Agriculture Programme was intended to reverse the impacts of drought but was implemented in a manner that aggravated land degradation.

The Community Based Natural Resources Management (CBNRM) strategy seeks to decentralize the management of local resources to communities and thereby increase local control of resources, increase local benefits and improve their management (Phutego and Chanda, 2004). Implementation of CBNRM has been hampered by lack of well-defined property rights, full empowerment of local institutions to control resources and management skills.

The Financial Assistance Policy (FAP), and the more recent Citizen Entrepreneurial Development Agency (CEDA), attempt to promote small business creation and employment to diversify the economy. In contrast to the agriculture sector programmes, FAP did not require financial contribution from beneficiaries. All these programmes, in addition to the recently introduced CEDA, aimed at diversifying the economy from the mining sector. However, in the majority of cases the initiatives survive only for as long as there is a government subsidy (MFDP, 1998).

The contribution of all these programmes towards enhancing economic development and therefore adaptation to multiple stresses in rural areas, including climate change, remain unsatisfactory. For example, despite all of the support, agriculture's contribution to GDP in the country has dropped from 40 per cent in 1966 to 2.6 per cent in 1999/2000 and the sector has grown to be less attractive to the young. The mushrooming of policies is evidence of a desperate and costly but futile effort on the part of the government to find a solution to address issues of low-income and capability poverty (CSO, 2003).

The failure of government interventions to enhance the capacity of the community to cope with the stresses of climate variability is due to numerous complex and interlinked factors. Execution of government policies and programmes was based on a top–down approach that did not engage local institutions in decision making or mobilization of local resources. They depended on financial inputs, mainly from internal but also from external sources. Most are conceived as temporary relief measures despite the fact that climate variability and drought are inherent features of the region. The necessary executive capacity is lacking or inadequate, leading to failure to effectively implement the programmes. Moreover, there has been limited emphasis on conducting feasibility studies prior to programme implementation to assess supporting infrastructure and manpower resource needs. Consequently, there has been insufficient attention to develop local skills and capacities for the programmes to succeed. There has also been a tendency to implement policies from different sectors, some of them contradictory and or with no clear linkages between them. For the rural poor, capacity poverty continues to constrain access to the available assistance.

The relief measures introduced during the 1980s drought were not withdrawn after the event, with the result that some of the programmes were incorporated into a burgeoning social welfare programme. It was obvious from this that there was an underlying problem of poverty in the country (Solway, 2002). Very poor female-headed households derived as much as 43 per cent of their total income from government programmes and just 18 per cent from their own initiatives, indicating a high level of dependence on social welfare assistance (BNA21, 2002).

Drought relief measures and agricultural subsidies implemented in the country continue to be at the expense of self-initiated indigenous social welfare systems (Solway, 1994; Sporton and Thomas, 2002; Thomas and Twyman, 2005). This is also acknowledged in government reports (Ministry of Finance and Development Planning, 1998). Programmes such as the Drought Relief Programme address income poverty and exclude capacity poverty; hence they do not lift individuals from poverty's vicious cycle. As a result of the foregoing, communities in rural areas are now more dependent on external intervention for both mild and severe stresses than was the case in the past. The increasing vulnerability to current climate variability is likely to make these communities more vulnerable to future climate change.

The Potential of Remnant Coping Strategies

Despite the social transformation experienced over the 20th century and the subsequent government interventions, communities still display a rich knowledge of past systems and innovative strategies for coping that include traditional institutional frameworks and use of veld products. These remnants of traditional institutions and practices have potential for incorporation into strategies for sustainable community adaptation to climate change.

Natural resources of the veld such as wood, medicinal plants and food products have a high potential to contribute to the livelihoods, incomes and food supplies of people in the Limpopo basin. By adding to and diversifying incomes and food sources, commercial development of veld products can play an important role in reducing the vulnerability of people who live in this semi-arid region. There are good examples outside Botswana, such as the commercialization of Rooiboos and Honeybush tea plants (Gleason, 2004; Nofal, 2004) and medicinal plants that used to be gathered from the veld but are now cultivated and support modern industries in South Africa with large international markets (Cunningham and Milton, 1987; Theron, 2001; Trutter, 2001; Kupka, 2001; Gleason, 2004). The potential of veld products is indicated by the high economic value of wild plants in southern Africa, estimated to be roughly US$270 per household per year, with even higher direct use values (Shackleton and Shackleton, 2000; Twine et al, 2003).

Examples of valuable natural products that are already commercialized in the Limpopo river basin include fuel wood and timber, food and medicinal plants, and insects (Phane caterpillar). These can be expanded to provide greater benefits and can also serve as models for commercializing other products of the veld.

Fuel wood trade, already common in the basin, could benefit from increasing concentrations in the atmosphere of carbon dioxide, one of the principal gases driving global climate change. Experimental results in semiarid regions show a potential for woody plants to out-compete grass species at higher concentrations of carbon dioxide (Bassiri et al, 1998; Bond and Midgley, 2000). Other studies have linked the currently observed increase in woody plants in southern Africa to an increase in anthropogenic CO_2 over the past 100 years (Bond et al, 2003; Joubert, 2004). On the basis of this hypothesis fuel wood trade might become an important economic activity in the future.

Prospects for developing a coordinated fuel wood trade, not yet explored in Botswana (Sekhwela, 1997), warrant investigation. Development of value-added products for purposes other than energy is also worth investigating. For example, fruits of the Morula tree (*Sclerocarya birrea*), found mainly in the Limpopo river basin, are already being used in the production of jam and oil by Kgetsi-ya-Tsie (KyT), a non-governmental organization with markets extending to the UK (*Lapologa Weekend Gazette*, 2003). With greater government support, such initiatives could be expanded to develop full fledged and diversified rural industries.

As already noted, trade in dried Phane caterpillar earns a few million US dollars per year for Botswana. Much of the economic benefit is captured by traders and processors. The potential for more income realization by local harvesters lies in development of local value-added processing, market development and harvester mobilization under a common umbrella, such as that offered by KyT, and direct access to markets by producers.

It is likely that the future availability of Phane would be affected by climate change. For example, late rainfall could result in caterpillars emerging when there are no Mophane leaves, resulting in a loss of the first (December) Phane emergence, while very wet conditions will result in high mortality of cocoons and young caterpillars. However, the development of the Phane industry now would enhance community rural industry skills that could be diversified to other products and broaden the income generation base, hence increasing resilience and adaptive capacity.

The potential of medicinal plants of the Limpopo basin have been developed to only a very limited extent. The Devil's Claw, one of the few examples of a commercially developed medicinal plant, thrives under desert to semiarid conditions and as a result has the potential to thrive under the future climate projected for Botswana. Many other plants could find commercial applications, and currently knowledge of the medicinal properties of plants is being plundered by scientists and pharmaceutical companies from developed countries (Moloi 2003a, b and c). This effectively denies communities their rightful potential income and benefits. For communities of the hardveld to benefit from commercial applications of medicinal compounds derived from local plants, changes are needed in intellectual property laws.

Development of cultural heritage tourism in rural areas is another potential source for income and employment creation if practised under appropriate policy framework (Dube and Moswete, 2003). Proper measures need to be established, first to make communities aware of their culture and other resources as a potential source of income under tourism. Second, there is a need to provide appropriate infrastructure and build capacity to facilitate such activities.

A culture of shared responsibility supported by available knowledge systems under defined traditional institutions such as the *Kgotla* and chieftainship is an intangible resource that could be developed as part of community adaptation strategies in the future. These traditional institutions for local administration and justice have persisted, albeit distorted, as the most viable basic community structure in rural settings, together with a wealth of human socio-cultural capital. Every Motswana is identified in legal police documents, applications for passport, government health cards and death notices in terms of their village, chieftainship and ward. The chieftainship structure is important for implementing government programmes and forms a communication link between rural communities and central government. The *Kgotla* still has roles to play in the administration of justice and serves as a platform for launching new programmes and the dissemination of information.

The chieftainship institution is now part of the local government system, recognized among other things by the national flag flying at *Kgotla*. Government endorses the selection and election of leaders of these institutions, and they are paid a formal monthly salary. But instead of labour regiments, councils and headmen of wards, the chief has to work with a Village Development Committee (VDC), which is elected every two years. Chiefs are now ranked, and paramount chiefs sit in the House of Chiefs.

Despite the institutional community structures described above, in reality, central government's new framework has defined power relations that inadvertently sidelined the community institutions in decision making. For example, the *Kgotla* is not part of decision making on natural resource use and management, which was one of its traditional core functions. The independence and community influence of such institutions have been severely minimized. There is growing reluctance to attend *Kgotla* meetings, which shows their growing loss of value in terms of impacts on people's livelihoods.

However, as a system of organization, the *Kgotla* still offers some important lessons that could be updated and used in the development of community adaptation strategies. This is crucial, as in addition to natural products, community coherence based on its traditional governing structures offers wealth in terms of cultural heritage – another potential resource for sustainable adaptation that has remained untapped in the Limpopo river basin and in the country as a whole.

Synthesis of Community Coping Strategies

The environmental and climatic conditions of the Limpopo river basin have shaped the adaptive lifestyles of rural communities. Drawing on accumulated knowledge and experiences with the local climate and resources, communities of the basin developed a strong institutional framework of collective decision making on resource use and management as well as innovative uses of natural products that provide a hedge against potential losses of cattle and crops. However, vulnerability to drought and other stresses increased as traditional institutions and practices were weakened by government policies both before and after independence, and by other social and economic changes. Attempts by the government to reduce rural poverty and vulnerability have largely been unsuccessful for reasons discussed earlier.

One of the critical factors has been the failure to engage communities in decisions that affect their livelihoods. Through the CBNRM strategy, the government has sought to transfer back the management of local resources to communities. But thus far the strategy has struggled because of a lack of local institutions that are accountable to the communities and that have the capacity to effect equitable access and use of resources and to represent local interests in the implementation of the national policy (Thomas and Twyman, 2005). Remnants of the traditional institutions of the *Kgotla* and chieftainship could be supported and revitalized to facilitate decentralization of resource management, giving back some of the responsibilities these institutions held before but with a capacity to operate under a modern system. This would provide a framework at the community level that could also be effective in promoting adaptation to climate change.

As a resource, indigenous knowledge has historically sustained precarious lifestyles in the shadow of a variable semiarid climate characterized by frequent droughts that have periodically reached catastrophic extremes, and this is evidence of its potential to further develop and enhance adaptation capacity to climate change. Some of the central issues to be addressed in providing a framework for successful utilization of indigenous resources and for implementing policies on decentralization of management of local resources include effective information dissemination, protection of intellectual property, development of modern rural industries where opportunities exist, cultivation of domestic and international markets, capacity building in entrepreneurship, and strategic provision of seed resources for community initiatives.

These can be achieved with an enabling policy of technical and other resource support systems operating as much as possible within the existing institutional frameworks in rural areas. The prevention of loss of indigenous knowledge systems through biopiracy could be a major intervention that would empower the communities for a sustainable future, based on the commercialization of their knowledge systems (Hansen and Van Fleet, 2003; Phuthego and Chanda, 2004). Another important aspect of the development of commercialization of veld products is the domestication, cultivation and sustainable production of the resources.

A series of stakeholder community workshops conducted in the Limpopo river basin showed that new programmes in rural areas need to take into account changes resulting from the interplay between traditional livelihood systems with Western institutions and values. Without the development of appropriate skills, income-generating community adaptation activities have a limited chance for success, as shown by the CBNRM initiatives (Phutego and Chanda, 2004). Many ongoing self-initiated women's groups on income-generation projects, such as weaving, dance groups and other projects faced problems of organizational, management and financial accounting skills but had no access to assistance in their localities. There is a need to investigate the kind of management that would be suitable at these levels, where trust, confidence and transparency can easily become elusive, with the resulting failure to achieve the high potential of such noble initiatives.

Also apparent from the stakeholder community workshops and the subsequent field surveys were the limited activities of non-governmental organizations (NGOs) in the Limpopo river basin rural areas, particularly in Bobirwa and the Northeast District. The categorization of Botswana as a middle-income country has resulted in less funding from donor agencies and a reduction in the activities of international NGOs in the country. NGOs play an important complementary role in the development of adaptation capacity in rural areas.

In the past, the *Kgotla* used to provide a selling point for cattle in villages under the supervision of the chief. Combined efforts involving government, NGOs and local initiatives, such as community-based organizations, are required to develop the modern *Kgotla* under the village development committees into rural trade hubs where guidance could be provided – within a familiar setup and in a language that would be understood by communities – on issues of entrepreneurship or possibilities to partner with the private sector for the development of local industries. The prospects for this happening are increasing with the general rise in education levels in the country, which has also yielded increasing numbers of well-educated chiefs.

It is worth noting that the economy of Botswana remains strongly driven by the government, because of its being almost wholly dependent on mineral exports. Government is the main source of revenues, and attempts to diversify have not been successful (BNA21, 2002). Industrial skills are limited in the country, and the role of the private sector is still emerging, partly explaining the failure of programmes such as the Financial Assistance Policy. Faced with issues of poverty and climate extremes, alleviation measures have had to be applied in the context of disaster management at the expense of productive initiatives. As a result, the spirit of self-reliance is destroyed and a society that expects nearly everything from the government is created. Under these circumstances, the main recommendation of this chapter, which is diversification of the economy through rural industries, remains a challenge within the framework of the current national economic structure.

An economy that is heavily dependent on export earnings from a single non-renewable product (diamonds in this case), that is heavily dependent for a high proportion of its livelihoods on agriculture and pastoralism in a semiarid, drought-prone climate and that has high rates of income and capability poverty is highly vulnerable. This vulnerability is heightened by signals that the future climate may become more arid and unpredictable, with consequences for the natural resource base. A proactive approach is required to reduce vulnerability to variable climate, variable productivity of natural resources and climate change. Reducing vulnerability needs to be supported by an appropriate policy framework that engages the private sector, civil society and local institutions to facilitate economic development and diversification in rural areas and build local capacity in managerial, entrepreneurial and technical skills for the sustainable management and use of natural resources. Traditional knowledge, practices and institutions are intangible but critical resources that should be fully utilized to build sustainable adaptation capacity in rural communities.

References

Bassiri, R., H. J. F. Reynolds, R. A. Virginia and M. H. Brunelle (1998) 'Growth and root NO_3 and PO_{43} uptake capacity of three desert species in response to atmospheric CO_2 enrichment', *Australian Journal of Plant Physiology*, vol 24, pp353–358

Bhalotra, Y. P. R. (1989) *Nature of Rainy Season in the Chobe District*, Department of Meteorological Services, Ministry of Works, Transport and Communication, Gaborone

BNA21 (2002) *Botswana National Report on the Implementation of Agenda 21 and Other Rio Earth Summit Decisions*, Botswana National Agenda 21 Coordinating Committee, Gaborone, Botswana

Bond, W. J. and G. F. Midgley (2000) 'A proposed CO_2-controlled mechanism of woody plants invasion in grasslands and savannas', *Global Change Biology*, vol 6, pp865–869

Bond, W. J., G. F. Midgley and F. I. Woodward (2003) 'The importance of low atmospheric CO_2 and fire in promoting the spread of grasslands and savannas', *Global Change Biology*, vol 9, pp973–982

Botswana Government (2002a) 'National Master Plan for Arable Agriculture and Diary Development (NAMPAADD)', Government White Paper No 1, Ministry of Agriculture, Gaborone

Botswana Government (2002b) 'Revised National Policy for Rural Development', Government White Paper No 3, Ministry of Finance and Development Planning, Gaborone

Botswana Government (2002c) 'Game Ranching Policy for Botswana', Government White Paper No 5, Ministry of Trade, Industry, Wildlife and Tourism, Gaborone

Botswana Government (2004) 'Agricultural Resources Conservation (Harvesting of Veld Products) Regulations', Statutory Instrument No 28 of 2004, Agricultural Resources Board, Ministry of Environment, Wildlife and Tourism, Gaborone

Buchanan-Smith, M. (1998) 'Food and nutrition insecurity: The impact of drought', in M. Mugabe, K. Gobotswang and G. Holmboe-Ottesen (eds) *From Food Security to Nutrition Security in Botswana*, National Institute of Research/Institute of General Practice and Community MedicineNIR/IASAM Collaborative Research Programme on Health, Population and Development, University of Botswana, Gaborone, Botswana and and University of Oslo, Oslo, Norway

Burton, I. (2004) 'Climate change and the adaptation deficit', Occasional Paper 1, Adaptation and Impacts Research Group (AIRG), Meteorological Service of Canada, Environment Canada, Toronto, Ontario

Campbell, A. C. (1978) 'The 1960s drought in Botswana', paper presented at the Drought Symposium, Botswana Society, Gaborone

Campbell, A. C. (1986) 'The use of wild food plants and drought in Botswana', *Journal of Arid Environments*, vol 11, pp81–91

Cunningham, A. B. and S. J. Milton (1987) 'Effects of basket-weaving industry on Mokola palm and dye plants in northwestern Botswana', *Economic Botany*, vol 41, pp386–402

Dube, O. P. (1984) 'Settlement pattern as a dynamic process: The Sebina Mathangwane area of North Central District, 1895–1959', unpublished thesis, University of Botswana, Gaborone

Dube, O. P. (1995) 'Monitoring vegetation and soil degradation in Botswana', Progress Report, Project No S010001, Australian Center for International Agricultural Research (ACIAR), Department of Geographical Sciences, University of Queensland, Brisbane, Australia

Dube, O. P. and R. M. Kwerepe (2000) 'Human induced change in the Kgalagadi sands: Beyond the year 2000', in S. Ringrose and R. Chanda (eds) *Towards Sustainable Management in the Kalahari Region—Some Essential Background and Critical Issues*', proceedings of the Botswana Global Change Committee – START Kalahari Transect Directorate of Research and Development, University of Botswana, Gaborone

Dube, O. P. and G. Pickup (2001) 'Effects of rainfall variability and communal and semi-commercial grazing on land cover in southern African rangelands', *Climate Research*, vol 17, pp195–208

Dube, O. P. and N. Moswete (2003) 'Tourism: Searching for adaptation options to climate change in southern Africa', AIACC Notes 2, pp6–7

Gichangi, K. K. and E. N. Toteng (2004) 'Policy and drought in the Kgatleng District of Botswana: Impact on communities' vulnerability and adaptation', Botswana First National Conference on Application of Science and Technology, 13–15 July, Botswana Environmental and Natural Resources Observation Network (BENRON), University of Botswana, Gaborone

Gleason, G. (2004) 'Cuttings are key to commercial cape honeybush production', *Farmer's Weekly*, 30 July, pp40–44

Hansen, S. A. and J. W. Van Fleet (2003) *Traditional Knowledge and Intellectual Property: A Handbook on Issues and Options for Traditional Knowledge Holders in Protecting their Intellectual Property and Maintaining Biological Diversity*, American Association for the Protection of Science, New York

Hulme, M. (2004) 'A change in the weather? Coming to terms with climate change', in F. Harris (ed) *Global Environmental Change Issues*, John Wiley and Sons, Chichester

Hulme, M., R. Dougherty, T. Ngara, M. New and D. Lister (2001) 'African climate change: 1900–2100', *Climate Research*, vol 17, pp195–208

Desanker, P. and C. Magadza (2001) 'Africa', in J. McCarthy, O. Canziani, N. Leary, D. Dokken and K. White (eds) *Climate Change 2001: Impacts, Adaptation and Vulnerability*, Contribution of Working Group II to the Third Assessment Report of the Intergovernmental Panel on Climate Change, Cambridge University Press, Cambridge, UK and New York

Jefferies, K. (1997) 'Poverty in Botswana', in D. Nteta, J. Hermans and P. Jeskova (eds) *Poverty and Plenty: The Botswana Experience*, Botswana Society, Gaborone

Joubert, L. (2004) 'CO_2-hungry super trees take over the savannas', *Farmer's Weekly*, 26 March, p34

Kates, R. W. (2000) 'Cautionary tales: Adaptation and the global poor', *Climate. Change*, vol 45, pp 5–17

Kgathi, D. L. (1984) 'Aspects of firewood trade between rural Kweneng and urban Botswana: A socio-economic perspective', NIR Working Paper No 46, National Institute of Development Research and Documentation, University of Botswana, Gaborone

Kgathi, D. L. (1989a) *Firewood Marketing in Urban Botswana*, Urban Household Energy Strategy Project, The World Bank, Washington, DC

Kgathi, D. L. (1989b) 'The grapple trade in Botswana', *Botswana Notes and Records*, vol 20, pp119–124

Khama, S. (1971) 'Traditional attitudes to land and management of property with special reference to cattle', *Botswana Notes and Records*, Special Edition No 1, Proceedings of the Conference on Sustained Production from Semi-arid Areas, October, Gaborone, pp57–61

Kupka, J. (2001) 'Growing profits from organic medicine', *Farmer's Weekly*, 3 May, pp4–5

Lapologa Weekend Gazette (2003) 'Kgetsi ya Tsie: Taking over the international cosmetics industry one cheru at a time', *Lapologa Weekend Gazette*, 31 January–8 February, pp1–2

Ministry of Finance and Development Planning (1998) *Study of Poverty and Poverty Alleviation in Botswana*, stakeholder views on the conclusions and recommendations of the study reports by the National Institute of Development Research and Documentation, Phase 1, vol 3, University of Botswana, Gaborone

Ministry of Finance and Development Planning (2002) 'Revised national policy for rural development', Government Paper No 3, Ministry of Finance and Development, Gaborone

Mirza, M. M. Q. (2003) 'Climate change and extreme weather events: Can developing countries adapt?' *Climate Policy*, vol 3, pp233–248

Moloi, E. (2003a) 'Traditional doctors enter the intellectual property fray', *The Botswana Guardian*, Friday 26 September, p13

Moloi, E. (2003b) 'USA linked to theft of Botswana herbs', *The Botswana Guardian*, Friday 5 September, p3

Moloi, E. (2003c) 'Traditional healers decry "half-hearted" government attitude', *The Botswana Guardian*, Friday 21 November, p7

Moruakgomo, M. B. W. (1996) 'Commercial utilization of Botswana's veld products: The economics and dimensions of Phane trade', in *Proceedings of the First Multidisciplinary Symposium on Phane*, Department of Biological Sciences, University of Botswana, Gaborone

Nofal, J. (2004) 'Honeybush hits the high road', *Farmer's Weekly*, 1 October, pp40–41

Ohiokpehai, O., J. Jagow, J. Jagwer and S. Maruapula (1998) 'Tsabana: Towards locally produced weaning foods in Botswana', in M. Mugabe, K. Gobotswang and G. Holmboe-Ottesen (eds) *From Food Security to Nutrition Security in Botswana*, National Institute of Development Research and Documentation, University of Botswana, and Department of General Practice and Community Medicine, University of Oslo, Oslo and Lentswe la Lesedi, Gaborone, pp179–195

Parida, B. P., D. B. Moalafhi and O. P. Dube (2005) 'Estimation of likely impact of climate variability on runoff coefficients from Limpopo basin using artificial neural network (ANN)', in *Proceedings of International Conference on Monitoring, Prediction and Mitigation of Water-Related Disasters*, Disaster Prevention Research Institute (DPRI), Kyoto University, Kyoto, Japan, pp443–449

Phuthego, T. C. and R. Chanda (2004) 'Traditional ecological knowledge and community-based natural resource management: Lessons from a Botswana wildlife management area', *Applied Geography*, vol 24, pp57–76

Schapera, I. (1951) *The Ethnic Composition of Tswana Tribes*, Monographs on Social Anthropology No 11, The London School of Economics and Political Science, London

Scholes, R. J. and R. Biggs (2004) *Ecosystem Services in Southern Africa: A Regional Assessment*, Millennium Ecosystem Assessment, Southern Africa Millennium Ecosystem Assessment, Council for Scientific and Industrial Research, Pretoria, South Africa

Sekhwela, M. B. M. (1997) 'Coordinated monitoring and management of the use of natural woodlands in Botswana: A strategy for woody biomass harvesting', *Journal of the Forestry Association of Botswana*, vol 1, pp32–44

Shackleton, C. M. and S. E. Shackleton (2000) 'Direct use values of secondary resources harvested from communal savannas in the Bushbuckridge Lowvel, South Africa', *Journal of Tropical Forest Products*, vol 6, pp28–47

Sharma, N., T. Damhang, E. Gilgan-Hunt, D. Grey, V. Okaru and D. Rothberg (1996) 'African water resources. Challenges and opportunities for sustainable development', Technical Paper No 33, African Technical Department Series, World Bank, Washington, DC

Sithole, B. (1993) 'Rethinking sustainable land management in southern Africa: The role of institutions', in F. Ganry and B. Campbell (eds) *Sustainable Land Management in African Semi-Arid and Subhumid Regions*, Proceedings of the SCOPE workshop, 15–19 November, French Agricultural Research Centre for International Development (CIRAD), Dakar

Solway, J. S. (1994) 'Drought as a revelatory crisis: An exploration of shifting entitlements and hierarchies in the Kalahari, Botswana', *Development and Change*, vol 25, pp471–495

Solway, J. S. (2002) 'Navigating the "neutral" state: Minority rights in Botswana', *Journal of Southern African Studies*, vol 28, pp711–729

Sporton, D. and D. S. G. Thomas (eds) (2002) *Sustainable Livelihoods in the Kalahari Environments: A Contribution to Global Debates*, Oxford University Press, Oxford, UK

Theron, K. (2001) 'Developing commercial products from wild plants', *Farmer's Weekly*, 7 December, pp30–32

Thomas, D. S. G. and C. Twyman (2005) 'Equity and justice in climate change adaptation amongst natural-resource-dependent societies', *Global Environment Change*, vol 15, pp115–124

Tlou, T. (1990) *A History of Ngamiland – 1750 to 1906: The Formation of an African State*, Macmillan Botswana Publishing Co. Pty, Gaborone

Trutter, M. (2001) 'Wild liquorice uplifts community', *Farmer's Weekly*, 1 June, pp52–53

Twine, W., D. Moshe, T. Netshiluvhi and V. Siphugu (2003) 'Consumption and direct-use values of savanna bio-resources used by rural households in Mametja, a semi-arid area of Limpopo Province, South Africa', *South African Journal of Science*, vol 99, pp467–473

United Nations Environment Programme (UNEP) (2001) *Vulnerability Indices: Climate Change Impacts and Adaptation*, UNEP Policy Series, UNEP, Nairobi, Kenya

Warren, A. (2005) 'The policy implications of Sahelian change', *Journal of Arid Environments*, vol 63, pp660–670

White, R. (1993) *Livestock Development and Pastoral Production on Communal Rangelands in Botswana*, Botswana Society, Gaborone

White, R. (1999) 'Fuelwood flow paths study in Francistown', final report, Energy Affairs Division, Ministry of Mineral Resources, Energy and Water Affairs, Gaborone

5

Community Development and Coping with Drought in Rural Sudan

Balgis Osman-Elasha, Nagmeldin Goutbi, Erika Spanger-Siegfried, Bill Dougherty, Ahmed Hanafi, Sumaya Zakieldeen, El-Amin Sanjak, Hassan A. Atti and Hashim M. Elhassan

Introduction

Persistent and widespread drought is a recurrent feature of the climate of Sudan. Drought and highly variable rainfall severely impact rural populations in Sudan and contribute to poverty, hunger, water scarcity, dislocation and even famine. Sudan's rural populations are also highly vulnerable to future climate change (Osman-Elasha and Sanjak, 2008). Still, there are examples of development strategies and resource management measures employed by rural populations of Sudan that have increased the resilience of communities to cope with drought and its effects. These examples provide models that can be applied to increase drought resilience more widely in Sudan. They also offer lessons that can guide the integration of climate change adaptation with community development strategies.

Development and resource management activities are examined in three rural, drought-prone areas of Sudan where traditional farming practices are common: Gireighikh Rural Council in Bara Province of North Kordofan State, Arbaat in the Red Sea State, and El Fashir Rural Council in North Darfur State. People in these rural communities face multiple threats that include climate hazards as well as poverty, resource scarcity, food insecurity, disease, land degradation and violence. A variety of coping and adaptive strategies have been developed and employed in the communities to address these threats. Many came into use autonomously within the studied communities, meaning without external intervention or planning. Others were adopted, or expanded in use, as a result of community development projects that were planned and supported by external agencies.

We examine the three cases to understand how development projects and resource management measures have affected community resilience to drought and other stresses, factors that enable or inhibit their effectiveness, and their potential as approaches to climate change adaptation. We briefly describe the framework and methods of the case studies, present and compare experiences from the different cases, and close with general lessons for climate change adaptation.

Sustainable Livelihoods and Adaptation

All of the case studies are sites of community development projects to enable and support sustainable livelihoods. Sustainable livelihoods is an approach to poverty reduction that focuses on the strategies and means by which people derive their livelihoods, the context of vulnerability to adverse outcomes in which they operate, and the assets needed to sustain and improve livelihoods, reduce vulnerability and move out of poverty (see, for example, DFID, 1999). Projects that apply a sustainable livelihoods strategy attack poverty and vulnerability holistically by increasing access to the assets needed to achieve positive livelihood outcomes. These livelihood assets include natural, physical, financial, human and social capital.

In the rural settings of the three case studies, livelihoods are based predominantly on traditional farming and pastoral activities. Exposure to drought and highly variable rainfall, coupled with highly constrained and unequal access to assets, are defining features of the vulnerability of people pursuing farming and pastoral livelihoods in the region. The development projects in the study areas, which were implemented partly in response to the 1980–1984 droughts that severely impacted much of the region, consist of multiple measures implemented together to increase access to the five different classes of livelihood assets. Each of the projects are considered by leaders and members of the communities where they were implemented to have been successful in increasing household and community assets that have generated multiple benefits, including greater resilience to drought. The measures that compose the community development projects therefore can be considered to be adaptation measures for reducing climate risks.

A variety of typologies have been applied to categorize adaptation measures (see Smit et al, 2001, for an overview of typologies). Adaptation measures have been categorized by their purposefulness (spontaneous or planned, anticipatory or reactive, and tactical or strategic; see Smit et al, 1996; Stakhiv, 1993), their function (reduce risks, diversify risks, spread risks and increase capacity to bear losses; see Burton et al, 1993), their form (technological, legislative, regulatory, financial, institutional and market-based instruments; see Bryant et al, 2000; Carter, 1996; Smithers and Smit, 1997) and implementing agent (individual, household, community, local government and national government; see Smit et al, 2000).

While the above attributes of adaptation are relevant to adaptation in Sudan, we found that within the context of our case studies, it was useful to apply the concepts and framework of sustainable livelihoods to categorize and assess adaptation measures. Our methodological approach, which is adapted from Springate-Baginski and Soussan (no date), is briefly outlined below and is described in greater detail in Osman-Elasha (2006).

Community members participated in identifying measures being used to increase drought resilience and in categorizing the measures according to the class of livelihood asset targeted by each measure. Structured interviews were used to gather information about climatic and other stresses and the adaptation measures and their effectiveness. Indicators of the effectiveness of adaptation measures were developed with community input and are based on the state of livelihood assets before and after implementation of the projects, as evaluated by community members. Changes in livelihood assets are evaluated in three dimensions: productivity, equity, and the sustainability of the measures and their benefits.

The specific indicators differ across the case studies due to differences in the environments, livelihoods and priorities of community members. In Bara Province, indicators of resource productivity changes focus on the area of rangelands rehabilitated, carrying capacity of lands and forage production. In Arbaat and Darfur indicators of resource productivity changes included crop yields per unit land area, volume of crop production, diversification of crops and livestock breeds, numbers of animals, quantity and reliability of water supply, and water conservation. Assessment of productivity changes also takes into account financial indicators such as income changes, diversity of income sources and access to credit.

The assessment of equity focused on the situation of minority groups and women. Indicators looked at changes in the access of women and minorities to land, water, social services and credit, their participation in training and production activities and their participation in decision making. Indicators of sustainability include improvement in environmental conditions, adoption of management practices that protect land and water, use of local knowledge, capabilities and technologies, formation and strengthening of community social institutions, raised awareness of climate hazards, and observed continuation of measures after discontinuation of externally funded projects.

Gireighikh Rural Council, Bara Province, North Kordofan State

Bara Province lies in the North Kordofan State of Western Sudan. The climate of Bara Province is semiarid, average rainfall is quite low at roughly 250–300mm per year, and seasonal and interannual variability of rainfall is high. The lands are marginal, with sandy, low fertility soils that are becoming increasingly degraded under combined human and climatic pressures. Most of the province is covered by desert scrub vegetation on undulating sand dunes. Agropastoral and transhumant livestock grazing are the predominant livelihoods of the study area.

The cumulative impact of recurring droughts, cultivation of marginal lands, fuel wood gathering and overstocking of livestock has drastically depleted the vegetation. As a result, soil erosion, desertification and dust storms have emerged as significant environmental challenges. The local resource base has been degraded, undermining livelihoods and leaving communities more vulnerable to adverse effects of future drought. The province was severely impacted by the 1980–1984 droughts that hit the entire Sahel, affecting family and tribal structures and their autonomous traditional practices of resource management, and leading to thousands of people migrating from their villages to refugee camps around the towns and cities.

In response to these devastating conditions, in 1992 the United Nations Development Program (UNDP) and the Global Environment Facility (GEF) initiated the project 'Community-Based Rangeland Rehabilitation (CBRR) for Carbon Sequestration'. The project sought to promote both climate change mitigation and adaptation goals by implementing community-based natural resource management strategies in 17 villages of the Gireighikh Rural Council in central Bara Province. The objectives included prevention of over-exploitation and degradation of marginal lands, rehabilitation of rangelands for the purpose of carbon sequestration, preservation of biodiversity, reduction of dust storms and reduction of risk of crop failure. The approach of the project was to increase livelihood opportunities and diversify local production systems; these are expected to have long-lasting benefits for improving socioeconomic conditions, decreasing out-migration and stabilizing the local population.

The focus of the CBRR project was on measures to improve the productivity and sustainability of the natural resources that support agropastoral and transhumant livelihoods. But achievement of this goal was reinforced by an approach that also simultaneously targeted the four other livelihood assets: physical, financial, human and social. The project emphasized community participation in decision making, training of members of the community to better manage natural resources, and greater participation of women in economic and social activities. Community development committees were established in the villages through which community members participated in decision making, implementation of resource management measures and organization of social services.

The measures implemented by the CBRR project, their benefits, actors who implemented them, the resources and capacities needed to implement them, and obstacles and risks impeding their implementation are listed in Table 5.1. Measures targeted at improving natural resource assets include land rehabilitation, controlling grazing pressures, introducing sheep as replacement for goats, planting trees and shrubs for shelterbelts, stabilizing sand dunes and developing women's gardens. Measures targeted at physical assets include installing and maintaining wells and water pumps, building grain milling and storage facilities, and changing building practices to conserve wood. Financial assets were increased by improving access to local and national markets, diversifying production activities (for example raising, fattening and marketing sheep), and providing greater access to credit through revolving credit funds. Human capital was increased through training of farmers and workers, including women, and providing health, education and other services. Social capital was enhanced by the formation of community development committees, the participation of community members in decision making, greater participation of women in production activities, and education, information exchange and networking activities.

Table 5.1 *Adaptation measures in Gireigbikb Rural Council, Bara Province, North Kordofan State*

Targeted Asset	Measures	Benefits	Actors	Needed resources and capacities	Obstacles and risks
Natural	• Rehabilitate rangeland • Shift from crop farming to livestock • Diversify livestock breeds • Water harvesting • Stabilize sand dunes • Construct and maintain windbreaks • Women's irrigated gardens • Provision of veterinary services	• Improve land productivity • Diversify production • Improve environmental conditions • Improve water supply and quality • Household food security • Food and financial security for women • Improve animal health and productivity	• Subsistence farmers • Female farmers • Project staff • Government officials	• Water source • Rangeland • Financial resources • Farming knowledge and skills • Training and extension	• Lack of financial resources • Out-migration of trained workers • Changing government policies • Competition among tribes • Conflict over resources
Physical	• Install and maintain wells and water pumps • Build grain storage and milling facilities • Conserve energy with improved stoves, other methods • Mud-walled houses to replace wooden huts	• More reliable water supply • Increase local processing of farm products • Increase local food reserves • Increase vegetable and fruit production with supplemental irrigation • Improve nutrition • Reduce clearing of vegetation for fuel wood and building materials	• Community Development Committees • Community members and elders	• Financial resources • Spare parts • Equipment for digging • Clay soil for building mud houses • Trained building workers • Trained mechanics • Informed pioneer women	• Lack of financial resources • Changing state level policies
Financial	• Expand and diversify on-farm income generating activities • Expand and diversify off-farm employment opportunities • Establish community revolving funds	• Access to credit • Greater and less variable financial income • Greater financial resources for bearing losses, recovering from losses, maintaining food security and maintaining and improving other livelihood assets	• Community development committees • Farmers, women and minority groups • Government development officers • Extension agents	• Financial resources • Trained workers • Training and extension facilities, services and materials • Rural development officers, extension officers • Trained community members	• Lack of financial resources • Out-migration of trained workers • Changing government policies

Table 5.1 *(continued)*

Targeted Asset	Measures	Benefits	Actors	Needed resources and capacities	Obstacles and risks
Human	• Build on traditional knowledge • Training of farmers and workers • Provide health, education and other services • Training of women	• Improve capabilities to manage natural resources • Improve resource productivity • Better human health and nutrition • Improve women's livelihoods	• Community Development Committees • Trained community members and elders • School teachers, extension workers, veterinary officers	• Schools • Training and extension facilities, services and materials • Clinics and health units • Veterinarian services • Teachers, trainers, extension workers and veterinarians	• Changing government policies • Migration of trained workers • Limited financial resources
Social	• Establish community development committees in local villages • Empower community development committees to make decisions, allocate resources, implement measures • Education, information exchange and networking	• Delivery of social services through a system in which the community has ownership • Participation of community members, including minorities and women, in decision making • Early preparedness for climate hazards through dissemination of climate information and forecasts • Equitable participation in project activities and equitable distribution of benefits	• Community Development Committees • Trained community members • School teachers, extension workers, veterinary officers • Women's groups	• Community leaders • Financial resources • Communication network • Updated climate information and forecasts • Extension materials • Teachers, community development officers	• Social conflicts • Changing government policies • Migration of trained workers • Diminished authority of traditional leaders and institutions

Survey data indicate that community members observed substantial improvements in land rehabilitation, carrying capacity and forage production as a result of the various measures implemented by the project. The improvement of rangeland also led to a significant increase in animal numbers, particularly sheep. High levels of participation of women in irrigated gardens and of women and marginalized groups in training, extension and social services were attained. Indicators for the sustainability of livelihood assets and adaptation measures showed improvement as a result of the project. Key to the sustainability, according to survey results, are the efficiency of the community development committees, efforts of a Sudanese non-governmental organization to support and continue measures, dissemination of information on rainfall, new production inputs and technologies, and prices, and high loan repayment rates to the community revolving funds.

Arbaat, Red Sea State

The Arbaat study area is the catchment of the Khor Arba'at, a small seasonal stream located in the Red Sea State of northeastern Sudan, about 50km north of Port Sudan, the state capital. The region is characterized by relative isolation, harsh terrain, highly variable rainfall, recurrent drought spells, small area of cultivable land and low population density. Surface runoff is the primary source of fresh water in the Red Sea area, where runoff rates are high due to the rocky and compact nature of soils, steep slope, high portion of rainfall from thunderstorms and the poor vegetation.

Animals represent the main means of economic and social mobility, recognition and survival of the *Beja* pastoralists of the region. Aware of their environment's vulnerability to drought and famine, they developed a primarily subsistence agropastoral system with a dispersed pattern of settlement and migration in pursuit of water, pasture and cultivable lands that helped maintain the carrying capacity of the land, reduce competition and conflict, and allow for recovery from shocks (Abdelatti, 2003). Adherence to *Beja* traditions of *salif*, a social code of conduct governing relations and resource use, helped to preserve land, animal and other resources. This included a strong social sanctioning system imposed by tribal leaders that helped to constrain overuse of land and wood, and rebuild of animal stocks after each drought cycle.

Frequent occurrences of drought in the Red Sea hills have been the norm during the 20th century. But the previous pattern of short-term recovery was shattered after the long drought and famine of the mid-1980s and the traditional agropastoral system of the *Beja* failed to re-configure (Abdelatti, 2003). Since then the Red Sea State has been in an almost constant state of emergency and relief operations that only vary in scale, length and location from one year to the next. Although the term emergency implies a short or limited duration, whereby people are temporarily in need of relief (ODI, 1995), the reality in the area is that most emergencies last for longer than one year. According to Abdelatti et al (2003), a 'relief culture' has become established in which the population and local authorities are heavily dependent on the central government and foreign aid organizations to address humanitarian crises. Under these conditions, long-term planning, including for building resilience to the impacts of drought, have become a low priority.

The Khor Arba'at Rehabilitation Project (KARP) was initiated in 1994 by SOS Sahel in response to the syndrome created by the Sahelian drought in the 1980s. The main objectives of the project were to improve the livelihoods and food security of the local population. Measures implemented by the project are identified in Table 5.2. Measures to improve the management of natural resources include micro-catchment water harvesting using contour bunds for planting trees (see Figure 5.1) and terracing, establishing home gardens and establishment of a system for equitable distribution of water. Physical assets were increased by digging wells for irrigation, installing and maintaining equipment for applying fertilizers and pesticides, and introducing improved seed varieties and new crops, including date palm. Financial assets were increased by diversifying farm production, expanding off-farm income opportunities, increasing access to credit through revolving funds and providing access to markets. Agricultural extension services, training, adult literacy, education for women and education about the use of credit added to human capacity. Social capacity was enhanced by the formation of community organizations, involvement of community members, including women, in decision making, and information exchange activities.

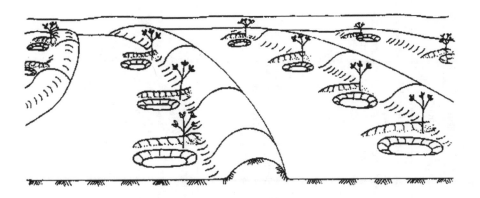

Figure 5.1 *Contour bunds for water harvesting and tree planting in Arbaat*

Benefits from the measures implemented by the KARP project include increased and less variable agricultural production, greater quantity, quality and reliability of water, enhanced livelihood opportunities, increased access to credit, reduced out-migration and stabilization of local population. Surveys also revealed that the project increased women's participation in public life, production activities and community development activities, and provided them greater access to resources.

Table 5.2 Adaptation measures in Arbaat, Red Sea State

Targeted Asset	Measures	Benefits	Actors	Needed resources and capacities	Obstacles and risks
Natural	• Earth bunds and terracing to harvest water • Increase recharge of ground water • Establish water distribution network for more equitable distribution • Establish home gardens	• More reliable and better quality water supply • Increase quantity and reliability of farm output and income • Diversify farm production • Stop out-migration • Improve household food security and nutrition	• Subsistence farmers • Women • Project staff • Government officials	• Water source (small stream or khor) • Agricultural land • Financial resources • Farming knowledge and skills • Technical capacity for irrigation • Training and Extension services	• Lack of financial resources • Government policies, including construction of dam
Physical	• Digging wells for irrigation • Install and maintain equipment for optimal application of fertilizers and pesticides • Introduce improved seed varieties, new crops and date palm	• Improve water availability • Increase quantity and reliability of farm output and income • Improve nutrition	• Community organizations • Farmers • Women	• Financial resources • Equipment for Digging • Spraying equipment • Fertilizers and insecticides • Spare parts • Trained irrigation workers • Trained mechanics • Informed pioneer women	• Lack of financial resources • Government policies, including construction of dam
Financial	• Expand and diversify on-farm income generating activities • Expand and diversify off-farm employment opportunities • Establish revolving funds	• Access to credit • Greater and less variable financial income • Greater financial resources for bearing losses, recovering from losses, maintaining food security and maintaining and improving other livelihood assets	• Community organizations • Women's groups • Extension agents	• Financial resources • Trained workers • Training and extension facilities, services and materials • Rural development officers, extension officers • Trained community members	• Government policies • Limited financial resources

Table 5.2 (*continued*)

Targeted Asset	Measures	Benefits	Actors	Needed resources and capacities	Obstacles and risks
Human	• Build on traditional knowledge • Training of farmers and workers • Training of community development committee members • Education of children • Adult literacy classes, especially for women • Access to information and ability to predict environmental changes	• Improve capabilities to manage natural resources • Increase productivity of farms and livestock • Improve women's livelihoods	• Community organizations • Traditional leaders and community elders • School teachers and extension officers	• Schools and literacy classes • Training and extension facilities, services and materials • Teachers, trainers and extension workers	• Limited financial resources • Migration of trained workers
Social	• Formation of community organizations • Promote women's participation in public activities, decision making and production process • Information exchange about climate hazards and forecasts	• Delivery of social services through a system in which the community has ownership • Participation of community members, including minorities and women, in decision making • Ability of women to participate in public life • Early preparedness for climate hazards through dissemination of climate information and forecasts • Equitable participation in project activities and equitable distribution of benefits	• Community Development Committees • Trained community members • School teachers and extension officers • Women's groups	• Community leaders • Financial resources • Communication network • Updated climate information and forecasts • Extension materials • Teachers, extension workers, community development officers	• Social conflicts • Changing government policies • Diminished authority of traditional leaders and institutions

Acquired skills and knowledge are considered to be the main sustainable benefits from the project. Elements of the project that have promoted sustainability of measures and benefits include government financial and technical support, establishment of community development organizations, continuous awareness campaigns, training programmes for different community groups, community participation in planning and contribution to the project costs and addressing community needs for health, education and other basic services.

El Fashir Rural Council, North Darfur State

North Darfur is situated on the northern transitional margin of the Intertropical Convergence Zone. Consequently, the state is one of the most drought affected regions of Sudan. Most of the area is deficient in water even in the wettest months of July to September, which account for 80 per cent of annual rainfall. The drought years of 1983–85 greatly affected the demographic and socioeconomic conditions of the area. Many people lost over half of their cattle, as well as large numbers of sheep, goats and camels as a result of the prolonged drought. Large numbers left their homes due to famine and the environmental impacts of desertification and drought, resulting in the growth of shantytowns. Seeking water and forage for their animals, transhumant pastoralists encroached on farmers' cultivated lands. Tribal conflicts and violence arose between pastoralists and farmers, and also between different groups of pastoralists. More recently, the region has been engulfed in civil war, the origins of which include conflicts over scarce resources in this harsh environment (Osman-Elasha and Sanjak, 2008).

In the two previous case studies, adaptive measures were introduced by projects that were externally promoted in response to a specific climatic event. In contrast to those cases, adaptation measures were developed autonomously by the local communities of El Fashir as a means of coping with variable resource productivity, recurrent drought and other stresses. These locally developed measures were later supported and expanded by an externally-funded food security project that was implemented in 1998 by the Intermediate Technology Development Group (ITDG). The ITDG food security project sought to build on indigenous knowledge of water-harvesting techniques with the involvement of local communities.

Autonomous adaptation measures that were expanded in El Fashir by the ITDG project, as well as new measures introduced by the project, are identified in Table 5.3. The greatest share of family food production has typically come from cultivation of sandy soils to grow millet, sesame and groundnuts. But in recent years, rain-fed farming on sandy soils has become increasingly risky and unable to produce enough food for the family. In response to this risk, *trus* cultivation was developed to grow sorghum, vegetables and tobacco on clay soils. First implemented in 1964 when a large *trus* or embankment was constructed across the *Wadi El Ku* to harvest water runoff, this practice has been promoted by the ITDG project to provide an important supplement to household food production and incomes, and to relieve pressure on cultivated sandy soils.

Magun cultivation is another practice that is indigenous to the area that has been promoted by the ITDG project. Adopted as a response to sand encroachment onto fertile soil, *magun* cultivation calls for digging holes 10–30cm in diameter, 5–15cm deep and spaced 40–70cm apart, loosening the soil in the centre of the holes and planting tobacco and melon seedlings. Terracing is used to harvest water and grow vegetables such as okra, eggplant and tomatoes that can be harvested up to 5 months after the rainy season. Home gardens, or *jubraka*, operated mostly by women, have been promoted for the growing of fast maturing crops and vegetables including okra, pumpkins and cucumbers. Gum trees (*Acacia senegal*) were planted to restore lands, increase vegetation cover and supply fuel wood.

Agricultural production was also improved by adding to physical assets with animal drawn ploughs, rental of tractors, expansion of traditional food storage facilities, construction of a central seed store, and use of improved seed varieties and new crop types. Financial assets were increased in El Fashir by diversifying and increasing agricultural production, improving access to credit and supporting marketing of farm products through unions. Training of farmers in techniques needed to diversify their production activities and improve resource management and the involvement of women in home gardening of vegetables has increased human capital. Social capital was increased through the formation of unions for traditional farmers, fruit growers and vegetable growers that have helped with organizing production, harvesting and marketing of products.

The benefits to the villages around El Fashir from the implemented measures include substantial increases in farm productivity, diversity of income sources, income levels and stability of incomes. Survey results indicate some improvement in the status of women, but there remain sharp distinctions between the roles of women and men. Women bear a disproportionately large share of the workload for crop farming, harvesting and vegetable growing, yet men make decisions about land use and farm planning. Sustainability of the measures and their benefits are aided by emphasis on indigenous water harvesting and cultivation techniques, and the prominent role of the traditional system of administration and traditional leaders in the management of resources and project implementation.

Adaptation Opportunities, Obstacles and Risks

Existing knowledge, skills, practices, resources and institutions that are indigenous to the case study sites provided opportunities to build on and increase resilience of the communities to drought and other hazards. But a variety of obstacles are identified by the case studies as impeding fuller adaptation. Furthermore, there are risks that threaten the sustainability of adaptation measures and benefits in each of the study areas.

Effective adaptation requires strong community institutions, leaders and innovators. The sustainable livelihood projects that were implemented in the case study areas worked with traditional administrative systems and leaders

Table 5.3 *Adaptation measures in El Fashir Rural Council, North Darfur State*

Targeted Asset	Measures	Benefits	Actors	Needed resources and capacities	Obstacles and risks
Natural	• Expansion of water harvesting for *trus* cultivation on clay soils • Construct terraces to grow vegetables • *Magun* cultivation to grow tobacco and melons • Restocking of gum trees (*Acacia Senegal*) • Use of crop residues to increase soil nutrients	• Increase quantity and reliability of farm output and income • Diversify farm output and incomes • Stable water supply • Reduce pressure on sandy soils • Improve household nutrition and food security	• Subsistence farmers • Women • Small commercial farmers • Merchants and urban farmers • Cooperative societies • Project staff • Government officials	• Agricultural land • Tree seedlings • Financial resources • Farming knowledge and skills • Indigenous knowledge of water harvesting • Technical capacity for irrigation • Training and extension services	• Conflicts over resources • Civil war • Changing Government policies • Financial resources
Physical	• Animal drawn ploughs for use in water harvesting • Rental of tractors • Traditional food storage technologies • Construction of central seed store • Use of improved seeds and new crop varieties	• Reliable water supply • Increased soil fertility and agricultural production • Diversification of crops • Food reserves for added food security	• Subsistence farmers • Cooperative societies • Intermediate Technology Development Group (ITDG)	• Domestic animals • Local material for building stores • Financial resources • Plant and animal residues • Farming knowledge and skills • Training and extension services	• Conflicts over resources • Civil war • Changing Government policies • Financial resources
Financial	• Provide diversified income opportunities from selling fruits and vegetable, gum garden, and employment opportunities. • Provide access to credit	• Greater and less variable financial income • Greater financial resources for bearing losses, recovering from losses, maintaining food security and maintaining and improving other livelihood assets	• Subsistence farmers • Women • Small commercial farmers • Cooperative societies • Farmers and trade unions • Extension workers	• Financial resources • Collateral assets • Farmers Union or Shail System • Extension services and facilities • Skilled farmers • Trained community members	• Inaccessible credit • Conflicts over resources • Civil war

Table 5.3 (*continued*)

Targeted Asset	Measures	Benefits	Actors	Needed resources and capacities	Obstacles and risks
Human	• Build on traditional knowledge • Training of farmers to manage their resources and diversify production • Involve women in production of vegetables	• Improve capabilities to manage natural resources • Increase productivity of farms and livestock • Better health and nutrition	• Community-based organizations • Women • Small commercial farmers • Traditional leaders and community elders • Cooperative societies • ITDG	• Training and extension facilities, services and materials • Community-based organizations • Farming knowledge and skills • Trainers, extension workers	• Conflicts over resources • Civil war • Migration of skilled labourers
Social	• Organize farmers unions • Organize women's groups • Organize social networks and cooperatives	• Mobilize resources for community projects • Participation of women in public life and production activities • Community participation in decision making	• Subsistence farmers • Women • Small commercial farmers • Merchants and urban farmers • Community-based organizations • Cooperative societies • Government officials	• Knowledgeable union members • Communication networks • Information sources • Community leaders • Development officers	• Conflicts over resources • Civil war • Diminished authority of traditional leaders and institutions

and local non-governmental organizations to gain community acceptance and ownership of the projects. The strengthening of institutions with training and resources, formation of new community institutions, empowering local institutions to plan and implement project activities, and promoting the participation of community members in the activities and decision-making processes of the institutions were central to the approach of the projects and are key factors contributing to their success. Local institutions have also been important for the continuation of development and adaptation activities after the termination of the externally funded projects.

However, the authority and coherence of traditional institutions have been weakened in Sudan by changes in government policies that have shifted authority to new government structures, social and economic changes, and displacement and migration of people. Weakening of local institutions is degrading social capital that is important for adaptation. It is also eroding the traditional systems for managing and accommodating migrations of herders and their livestock and for resolving tribal disputes.

A planned heightening of the Khor Arba'at Dam is another example of government policy that is jeopardizing adaptation gains. Heightening of the dam is expected to divert more water for urban use in the capital city of Port Sudan to alleviate severe water shortage during summer time. But increasing water storage in the dam and diverting more water to Port Sudan will reduce the volume of water spillover that supplies the Arbaat community. Potential adverse effects of reduced water supply to Arbaat include reduced cultivated area, displacement of families, spread and invasion of aggressive mesquite trees into fertile agricultural land, and reduced production of food for subsistence and marketing to urban dwellers in Port Sudan. Members of the Arbaat community are petitioning the government through traditional and religious leaders to either take necessary measures to mitigate the adverse impacts of the project or, preferably, to drop the idea.

Lack of natural and physical resources of all kinds is an obstacle to adaptation in rural communities of this semiarid region. Lack of water and fertile lands are key constraints on livelihoods and most of the adaptation strategies are aimed at improving water supply and management, and improving the fertility and productivity of land. Shortage of financial resources is identified in all three case studies as threatening the sustainability of adaptation measures. Credit for the purchase of seed is particularly needed. Other resources repeatedly identified in the case studies as essential for adaptation and in short supply include facilities and material for education, health, marketing and agricultural extension services, food storage facilities, equipment and spare parts.

Human capital is scarce and enhancement of human capital is an important component of the development projects. While most of the technologies promoted by the projects to improve the use and management of land and water, and to diversify production activities and incomes are indigenous to the areas, not all have the knowledge and skills needed to implement them effectively. Training and agricultural extension services are needed to disseminate the needed knowledge and skills. This, in turn, requires skilled rural development officers, extension workers and teachers. An important threat highlighted in two of the three case studies is the out-migration of skilled workers and technicians. In Bara, in-migration to the region is also a threat, as people are being attracted by the improved condition of rangelands and availability of water and may come into conflict with residents of the region.

Our case study of El Fashir Rural Council began prior to the current conflict and violence in Darfur, which pose substantial risks to the sustainability of adaptation efforts, as well as to the very survival of the people and communities of the region. Historically, conflicts in Darfur have arisen in response to scarcity of natural resources as farmers and pastoralists have clashed over water and land. The current crisis, which has reached a state of civil war, also has its origins in conflict over resources that have been exacerbated by the weakened state of local institutions. The context of vulnerability of the people of Darfur has deteriorated severely as a result.

But despite the violence and displacement of large numbers of people in Darfur, within the case study villages of El Fashir Rural Council water harvesting and associated agricultural activities continue without external support, cultivated area has increased, crop productivity is up from 2005, and livestock production has increased such that the villages are the main meat suppliers to the town of El Fashir.[1] Whether these activities and benefits can be sustained in El Fashir is doubtful as the crisis in Darfur continues. Extension of them to wider areas is impossible without cessation of hostilities and establishment of security in the region. The current conflict in Darfur deserves detailed assessment and in-depth analysis that go beyond the scope of this paper.

Lessons for Adaptation Processes

In the three Sudanese case studies, strategic approaches were taken towards improving livelihoods, increasing food security and reducing vulnerability to multiple sources of risk, including but not limited to climatic risks. The community development projects each implemented multiple measures aimed at these goals. Some measures were motivated explicitly by the desire to reduce climate risks (for example, water harvesting), some to expand livelihood options (for example, introduction of new crops and types of livestock), some to improve and sustain the productivity of land (for example, rehabilitating rangelands, controlling access and use of rangelands) and some to increase individuals' skills (training and extension services). Others were introduced to expand participation in decision making (for example, formation and engagement of village committees and women's groups), empower marginalized persons (for example, literacy programmes and irrigated gardens for women) and improve financial conditions (for example, revolving credit fund and marketing assistance). Taken together, they had the effect of increasing resilience of the communities to recurrent drought.

The strength of the development projects lay not in their individual measures for responding to drought, but in the holistic way in which the measures were planned and implemented to initiate and support adaptive processes and to build capacity to sustain the processes. They represent experiments in the integration of adaptation with community development. The result is an enhanced capacity for adaptation, or broader ability of the communities to manage climate and other risks to livelihoods and food security. The experiences from the studied communities suggest that, consistent with the findings of Smit et al (2001), future adaptation should focus on enhancing adaptive capacity, beginning with developing flexible management approaches and increasing current resilience to drought and other stresses. In the context of sustainable livelihoods, this takes the form of increasing access to natural, physical, financial, human and social assets that can be used to respond to climatic and other stresses.

Community institutions played an important role in the development projects and implementation of adaptation measures. The sustainable livelihood projects that were implemented in the case study areas worked with traditional administrative systems and leaders and local non-governmental organizations to gain community acceptance and ownership of the projects. The strengthening of institutions with training and resources, formation of new community institutions, empowering local institutions to plan and implement project activities, and promoting the participation of community members in the activities and decision-making processes of the institutions were central to the approach of the projects and are key factors contributing to their success. Local institutions have also been important to the continuation of development and adaptation activities after the termination of the externally-funded projects.

Each of the development projects relied on local institutions and local leaders to organize and implement their programmes. These institutions engaged persons at risk in decision processes, including farmers, herders, women and minorities. Using this bottom–up approach enables a community to have a better understanding of its own vulnerability, priorities and adaptation needs to shape the project's objectives, design and implementation. It also facilitates cooperation within the community, mobilization of local resources and access to indigenous knowledge to achieve project objectives. Integration of indigenous knowledge and experience with climate variability and traditional practices for managing the effects of variability proved to be important in each of the case study sites.

Human and social capacity building were critical to the effectiveness of the development projects in Bara, Arbaat and El Fashir, and appear to have produced long-lasting benefits. Each of the projects invested in creating and strengthening local institutions and providing training to community members on managing resources, diversifying livelihoods and responding to risks. The human and social capacities built by the projects are the mechanisms by which the adaptive processes in these communities can continue and become more effective.

Looking to the future, climate change is likely to alter patterns of climate variability and extremes from those experienced in the past, possibly in ways that would exceed the existing tolerance ranges of Sudan's rural communities. Traditional knowledge based on historical observations and experience will be less reliable as a guide for managing climate risks. While rural communities have adapted their livelihoods to variable climate conditions and developed coping strategies to deal with local risks, events that are unusually severe, persistent or frequent can invalidate those strategies and increase risk (Pelling et al, 2004). Economic globalization and other economic and social changes add to the uncertainty about the viability of current livelihood strategies and adaptation measures. In this context of change and uncertainty, the advantages of building adaptive capacity, in contrast to focusing on implementation of specific adaptation measures, and integrating adaptation with holistic community development programmes are all the more compelling.

Note

1   Personal communication to N. Goutbi from Practical Action, 6 September 2006.

References

Abdelatti, H. (2003) 'Khor Arba'at livelihoods and climate variability', AIACC-AF14 case study report, Higher Council for Environment and Natural Resources, Khartoum

Bryant, C.R., B. Smit, M. Brklacich, T. Johnston, J. Smithers, Q. Chiotti and B. Singh (2000) 'Adaptation in Canadian agriculture to climatic variability and change', *Climatic Change*, vol 45, no 1, pp181–201

Burton, I., R. Kates and G. F. White (1993) *The Environment as Hazard*, Guilford Press, New York

Carter, T. R. (1996) 'Assessing climate change adaptations: The IPCC guidelines', in J. Smith, N. Bhatti, G. Menzhulin, R. Benioff, M. I. Budyko, M. Campos, B. Jallow and F. Rijsberman (eds) *Adapting to Climate Change: An International Perspective*, Springer-Verlag, New York, pp27–43

DFID (1999) 'Sustainable livelihoods guidance sheets', Department for International Development, UK, available at www.livelihoods.org/info/info_guidancesheets.html

Ministry of Environment and Physical Development (2003) 'Sudan's first national communication under the United Nations Framework Convention on Climate Change', available at http://unfccc.int/resource/docs/natc

Overseas Development Institute (ODI) (1995) 'General food distribution in emergencies: From nutritional needs to political priorities', *Good Practice Review*, vol 3

Osman-Elasha, B. (2006) 'Human resilience to climate change: Lessons for eastern and northern Africa', final report of AIACC Project No AF 14, International START Secretariat, Washington, DC, available at www.aiaccproject.org

Osman-Elasha, B. and E-A. Sanjak (2008) 'Livelihoods and drought in Sudan', in N. Leary, C. Conde, J. Kulkarni, A. Nyong and J. Pulhin (eds) *Climate Change and Vulnerability*, Earthscan, London

Pelling, M., A. Maskrey, P. Ruiz and L. Hall (2004) *Reducing Disaster Risk: A Challenge for Development*, United Nations Development Programme, Bureau for Crisis Prevention and Recovery, New York, www.undp.org/bcpr

Smit, B., E. Harvey and C. Smithers (2000) 'How is climate change relevant to farmers?' in D. Scott, B. Jones, J. Andrey, L. Mortsch and K. Warriner (eds) *Climate Change Communication*, proceedings of international conference, 22–24 June, Kitchener-Waterloo, Environment Canada, Toronto

Smit, B., D. McNabb and J. Smithers (1996) 'Agricultural adaptation to climatic variation', *Climatic Change*, vol 33, pp7–29

Smit, B., O. Pilifosova, I. Burton, B. Challenger, S. Huq, R. Klein and G. Yohe (2001) 'Adaptation to climate change in the context of sustainable development and equity', in J. J. McCarthy, O. F. Canziani, N. A. Leary, D. J. Dokken and K. S. White (eds) *Climate Change 2001: Impacts, Adaptation and Vulnerability*, contribution of Working Group II to the Third Assessment Report of the Intergovernmental Panel on Climate Change, Cambridge University Press, Cambridge, UK and New York

Smithers, J. and B. Smit (1997) 'Human adaptation to climatic variability and change', *Global Environmental Change*, vol 7, no 2, pp129–146

Springate-Baginski, O. and J. Soussan (no date) 'A methodology for policy process analysis: Improving policy–livelihood relationships in South Asia', Working Paper 9, University of Leeds, Leeds, UK, www.geog.leeds.ac.uk/projects/prp

Stakhiv, E. (1993) *Evaluation of IPCC Adaptation Strategies*, Institute for Water Resources, US Army Corps of Engineers, Fort Belvoir, VA

6

Climate, Malaria and Cholera in the Lake Victoria Region: Adapting to Changing Risks

Pius Yanda, Shem Wandiga, Richard Kangalawe, Maggie Opondo, Dan Olago, Andrew Githeko, Tim Downs, Robert Kabumbuli, Alfred Opere, Faith Githui, James Kathuri, Lydia Olaka, Eugene Apindi, Michael Marshall, Laban Ogallo, Paul Mugambi, Edward Kirumira, Robinah Nanyunja, Timothy Baguma, Rehema Sigalla and Pius Achola

Introduction

Malaria and cholera are potentially fatal diseases that affect millions of people every year and result in more than a million mortalities, particularly in the developing countries. Malaria itself causes about one million deaths per year globally, of which more than 90 per cent occur in Africa (WHO/UNICEF, 2003). Global pandemics of cholera have been recorded since the beginning of the 19th century and the 7th pandemic, beginning in 1961, affected Asia, Europe, Africa and Latin America, causing thousands of fatalities (WHO, no date).

In the East African countries, malaria is ranked as the primary cause of morbidity and mortality in both children and adults. It causes about 40,000 infant deaths in Kenya each year; in Uganda annual cases of malaria range between 6 to 7 million, with 6500 to 8500 fatalities, and in Tanzania the annual death toll is between 70,000 and 125,000 and accounts for 19 per cent of health expenditure (De Savigny et al, 2004a and b). In the case of cholera, the first epidemic in Africa was reported as far back as 1836 (Rees, 2000). Major outbreaks were next reported in 1970 and affected West Africa (Guinea), the horn of Africa (Ethiopia, Somalia and Sudan) and Kenya (Waiyaki, 1996). The most severe cholera outbreak on the African continent was in 1998, accounting for more than 72 per cent of the global total number of cholera cases and acutely affecting the Democratic Republic of Congo, Kenya, Mozambique, Uganda and the United Republic of Tanzania. Cholera outbreaks in East Africa have been reported to the World Health Organization (WHO) since 1972. In the Lake Victoria region of East Africa both malaria and cholera are common, with malaria endemic in the lowlands and epidemic in the highland areas and cholera endemic in the basin since the early 1970s (Rees, 2000).

Recently, there has been a marked increase in the severity and intensity of malaria outbreaks in the highlands and cholera outbreaks in the basin. Scientific studies have linked this resurgence in disease episodes to recent climatic anomalies such as above average temperatures and rainfall (Wandiga et al, 2006 and Olago et al, 2007), which in turn has raised serious concerns about the impact of climate change on human health. Further, research studies conducted in this area also suggest that incidences of highland malaria and cholera are likely to increase with future changes in climate (Githeko et al, 2000). In the light of the fact that the present capacity with respect to services, programmes and infrastructure is quite inadequate, the above observations and findings have only created an added uncertainty about the ability of this region to cope with increased disease occurrences in the future.

This vulnerability of the Lake Victoria communities is further compounded by other pre-existing social and economic issues such as poverty, increasing population, lack of adequate infrastructure, human stresses on the local environment and the presence of other diseases (HIV/AIDS, diarrhoeal diseases and respiratory diseases). Non-health impacts of climate change on the environment, ecosystems, agriculture and livelihoods and infrastructure could also pose as additional stressors (see Confalonieri and Menne, 2007). Moreover, the absence of any policy interventions for implementing effective adaptive strategies to manage the increasing regional vulnerability to these diseases further reduces the ability of people to cope.

This research study was therefore undertaken to attempt to address these climate change-related human health concerns in the Lake Victoria region. Our particular objective is to assess the existing vulnerability to highland malaria and cholera, and to evaluate the present adaptive capacity in terms of its strengths and weaknesses, using specific case studies from Tanzania, Kenya and Uganda.[1] It is hoped that the results of this analysis will help to draw out important lessons that can inform the development and institution of appropriate policies and programmes to address the excess risks to communities in the Lake Victoria region of East Africa from future climate change.

Methodology for Assessment

Various tools such as the review of published and unpublished material, participatory assessments through focus group discussions, household and key informant interviews, and field observations were employed for the collection and analyses of data for this study. Published and unpublished material was referenced to gather secondary data to facilitate the selection of study sites and to establish disease patterns. This included records from hospitals and health centres, government statistical abstracts, ministerial offices, local institutions and policy documents and the *Weekly Epidemiological Review*[2] (WER-WHO).

Climate, hydrology and disease outbreak data were obtained from malaria and cholera study sites in Kenya, Uganda and Tanzania. The malaria study sites selected were Kericho (Kenya), Kabale (Uganda) and Muleba (Tanzania), all of which were highland sites (above 1100m above sea level) (see Figure 6.1). These sites have reported resurgence in malaria epidemics in the last two decades and have also experienced changes in climate since the turn of the 20th century. The cholera study sites in the Lake Victoria basin included Kisumu (Kenya), Kampala (Uganda) and Biharamulo (Tanzania) (see Figure 6.2). These sites also reported a history of cholera epidemics and recent changes in climate.

It should be noted that it was important to include households from different elevations because previous studies (Githeko et al, 2006) have shown that prevalence of highland malaria is differentiated by elevation, with 70 per cent, 40 per cent and 30 per cent malaria prevalence among households in valley bottom, hillside and hilltop respectively. Other factors considered in the selection of sites were proximity to a hospital and a meteorological station with reliable data.

Informal and formal interviews with individual households and key informants and focus group discussions served as primary data sources for this study. Both qualitative and quantitative methods were used and direct observations were made to supplement interview data (for detailed methodology see Chambers, 1992; Mettrick, 1993; Mikkelsen, 1995). Qualitative data were obtained from group discussions in selected villages, with participants drawn from all the sub-villages and representing various social groups. The aim was to capture the indigenous knowledge base on local vulnerability to climate impacts and adaptation mechanisms (curative and preventive) for malaria and cholera. For the quantitative data, a random sample of 900 households were interviewed, 150 from each of the 6 study sites. Information was collected on socioeconomic characteristics; infrastructure and services; climate-related human health problems and management strategies at the household level; exposure to diseases; and accessibility and availability of health services.

Stakeholder workshops were conducted separately in Kenya, Tanzania and Uganda to present the study and research findings, and to obtain participant feedback.

Research Findings

Malaria

Malaria is caused by the *Plasmodium* parasite and is transmitted via the bite of the female Anopheles mosquito, which transfers parasite infected blood from a sick person to a healthy person (TDR, no date). Malaria causes more than a million deaths in Africa every year and in the highlands of the Lake Victoria region in East Africa it is known to be highly unstable and epidemic in nature (see Figure 6.1). Such zones of unstable malaria are shown to be more sensitive to climate variability and environmental changes than those where the disease is endemic (Mouchet et al, 1998).

Location of the Malaria Study Sites

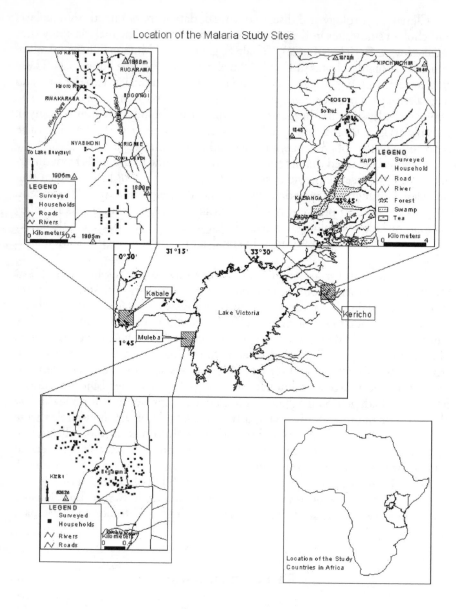

Figure 6.1 *Map showing the Lake Victoria highland malaria region and the studied villages*

A significant increase in the severity and frequency of highland malaria epidemics has been observed in recent decades in comparison to the early episodes of the 1920s and 1950s (Garnham, 1945; Roberts, 1964), with virtually no recorded epidemics from the 1960s to the early 1980s. Scientific studies

Location of the Cholera Study Sites

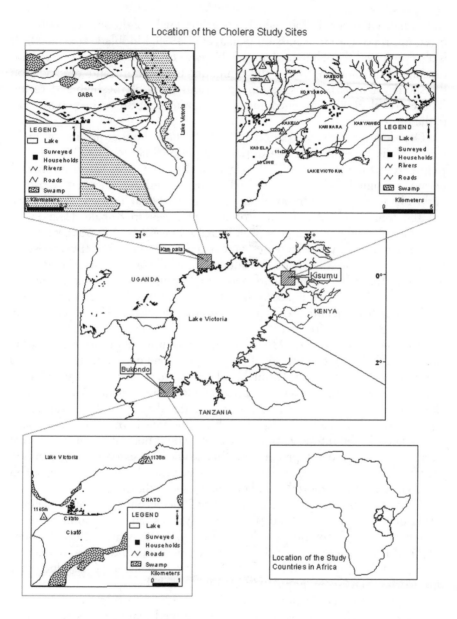

Figure 6.2 *Map showing the Lake Victoria cholera region and the studied villages*

have closely linked this resurgence of highland malaria epidemics with climate variability (Lepers et al, 1988; Khaemba et al, 1994; Lindsay and Martens, 1998; Malakooti et al, 1998; Mouchet et al, 1998; Some, 1994; Matola et al, 1987; Fowler et al, 1993) and El Niño events that lead to elevated temperatures and enhanced precipitation, which create optimal conditions for mosquito breeding and disease transmission (Kilian et al, 1999; Lindblade et al, 2000). It is also projected that climate change is likely to cause an increase in temperature and precipitation, above the minimum temperature and precipitation thresholds of malaria transmission, in various parts of the highland region (Githeko et al, 2000), thus encouraging greater incidences of disease outbreaks in the future.[3]

Our data for the study period 1996–2001 show the first upsurge in highland malaria cases in Tanzania during the period May to July 1997, and in Kenya from June to July 1997. In Uganda, the number of cases during this period remained below normal. The most significant surge in seasonal outbreaks was observed from January to March 1998 in Tanzania and Kenya, but the trends extended to May of the same year in Kabale, Uganda. In Tanzania the epidemic caused a peak increase in cases by 146 per cent, while in Kenya and Uganda the increases were 630 per cent and 256 per cent respectively. The peak month for hospital admissions in all countries was March. It is likely that the increases in malaria cases in Tanzania and Uganda reflect the true trends (Wandiga et al, 2006) since the Kenyan hospital used in this study is a Mission hospital and also includes cases that should have been admitted to the Kericho District Hospital.

This period during 1997–98 when the malaria outbreaks occurred was also an El Niño period during which significantly higher than normal temperature and precipitation were recorded. The highland malaria outbreaks in the Lake Victoria region followed on the heels of this phase of anomalous climate. Data show that climate change is already affecting this area and that the highlands are warming at a greater rate than the lowlands, which has important implications for malaria transmission in the future. A warmer climate would not only reduce mosquito larvae and parasite development times but would also extend the duration of epidemics due to the longer duration of favourable climatic conditions. This could greatly increase mortality and morbidity due to the disease.

Other factors, such as environmental and socioeconomic change, deterioration of health care and food production systems, and the modification of microbial/vector adaptation (McMichael et al, 1996; Morse, 1995; Epstein, 1992 and 1995), act as co-contributors in the emergence and spread of malaria. In the East African highlands, human exposure to malaria has also increased due to the increase in population density, which has significantly stressed limited productive land (Lindsay and Martens, 1998), forcing farmers to clear forests and reclaim swamps. Puddles and elevated temperatures result from lost tree and ground cover, providing ideal breeding sites for mosquitoes (Walsh et al, 1993). Even small increases in temperature can greatly hasten mosquito development, resulting in a greater density of mosquito population (Lindblade et al, 2000).[4]

The vulnerability of highland households is also significantly affected by the scarcity of economic resources, which prevents them from investing in health-coping mechanisms. Wealthy households can afford medical/health and other social services and are less susceptible to disease outbreaks in contrast to poorer households. For example, in Bugarama village in Tanzania, the richer people, locally known as *Washongole* can better afford to meet health-related costs compared to the poor group, locally know as *Abworu* (Yanda et al, 2005).

Children, pregnant women and the elderly are reported to be most likely to succumb to the disease (Greenwood and Mutabingwa, 2002; McMichael et al, 1996). Trends in malaria cases in our study sites show that children under five years of age display a higher vulnerability to malaria attacks compared to older individuals (see Figures 6.3 and 6.4). This is consistent with the fact that young children have lower immunity (Wandiga et al, 2006; Yanda et al, 2005). Household interviews at the Tanzanian study sites revealed that about 15 per cent of households had lost at least one member due to malaria, with 12 per cent of households reporting to have lost a child member and 3 per cent having lost an adult member (Yanda et al, 2005).

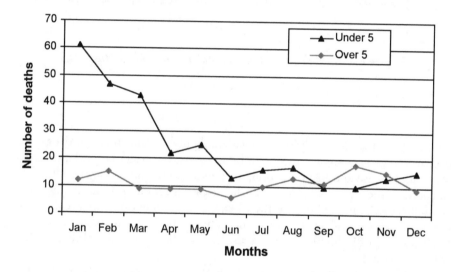

Figure 6.3 *Total death toll due to malaria for Ndolage Hospital, Tanzania, in 2001*

Source: Yanda et al (2005).

Other contributors that increase vulnerability to malaria in this region include drug resistance, home treatment of malaria and changes in vector biting behaviour (Mbooera and Kitua, 2001). The locals also harbour certain traditional beliefs about malaria, for example, the belief in Muleba, Tanzania, that eating maize meal instead of bananas causes the disease (Mbooera and Kitua, 2001).[5]

Existing capacity to cope

Preparedness in terms of preventing and treating malaria was observed to be substantially better in the lowlands, where malaria is endemic, in comparison to our study sites in the highlands of the Lake Victoria region. In the lowlands

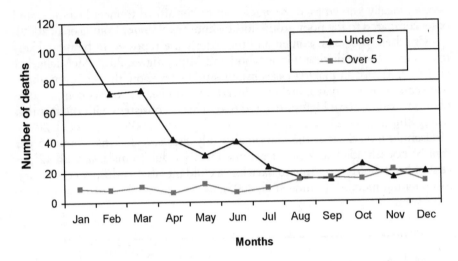

Figure 6.4 *Total death toll due to malaria for Rubya Hospital, Tanzania, in 2001*

Source: Yanda et al (2005).

the number of health facilities is considerably higher; anti-malarial drugs and bed nets to protect against mosquito bites are readily available; and awareness about the disease and its prevention, diagnosis and treatment is generally high. In contrast, the capacity to deal with the disease is much lower in the highland areas surveyed, largely because the disease is not common to this region and there is a lack of local awareness about its prevention and treatment. The relative poverty in the highland region also further hampered the ability to deal with the disease.

We identified some techniques commonly used by households in our study sites to overcome mosquitoes as an adaptation strategy for malaria. These include (number in brackets indicates percentage of respondents who used the particular technique):

- sleeping by a fire inside the house to avoid mosquitoes – burning of eucalyptus leaves and other herbs was reported to be very effective in chasing away mosquitoes (4 per cent);
- using mosquito coils (21 per cent);
- clearing of bushes around homesteads to destroy potential breeding grounds (18 per cent);
- house screening (16 per cent);
- draining stagnant water (15 per cent);
- treating bed nets with insecticides (mainly with Ngao), at most twice a year (15 per cent); and
- spraying insecticides inside the house (11 per cent).

In Tanzania, an increasing use of bed nets and insecticide-treated bed nets was observed, although most people could not afford to buy bed nets for the entire household. On average, each household of six or seven persons owned 1.5 bed nets, which were sufficient for only about 2.4 persons per household (Yanda et al, 2005). In Kenya, few people had access to bed nets, and the proportion of people using bed nets treated with insecticides was even lower. A household could have as many as 16 persons, the average size being 3.7 persons, and the number of bed nets per household ranged from 1 to 6. Only about 28 per cent of the households that use bed nets treat them with insecticides, that too at most twice a year, when the recommended frequency is four times per year (although some of the newer insecticides require only one annual treatment). The lack of use of bed nets, especially insecticide-treated bed nets tends to greatly reduce the efficacy of programmes designed to address malaria, for example, the World Health Organization's 'Roll Back Malaria' campaign, which specifically promotes the use of insecticide treated bed nets for the prevention of malaria as a low-cost and effective measure.[6]

We also noted that many people in the study sites preferred traditional curative measures (local herbs) to treat malaria rather than visiting health facilities.[7] Participants in the stakeholder workshop in Muleba town, Tanzania, estimated that about two-thirds of the malaria patients get cured with traditional medicines. A high reliance on herbal medicines was also reported at the other study sites in Tanzania (in Bukoba Rural District see Mwisongo and Borg (2002) and Bugarama village), with some people relying entirely on herbal remedies. According to Mwisongo and Borg (2002), more than 80 per cent of rural Tanzanians depend on herbal remedies for their primary healthcare.

The preference for local herbs for the treatment of malaria over clinical medicines was found to be because (1) they were common, well known and familiar to most people; (2) they were easily available, less expensive, and effective as a first aid before taking the patient to a hospital or health centre; and (3) pregnant women using these herbs against malaria did not encounter problems during delivery, largely due to the fact that these herbs are also effective in reducing complications during pregnancy (Yanda et al, 2005). The herbs are usually prescribed by traditional healers who are reported to be familiar with the symptoms of malaria.

Researchers from the National Institute for Medical Research (NIMR) have confirmed that traditional healers, like other cadres in medicine, do possess the knowledge and skills for malaria disease management (diagnosis, treatment and prevention). Laboratory analyses of the traditional herbs by the NIMR revealed that the majority were antimalarial in nature (with variable potency) and could treat other diseases as well (Mwisongo and Borg, 2002).

Clinical treatment for malaria is available at the village health facility in every village. An annual contribution of 1500 shillings per household to the health facility ensures medical services (including medication). Treatment is available to non-contributors only under 'emergency' situations subject to the condition that they pay their dues on recovery. Only the extremely poor receive treatment free of charge. The receipt of contribution to the village health facility also serves as a guarantee for receiving treatment at government hospitals, such as Rubya, and health centres such as Kabare (in Biirabo village) and Chato (in Chato village).

Data collected on the preferred means of treating malaria at the household level and analysed on the basis of the level of education of the surveyed people (Table 6.1) reveal that a majority of the people (78 per cent) do consider modern medicine to be the ultimate cure for malaria, despite the high reliance on traditional medicines. No variations in malaria treatment practices at household level could be discerned on the basis of education. A combination of modern and herbal/traditional medicine is locally perceived by some (about 16 per cent of respondent households) to be an important means of combating malaria. Only a small group of people with primary education (about 3 per cent of the total) considered herbal medicine alone to be sufficient for treating malarial disease. Given that the majority believe in the efficacy of modern medicine, the observed high reliance on indigenous medicines is likely to be due to financial barriers.

Table 6.1 *Percentage responses of how malaria is treated at the household level by people with different levels of education*

Level of Education	How Malaria is Treated					Total
	Modern Medicine	Herbal Medicine	Modern Medicine/ Tepid Sponging	Modern Medicine/ Herbal Medicine	Modern Medicine/ Prayer	
None	12.0	0	0.3	3.3	0.7	16.3
Primary	56.3	3.0	0.7	11.7	1.4	73.0
Secondary	6.0	0	0	0.3	0	6.3
Tertiary	2.3	0.3	0	0	0	2.7
Others	1.4	0	0	0.3	0	1.7
Total	78.0	3.3	1.0	15.6	2.1	100

Local capacity to adapt is also constrained by the absence of adequate early warning mechanisms that could forewarn vulnerable communities about impending malaria epidemics. One such model linking temperature anomalies with malaria outbreaks in the highlands of East Africa has been developed by Githeko and Ndegwa (2001) and has been used with some success in Kenya. However, the current unavailability of such warning systems for the entire region means response to epidemics in most areas is typically reactive and often delayed.

Cholera

Cholera is caused by the bacteria *Vibrio cholerae* (Waiyaki, 1996) and is transmitted through the faecal-oral route largely due to (1) the ingestion of contaminated water and food and (2) a lack of scrupulous attention to personal cleanliness (Snow, summarized in Waiyaki, 1996). Cholera causes thousands of fatalities in Africa every year and is endemic to the Lake Victoria basin. It is found to be more common among those living in villages bordering the lake compared to those who live in the hinterland (Shapiro et al, 1999). Specific risk factors associated with cholera outbreaks in the Lake Victoria basin include drinking water from Lake Victoria or from a stream, sharing food with a person with watery diarrhoea, and attending funeral feasts (Shapiro et al, 1999).

Several environmental, social and demographic factors are believed to contribute to increasing the risk of cholera outbreaks in this region. Environmental factors include heavy rains and subsequent floods (Desanker and Magadza, 2001; Ulisses and Menne, 2007) and more recently, increased sea surface temperatures have also been implicated as a contributing cause (Desanker and Magadza, 2001; Rodo et al, 2002; Pascual et al, 2000; Colwell, 1996). According to Colwell (1996), increased precipitation increases nutrient laden discharge from rivers and streams into the sea. In combination with warmer sea-surface temperatures, this creates optimum conditions for phytoplankton growth, which subsequently leads to an increase in zooplanktons that feed on the phytoplanktons (Colwell, 1996; Desanker and Magadza, 2001; Cruz et al, 2007; Ulisses and Menne, 2007). Zooplanktons are the preferred host for *Vibrio cholerae*, and when zooplankton populations increase, so do *Vibrio cholerae* populations (Colwell, 1996; Patz, 2002; Desanker and Magadza, 2001; Ulisses and Menne, 2007). Contamination of coastal drinking water supplies by sea water containing zooplanktons (due to flood events) leads to the spread of the disease among humans (Colwell, 1996). Similar effects are also thought to be possible for lakes and other inland water bodies (Desanker and Magadza, 2001). Heavy rains and increased sea-surface temperatures are also associated with El Niño, and studies undertaken in Bangladesh have specifically correlated cholera epidemics with El Niño events (Rodo et al, 2002; Pascual et al, 2000).[8]

In the Lake Victoria basin cholera epidemics recorded during the El Niño years in 1982/83 and 1997/98 were found to coincide with high stream flows and well above normal temperatures. Above normal precipitation and flooding alone, without the above normal temperatures, did not appear to trigger cholera epidemics. Non-epidemic (hygienic) cholera outbreaks were found to be associated with the rainy season when there are above normal rainfall and temperatures, but not as intense as those experienced during El Niño. The morbidity due to such hygienic cholera outbreaks is several orders of magnitude lower than that due to epidemics. Typically cholera epidemics/outbreaks tend to occur anytime between April and December (Wandiga et al, 2006 and Olago et al, 2007), following periods of mainly sustained anomalous high temperatures in the months of January, February and March, and heavy rains.

Of the socioeconomic and demographic factors that also play a critical role in encouraging cholera outbreaks in the Lake Victoria basin, poverty is the most significant. Most of the communities here rely predominantly on either farm earnings or self-employment, with very few people having a source of steady income. Poverty therefore proves to be a restrictive factor in the prevention of cholera outbreaks, for example, in the inability of local communities to construct durable sanitary facilities due to a lack of adequate financial resources.

In the case of Chato village in Tanzania (see Figure 6.2), most people have access to toilet facilities, including pit latrines and a few flush toilets, although the extent of use of these facilities is hard to determine. Moreover, like many other rural areas in the country, there is no sewage removal system. This makes waste disposal difficult and is locally believed to contribute to cholera outbreaks (Yanda et al, 2005) due to the contamination of water supplies.

We found that a good number of households in the study area used various treatments for drinking water to prevent cholera incidences: (1) boiling drinking water (56.7 per cent), (2) filtering drinking water (50 per cent), and (3) treating drinking water with chemicals (3.3 per cent). However, in certain villages, particularly those around the lake, the proportion of households drinking untreated water can be as high as 20 per cent. This is due to the perception that piped water or water from the pumped wells is safe for drinking (Yanda et al, 2005). The safety of this water depends, however, on the level of treatment at the source, which may not be adequate, and consuming untreated water greatly increases vulnerability to cholera. Moreover, field observations show that the tap water and pumped water supply is often unreliable; in such situations people are forced to collect water from the lake for drinking and other domestic purposes. This potentially exposes users to cholera pathogens in the lake water, as happened during the cholera outbreak of the 1980s. Table 6.2 provides a list of explanations offered by the surveyed households for not boiling or treating drinking water.

Table 6.2 *Percentage of reasons/explanations for not treating/boiling drinking water in Chato village*

Reason/Explanation	Percentage of Responses
Tap water considered safe	63.8
Not used to boiling drinking water	15.9
Lack of fuelwood for boiling water (boiling water is too costly)	10.1
No utensils for boiling water	5.8
Boiling water is tiresome	2.9
Fear of losing the taste of water	1.5
Total	100

Finally, it must be noted that currently there are no traditional medicines/cures for cholera in the region. The disease is relatively new to the area and traditional herbalists have yet to invent any curative or preventive medication.

Existing capacity to cope

Mortality due to cholera is generally observed to be lower compared to malaria in the Lake Victoria region. One possible reason is the relatively higher level of community awareness about the disease. This can be attributed to the awareness programmes instituted by the local administration largely to compensate for the insufficient capacity of public health facilities to handle epidemics. Local civil society organizations are also found to play an important role in helping construct sanitary facilities and water supply sources and in providing free medication during epidemics.

According to our survey results, a significant proportion of the respondents (86.6 per cent, 67.2 per cent and 52.7 per cent in Kisumu, Kampala and Biharamulo respectively) exhibited awareness about the influence of weather conditions on the health of household members (Olago et al, 2007). They observed that cholera outbreaks occurred mostly during wet weather and during periods of low water supply associated with dry seasons. According to health workers from Biharamulo, this could be blamed on the sandy nature of the soil pit latrines that often collapsed during the rainy season and contaminated water supplies. During periods of low water supply there is increased dependency on water from the lake, which may be contaminated with cholera pathogens. Periods of low water supply may also lead to a reduced level of sanitary practices, such as hand washing. This knowledge about the association of specific weather patterns with the occurrence of the disease can, however, serve to warn people about possible outbreaks and enable them to be better prepared.

Awareness about the need to treat or boil drinking water was also found to be high, despite the fact that some people drink untreated water. Ninety-four per cent of respondents reported that they were knowledgeable about cholera and the consequences of drinking untreated water. Ninety-one per cent knew that drinking untreated water could cause diarrhoeal diseases, such as cholera, and 49 per cent reported to have had at least one member of the household suffering from diarrhoeal diseases during the last 5 years (Yanda et al, 2005) as a result of having consuming untreated water from the lake. This awareness was attributed to various sources such as health service providers, formal sources such as schools, informal networks, media and community awareness campaigns (see Figure 6.5).

Most households appeared to be sufficiently informed about medical treatment options for cholera such as the use of antibiotics and oral rehydration salts. However, costs of medical treatment are a barrier and the locals often have to rely on the free medications distributed during epidemics by civil society organizations. Then again, because cholera does not have any local/indigenous treatment options, people do report to health facilities immediately when there are signs of the disease. This greatly contributes to reducing mortality in comparison to malaria, where many people attempt home treatment before reporting to health facilities.

Viewpoints expressed by Chato village (Tanzania) residents on measures that are or should be taken to prevent or treat incidences of cholera include:

1   general cleanliness;

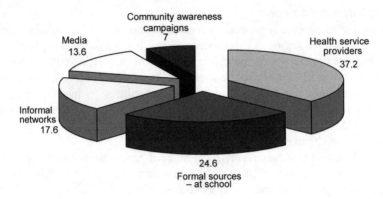

Figure 6.5 *Sources of information on consequence of cholera*

Source: Yanda et al (2005).

2   washing hands with soap;
3   boiling drinking water; and
4   using health facilities when sick.

Participants in the stakeholder workshops also identified and suggested strategies to reduce the vulnerability of the community to cholera epidemics,
distinguishing between the allocation of responsibility at the village and district
levels (Table 6.3). Some measures such as observing proper hygiene and protection of and proper management of water sources were identified for
implementation on a routine basis, while other measures would have to be
implemented specifically during cholera outbreaks, for example, promptness
in reporting cholera outbreaks and sending sick people to health centres and
hospitals for treatment.[9]

In the case of cholera, no early warning mechanisms about potential disease
outbreaks have so far been instituted to allow residents to be better prepared to
cope. Lipp et al (2002) have put forth the suggestion that the spatial and
temporal correlation of *Vibrio cholerae* with El Niño, or its proxies and
predictors, could help to create such a forewarning mechanism and provide an
effective way to prevent exposure to cholera in vulnerable regions. However,
any such system has yet to be developed and implemented on the ground.

Conclusions and Recommendations

The findings of our assessment indicate that communities in the Lake Victoria
region of Africa display a significantly high vulnerability to climate change/
climate variability-induced diseases such as malaria and cholera. The capacity
of these communities for the prevention and treatment of such diseases is in
contrast quite low, as a result of which mortality and morbidity in the event of
disease outbreaks tends to be high.

Table 6.3 *Cholera control strategies suggested by stakeholders*

Village Level	District Level
Construction and use of improved toilets	Awareness campaigns on how to prevent cholera outbreak
Use of clean and safe water (boiled)	Outbreak preparedness (districts need to have plans for controlling cholera in the event of outbreaks)
Use of clean and safe water (boiling cooking and drinking water)	Outbreak preparedness
Proper collection and disposal of wastes • Collecting solid wastes in pits and burying the pits when they fill up • Burning the wastes, whenever possible	• Planning for cholera control strategies in cooperation with community leaders • Providing equipment necessary to keep the environment clean and improve the hygienic conditions
Protection and proper management of water sources	Recruit more health staff
Cost-sharing in the management of water sources	Undertake environmental assessment to ascertain causes of problems and how to control the situation
Washing hands before taking any food	
Washing hands after every use of the toilet	
Cleanness of household utensils	Establish temporary camps for patients during cholera outbreaks
Community to report promptly when there is a cholera outbreak	Ensure prompt response to cholera outbreak
Sick people to report promptly at health centres and hospitals for treatment	Undertake laboratory analysis to confirm outbreak

Communities in this region are found to display a greater vulnerability to climate-induced highland malaria outbreaks in comparison to cholera outbreaks in the basin. This is largely due to a relative lack of awareness about the disease and a relative lack of preparedness in terms of availability of health care facilities and medication. The predominant reliance of communities on local herbs to treat malaria as a first and often only recourse (as opposed to clinical treatment) also contributes to increased mortality and morbidity, especially due to the herbs' variable efficacy. In comparison, greater local knowledge about the prevention and treatment of cholera and the absence of any indigenous treatment options tends to ensure more prompt clinical attention.

Vulnerability to malaria is also observed to be higher in the highland region, where the disease is epidemic, in comparison to the lowlands, where the disease is endemic. The endemicity of malaria in the lowlands has ensured greater awareness and better availability of health care services and medications, while in the highlands there is a lack of experience in dealing with the disease due to its recent resurgence, which translates into a lack of knowledge about the disease and a lack of accessibility to adequate health care and treatment. The poverty of the population in this region in comparison to the lowlands is also a critical factor that greatly impedes its capacity to invest in disease prevention (for example, use of insecticide-treated nets) or in disease treatment. The inability to pay also creates a high dependence on indigenous medicine despite the strong belief in the efficacy of modern medicine.

In the case of cholera, although awareness about the disease and its treatment is relatively high and mortality is subsequently lower, many still tend to follow unsanitary practices and drink untreated water from the lake. Poverty is once again observed to be an important factor in the inability of the people to invest in better sanitation and appropriate medication to treat the disease. Many rural areas also do not have adequate sanitary services such as sewage removal services, which can cause contamination of drinking water supplies. Ignorance about the safety of water supplies among the rural population and the scarcity of well-equipped health centres tends to compromise the health of the people in the event of outbreaks and epidemics.

Common to both highland malaria and cholera epidemics is the need for stronger and better public information campaigns to ensure that the rural population in this area is sufficiently informed about strategies for the prevention and treatment of these diseases. Second, better infrastructure in terms of sanitation and living conditions, access to basic services such as clean water supplies, and access to adequate health services and medication are critical for preventing and addressing disease epidemics in the region. Third, a better understanding of the correlation of climate phenomena with disease outbreaks would help to establish early warning mechanisms in order to allow for advance preparedness in the community in terms of strategies for prevention and treatment. Githeko and Ndegwa (2001) have initiated this process to some extent by applying such a model for malaria with some effectiveness in Kenya. The expansion of this model to other areas in the highland region could potentially help to greatly reduce the severity of malaria outbreaks in the future. No such system exists for cholera at the moment. Finally, strategies for poverty reduction, by developing and providing livelihood opportunities, could greatly help communities to invest in better health coping mechanisms, for example, the use of insecticide-treated nets in the case of malaria and effective sanitary facilities in the case of cholera. Better income security would also enable communities to seek prompt clinical treatment and help reduce mortality due to these diseases.

Important lessons can be learnt from existing adaptation strategies that have been found to work. For example, the better adaptation to malaria in the lowlands of East Africa could instruct the development of better adaptation strategies in the highlands. In the case of cholera, community awareness campaigns can be further strengthened to ensure adequate sanitary precautions. Additionally, success stories from programmes dealing with other diseases in the region could help inform strategies for adaptation to climate-related malaria and cholera outbreaks. In designing future adaptation programmes a range of socioeconomic factors and demographic trends would also have to be accounted for, because they influence social responses to climatic stresses and would be critical in the determination of strategies to cope with the impacts.

National and local governments have a vital role to play in increasing the adaptive capacity of communities in the Lake Victoria region to climate related disease epidemics and outbreaks. In addition, various civil society organizations and relevant regional, national and international institutions and agencies could potentially make important contributions to this process. There are presently no comprehensive government programmes or fiscal facilities in place for climate-related disaster preparedness. Many of the current preventive and curative programmes for malaria and cholera that are run by governments or civil society organizations often predominantly rely on external sources of assistance, whose long-term sustainability is not always guaranteed. Therefore, finding ways to develop, institute and sustain programmes that would strengthen local capacity to adapt to the health impacts of climate change or climate variability is critical.

In short, more effective adaptation to the health impacts of climate variability and future climate change in the Lake Victoria region would entail a combination of a multitude of strategies, including increased awareness, better health care, improved sanitation, adequate infrastructure, development of early warning mechanisms, creation of livelihood opportunities, and increased support from national and international governments and institutions.

Notes

1   Details of this study have been described in papers by Wandiga et al (2006), Olago et al (2007) and Yanda et al (2005).
2   International health regulations require national health administrators to report the number of indigenous and imported cases of cholera and deaths to the World Health Organization (WHO) within 24 hours of receiving such information. This cholera data are then reported in the *Weekly Epidemiological Review* (WER) detailing the date and geographical location.
3   Hay et al (2002) have disputed the association of malaria outbreaks with climate variability and change. They analysed climate data for the diurnal temperature range spanning the 1950–1959 period and found no significant changes in temperature or vapour pressure at any of the highland sites that had reported high malaria incidences. However, these findings have been challenged by Patz et al (2002), who claim that the use of a downscaled gridded climate data set by interpolating it to specific sites by Hay et al (2002) ignores climate dependencies on local elevation, which compromises the accuracy of the results. In opposition to Hay et al's (2002) findings, Patz et al (2002) have reported a warming trend at highland sites in East Africa coinciding with an increasing trend in the incidence of malaria at those sites, using data obtained from the specific locations over the specific time period. Similar associations have also been reported by other scientists (Ulisses and Menne, 2007).
4   In contrast to reclaimed swamps, the natural swamps in the valley bottoms contain Papyrus, which, due to its cooling properties and natural oil secretions, is believed to inhibit anopheles mosquito development (Lindblade et al, 2000; Reiter, 2001).

Papyrus and other naturally occurring grasses tend to bring down temperatures in the swamp, which slows down mosquito development (Lindblade et al, 2000). Papyrus also produces natural oils, which are believed to inhibit anopheles mosquito development (see Reiter, 2001).

5   In Muleba, Tanzania, maize meal is usually eaten only during periods of food shortages that occur due to above and/or below average rains (for example, the El Niño rains or La Niña droughts). Such climatic conditions are also conducive to the occurrence of malaria, which is therefore more rampant at such times. Additionally, poor nutrition due to the food shortage further makes people more susceptible to malaria, especially children who tend to become anaemic due to nutritional shortages (Mwisongo and Borg, 2002). As a result the traditional understanding that eating maize meal causes malaria has developed. In Kenya, the consumption of an edible oil called *chipsy* is associated with the occurrence of malaria simply because this oil was first introduced in Kenya in 1990, which was also the year of the El Niño rains and malaria episodes. Similarly, in Uganda, supernatural forces are considered to cause the convulsions that are associated with malarial complications and traditional medicines are considered to be the only solution (Nuwaha, 2002). This often leads to either no or delayed medical care and increases morbidity and mortality due to the disease.

6   The World Health Organization's (WHO) programme 'Roll Back Malaria' has been adopted by most countries in Africa. The three East African governments actively promote this programme, whose objectives towards malaria eradication are to increase the use of ITNs, early diagnosis and treatment of malaria, and the use of effective anti-malarial drugs. This programme has attracted several local and international civil societies. One such non-governmental organization active in East Africa is Population Services International (PSI), which receives financial support from both the British and American Governments. Its stated objective is to increase the use, ownership and availability of ITNs in Kenya, Uganda and Tanzania within 15 minutes' walk of malaria endemic areas. Promotions of ITNs are prevalent in most market centres in East Africa. However, the cost of a subsidized ITN is still US$1.50, putting it beyond the reach of households living below the poverty line.

7   The plants primarily used for this purpose include (using Haya names) *Mbilizi*, *Kajule*, *Nkaka*, *Ikintuntumwa* and *Mwarobaini* (Yanda et al, 2005), although the level of success may vary among the different varieties.

8   A study in Chesapeake Bay though found the link between temperature and cholera in suboptimal environments (freshwater or high salinity) to be weak (Louis et al, 2003). Lake Victoria, which is a freshwater lake, has a salinity ranging between 3.9 and 7.0, much lower than the optimal ocean salinity. However, these research findings were based on a limited time record (2 years) and did not account for indicators associated with nutrient load and zooplanktons (for example, discharge and precipitation) (Louis et al, 2003).

9   The Chato Health Center that caters for the entire Chato Division (and other neighbouring divisions) is conveniently located right within Chato village.

References

Chambers, R. (1992) 'Rural appraisal: Rapid, relaxed and participatory', Discussion Paper No 311, Institute of Development Studies, Brighton, Sussex, UK

Checkley, W., L. D. Epstein, R. H. Gilman, D. Figueroa, R. I. Cama, J. A. Patz and R. E. Black (2000) 'Effects of El Niño and ambient temperature on hospital admissions for diarrhoeal diseases in Peruvian children', *The Lancet*, vol 355, pp442–450

Colwell, R. (1996) 'Global climate and infectious disease: The cholera paradigm', *Science*, vol 274, pp2025–2032

Confalonieri, U. and B. Menne (2007) 'Human health', in S. Solomon and D. Qin (eds) *Climate Change 2007: Climate Change Impacts, Adaptation and Vulnerability*, Contribution of Working Group II to the Fourth Assessment Report, Intergovernmental Panel on Climate Change (IPCC), Cambridge University Press, Cambridge, UK

Cruz, R. V., H. Harasawa, M. Lal and W. Shaohong (2007) 'Asia', in S. Solomon and D. Qin (eds) *Climate Change 2007: Climate Change Impacts, Adaptation and Vulnerability*, Contribution of Working Group II to the Fourth Assessment Report, Intergovernmental Panel on Climate Change (IPCC), Cambridge University Press, Cambridge, UK

De Savigny, D., E. Mewageni, C. Mayombana, H. Masanja, A. Minhaji, D. Momburi, Y. Mkilindi, C. Mbuya, H. Kasale, H. Reid and H. Mshinda (2004a) *Care Seeking Patterns in Fatal Malaria: Evidence from Tanzania*, Tanzania Essential Health Interventions Project (TEHIP), Rufiji Demographic Surveillance System, Tanzania, Ifakara Health Research and Development Centre, Tanzania, Tanzania Ministry of Health and International Development Research Centre, Canada

De Savigny, D., E. Mewageni, C. Mayombana, H. Masanja, A. Minhaji, D. Momburi, Y. Mkilindi, C. Mbuya, H. Kasale, H. Reid and H. Mshinda (2004b) 'Highland malaria in Uganda: Prospective analysis of an epidemic associated with El Niño', *Transactions of the Royal Society of Tropical Medicine and Hygiene*, vol 93, pp480–487

Desanker, P. and C. Magadza (2001) 'Africa', in J. J. McCarthy, O. F. Canziani, N. A. Leary, D. J. Dokken and K. S. White (eds) *Climate Change 2001: Impacts, Adaptation and Vulnerability*, Contribution of Working Group II to the Third Assessment Report, Intergovernmental Panel on Climate Change (IPCC), Cambridge University Press, Cambridge ,UK

Epstein, P. R. (1992) 'Cholera and environment', *Lancet*, vol 339, pp1167–1168.

Epstein, P. R. (1995) 'Emerging diseases and ecosystem instability: New threats to public health', *American Journal of Public Health*, vol 85, pp168–172, available at http://extdr/offrep/afr

Fowler, V. G. Jr., M. Lemnge, S. G. Irare, E. Malecela, J. Mhina, S. Mtui, M. Mashaka and R. Mtoi (1993) 'Efficacy of chloroquine on *Plasmodium falciparum* transmitted at Amani, eastern Usambara mountains, northeast Tanzania: An area where malaria has recently become endemic', *Journal of Tropical Medicine and Hygiene*, vol 6, pp337–345

Garnham, P. C. C. (1945) 'Malaria epidemics at exceptionally high altitudes in Kenya', *British Medical Journal*, vol 11, pp45–47

Githeko, A. K., S. W. Lindsay, U. E. Confaloniero and J. A. Patz (2000) 'Climate change and vector-borne disease: A regional analysis', *Bulletin of the World Health Organization*, vol 78, pp1136–1147

Githeko, A. K., J. M. Ayisi, P. K. Odada, F. K. Atieli, B.A. Ndenga, I. J. Githure and G. Yan (2006) 'Topography and malaria transmission heterogeneity in the western Kenya highlands: Prospects for focal vector control', *Malaria Journal*, vol 5, p107

Greenwood, B. and T. Mutabingwa (2002) 'Malaria in 2002', *Nature*, vol 415, pp670–672

Hay, S. I., M. Simba, M. Busolo, A. M. Noor, H. L. Guyatt, S. A. Ochola and R. W Snow (2002) 'Defining and detecting malaria epidemics in the highlands of western Kenya', *Emerging Infectious Diseases*, vol 8, pp555–562

Hulme, M. (1996) 'Recent climatic change in the world's drylands', *Geographical Research Letters*, vol 23, pp61–64

Intergovernmental Panel on Climate Change (IPCC) (2001). *Climate Change 2001: Impacts, Adaptation and Vulnerability*, J. McCarthy, O. F. Canziani, N. Leary, D.

Dokken and K. S. White (eds) contribution of Working Group II to the Third Assessment Report, Cambridge University Press, Cambridge, UK

Khaemba, B. M., A. Mutani and M. K. Bett (1994) 'Studies of anopheline mosquitoes transmitting malaria in a newly developed highland urban area: A case study of Moi University and its environs', *East African Medical Journal*, vol 3, pp159–164

Kilian, A. H. D., P. Langi, A. Talisuna and G. Kabagambe (1999) 'Rainfall pattern, El Niño and malaria in Uganda', *Transactions of the Royal Society of Tropical Medicine and Hygiene*, vol 93, pp22–23

Lepers, J. P., P. Deloron, D. Fontenille and P. Coulanges (1988) 'Reappearance of Falciparum malaria in central highland plateaux of Madagascar', *Lancet*, vol 12, pp585–586

Lindblade, K. A., E. D. Walker, A. W.Onapa, J. Katunge and M. L. Wilson (2000) 'Land use change alters malaria transmission parameters by modifying temperatures in a highland area of Uganda', *Tropical Medicine and International Health*, vol 5, pp263–274

Lindsay, S. W. and W. J. M. Martens (1998) 'Malaria in the African highlands: Past, present and future', *Bulletin of the World Health Organization*, vol 76, pp33–45

Lipp, E. K., A. Huq and R. R. Colwell (2002) 'Effects of global climate on infectious diseases: The cholera model', *Clinical Microbiology Review*, vol 15, pp757–770

Louis, V. R., E. Russek-Cohen, N. Choopun, I. N. Rivera, B. Gangle, S. C. Jiang, A. Rubin, J. A. Patz, A. Huq and R. Colwell (2003) 'Predictability of *Vibrio cholerae* in Chesapeake Bay', *Applied and Environmental Microbiology*, vol 69, pp2773–2785

Malakooti, M. A., K. Biomndo and G. D. Shanks (1998) 'Re-emergence of epidemic malaria in the highlands of western Kenya', *Emerging Infectious Diseases*, vol 4, pp671–676

Matola, Y. G., G. B. White and S. A. Magayuka (1987) 'The changed pattern of malaria endemicity and transmission at Amani in the eastern Usambara mountains, northeastern Tanzania', *Journal of Tropical Medicine and Hygiene*, vol 3, pp127–134

Mbooera, L. E. G. and A.Y. Kitua (2001) 'Malaria epidemics in Tanzania: An overview', *African Health Sciences*, vol 8, pp17–23

McMichael, A. and A. Githeko (2001) 'Human health', in J. J. McCarthy, O. F. Canziani, N. A. Leary, D. J. Docken and K. S. White (eds) *Climate Change 2001: Impacts, Adaptation and Vulnerability*, contribution of Working Group II to the Third Assessment Report of the Intergovernmental Panel on Climate Change (IPCC), Cambridge University Press, Cambridge, UK

McMichael, A. J., A. Hames, R. Scooffand and S. Covats (eds) (1996) 'Climate change and human health: An assessment', report prepared by task group on behalf of the World Health Organization, the World Meteorological Organization and the United Nations Environment Programme, Geneva, Switzerland

Mettrick, H. (1993) *Development Oriented Research in Agriculture: ICRA Textbook*, The Centre for Development Oriented Research in Agriculture, Wageningen, The Netherlands

Mikkelsen, B. (1995) *Methods for Development Work and Research: A Guide for Practitioners*, Sage Publications, New Delhi

Mocumbi, P. (2004) 'Plague of my people', *Nature*, vol 430, p925

Morse, S. S. (1995) 'Factors in the emergence of infectious diseases', *Emerging Infectious Diseases*, vol 1, pp7–15

Mouchet, J., S. Manuin, S. Sircoulon, S. Laventure, O. Faye, A. W. Onapa, P. Carnavale, J. Julvez and D. Fontenille (1998) 'Evolution of malaria for the past 40 years: Impact of climate and human factors', *Journal of the American Mosquito Control Association*, vol 14, pp121–130

Mwisongo, A. and J. Borg (eds) (2002) *Proceedings of the Kagera Health Sector Reform Laboratory 2nd Annual Conference*, Ministry of Health, Dar es Salaam, United Republic of Tanzania

Olago, D., M. Marshall, S. O. Wandiga, M. Opondo, P. Z. Yanda, R. Kangalawe, A. Githeko, T. Downs, A. Opere, R. Kabumbuli, E. Kirumira, L. Ogallo, P. Mugambi, E. Apindi, F. Githui, J. Kathuri, L. Olaka, R. Sigalla, R. Nanyunja, T. Baguma and P. Achola (2007) 'Climatic, socio-economic and health factors affecting human vulnerability to cholera in the Lake Victoria basin, East Africa', *Ambio*, vol 36, no 4, pp350–358

Pascual, M., R. Xavier, S. P. Ellner, R. Colwell and M. J. Bouma (2000) 'Cholera dynamics and El Niño – Southern oscillation', *Science*, vol 289, pp1766–1769

Patz, J. (2002) 'A human disease indicator for the effects of recent global climate change', *Proceedings of the National Academy of Sciences*, vol 99, pp12,506–12,508

Patz, J. A., K. Strzepek, S. Lele, M. Hedden, S. Greene, B. Noden, S. I. Hay, L. Kalkstein and J. C. Beier (1998a) 'Predicting key malaria transmission factors, biting and entomological inoculation rates, using modelled soil moisture in Kenya', *Tropical Medicine and International Health*, vol 3, pp818–827

Patz, J. A., K. Strzepek, S. Lele, M. Hedden, S. Greene, B. Noden, S. I. Hay, L. Kalkstein and J. C. Beier (1998b) 'Predicting key malaria transmission present and future', *Bulletin of the World Health Organization*, vol 76, pp33–45

Patz, J. A., M. Hulme, C. Rosenzweig, T. D. Mitchell, R. A. Goldberg, A. K. Githeko, S. Lele, A. J. McMichael and D. Le Sueur (2002) 'Regional warming and malaria resurgence. Brief communications', *Nature*, vol 420, pp627–628

Rees, P. H. (2000) 'Editorial – Cholera', *East African Medical Journal*, vol 77, pp345–346

Reiter, P. (2001) 'Climate change and mosquito-borne diseases', *Environmental Health Perspectives*, vol 109, Supplement 1, pp141–161

Roberts, J. M. D. (1964) 'Control of epidemic malaria in the highlands of western Kenya, Part I: Before the campaign', *Journal of Tropical Medicine and Hygiene*, vol 61, pp161–168

Sachs, J. and P. Malaney (2002) 'The economic and social burden of malaria', *Nature*, vol 415, pp680–685

Shapiro, R. L., M. R. Otieno, P. M. Adcock, P. A. Phillips-Howard, W. A. Hawley, L. Kumar, P. Waiyaki, B. L. Nahlen and L. Slutsker (1999) 'Transmission of epidemic *Vibrio cholerae* in rural western Kenya associated with drinking water from Lake Victoria: An environmental reservoir for cholera?' *American Journal of Tropical Medicine and Hygiene*, vol 60, pp271–276

Some, E. S. (1994) 'Effects and control of highland malaria epidemics in Uasin Gishu District, Kenya', *East African Medical Journal*, vol 71, pp2–8

TDR (Special Programme for Research and Training in Tropical Diseases) (no date) 'Malaria: Disease information', available at www.who.int/tdr/diseases/malaria/diseaseinfo.htm, accessed 31 January 2007

Waiyaki, P. G. (1996) 'Cholera: Its story in Africa with special reference to Kenya and other East African countries', *East African Medical Journal*, vol 73, pp40–43

Walsh, J. F., D. H. Molyneux and M. H. Birley (1993) 'Deforestation: Effects on vector-borne disease', *Parasitology*, vol 106, ppS55–S75.

Wandiga, S. O., M. Opondo, D. Olago, A. Githeko, F. Githui, A. Opere, P. Z. Yanda, R. Kangalawe, R. Kabumbuli, E. Kiramura, J. Kathuri, E. Apindi, L. Olaka, L. Ogallo, P. Mugambi, R. Sigalla, R. Nanyunja, T. Baguma, P. Achola, M. Marshall and T. Downs (2006) 'Final report, Project AF91, Assessments of Impacts and Adaptations to Climate Change', The International START Secretariat, Washington, DC

WHO (World Health Organization) (no date) 'Global epidemics and impact of cholera', available at www.who.int/topics/cholera/impact/en/index.html, accessed 3 April 2007

WHO/UNICEF (2003) 'Africa malaria report', WHO/UNICEF, Geneva, Switzerland, available at www.rbm.who.int/amd2003/amr2003/amr_toc.htm

Yanda, P. Z., R. Y. M. Kangalawe and R. J. Sigalla (2005) 'Climatic and socio-economic influences vulnerability on cholera in the Lake Victoria region', AIACC Working Paper No 12, International START Secretariat, Washington, DC

7

Making Economic Sense of Adaptation in Upland Cereal Production Systems in The Gambia

Momodou Njie, Bernard E. Gomez, Molly E. Hellmuth, John M. Callaway, Bubu P. Jallow and Peter Droogers

Introduction

The Gambia lies in the Sahel region, where rainfall is directly linked to the zonal position of the Inter-Tropical Discontinuity and is highly sensitive to perturbations of the global monsoon circulation. To cope with seasonal variability associated with such perturbations, Gambian farmers have traditionally used a number of strategies. But how successful these are is open to debate, considering rural–urban migration trends in the past three decades. In the face of imminent threats from climate change, adaptation strategies inspired and informed by past and current coping strategies are reported in The Gambia's Initial National Communication (GOTG, 2003), but their performance remains to be evaluated in terms of economic viability or impact on national food security.

In this chapter, the SWAP-WOFOST model is used in combination with CEREBAL to investigate the impact of climate change on The Gambia's cereal balance under different management options. Most significantly, the analysis looks into the economic efficiency of specific management options, subject to social and political acceptability.

Before and immediately after independence in 1965, agricultural policy in The Gambia was primarily driven by the need to generate foreign exchange to pay for goods and services required for economic development. Over the years, this paradigm was reinforced by buoyant world market prices for groundnuts and cheap food prices (Carney, 1986) and only began to lose ground after protracted drought, and economic hardship experienced by farmers in the last two decades. The gradual move away from cash to cereal crops is clearly shown in agricultural statistics (Department of Planning, 2001, 2003 and 2005). Cereal production is mainly for consumption, but surplus production by individual farming households, or *dabada*, is sold off in local grain markets. Cereals grown include rice (*Oryza sativa*), millet (*Pennisetum typhoides*), sorghum (*Sorghum bicolor*) and maize (*Zea mays*), with millet accounting for nearly 60 per cent of area planted and slightly more than 50 per cent of total cereal production (Department of Planning, 2005).

The River Gambia divides the country into two strips of land no wider than 30km at any transect. Over 48 per cent of the total land area of The Gambia is below 20m above mean sea level, and nearly one-third of the country is at or below 10m above sea level. In general, elevation increases with axial distance from the river. Geomorphological units are described as lowland and uplands. Weathered tropical soils found in the uplands are not very fertile but are well drained. In contrast, the fine-textured soils of the lowlands are poorly drained. This juxtaposition of topography, pedology and hydrology leads to spatial differentiation of cereal cultivation areas. In general, the River Gambia valley and adjacent swamps are used to cultivate rice, whereas the plateau is the center for millet cultivation.

The traditional agricultural system depends on extensive land use, using little agricultural input. To this effect, successful crop production is dependent on rainfall and favourable environmental conditions. Farmers' vulnerability is systemic and inextricably linked to climate variability, natural soil fertility and economy-wide policy framework.

A sharp and significant drop in average rainfall in The Gambia since the late 1960s has put tremendous pressure on crop production. Lowland rice production, in particular, is under heavy pressure from reduced flood duration and frequency, saline intrusion, soil acidification and deposition of sediments eroded from uplands. To fix some ideas, protracted drought and saltwater intrusion in cropland areas have resulted in a 50 per cent decline in the area under rice cultivation (Department of Planning, 2001). Declining soil fertility in uplands is forcing fundamental changes in production such as the use of marginal land, reduced fallows and deforestation to compensate for low productivity. Although tidal irrigation in lowlands and introduction of improved rice cultivars represent opportunities for increasing total cereal production, water still remains the limiting factor to expansion. Njie (2002a) and, more recently, Verkerk and van Rens (2005), demonstrate the environmental and economic risks of expanding irrigation schemes under natural flow conditions in the River Gambia.

Adaptation Strategies

In response to climate hazards, farmers have experimented and adopted a raft of strategies to cope with erratic rainfall patterns. An insightful study by Jallow (1995) divides these into risk-aversion and risk-management strategies. The first category includes crop diversification, crop selection and plot dispersal. If these fail to provide adequate insurance, farmers, depending on their circumstances, sell off assets, use kinship networks, receive government assistance in the form of food aid and harvest natural forest food to get over the period of hardship.

Cole et al (2005) report adaptation strategies at government level as including sustaining support for farmer risk-aversion strategies through appropriation and dissemination of high-yield cultivars, providing engineering and technical leadership in land rehabilitation/conservation and water control, providing scientific advice in the form of seasonal rainfall forecasting and, when all these fail, providing disaster relief with the assistance of relevant United Nations agencies, multilateral institutions and non-governmental organizations.

Although these strategies worked sufficiently well in the past, a critical evaluation in the context of climate change and socioeconomic trends casts doubts on the prospects of some of them. The subsisting problem is how to maintain or increase production under adverse conditions. Limiting constraints include further decline in rainfall, per capita availability of land, land degradation, widespread poverty and social mutations.

Our point of departure in this study is the understanding that crop production may be increased by increasing cultivated area and/or increasing crop yields. Through screening and integration of previously proposed adaptation options (Jallow, 1995; GOTG, 2003), we identify (1) crop breeding and selection, (2) crop fertilization and (3) irrigation as the most comprehensive, no-regrets, flexible strategies to improve crop yields. The main argument in favour of crop breeding and selection is that of probable decline in rainfall and increased variability. On the other hand, promotion of crop fertilization as an adaptation strategy is influenced by continuous decrease in available prime land and concurrent degradation of arable land. Moreover, land that requires some amendments represents the second largest class of agriculturally suitable land (National Environment Agency, 1997). Irrigation provides a sorely needed means of mitigating impacts of spatial and temporal variability of rainfall and offers the potential for extending the growing season and expanding total cultivated area. Except perhaps for irrigation of upland cereals, these strategies are not entirely new. What is novel about our restatement of these strategies is the systematic, rigorous and quantitative approach used in this study.

Analytical Framework

The analytical framework used in this study (see Figure 7.1) is built around two key components: (1) crop modelling and (2) economic feasibility analyses using SRES A2 forcing and outputs from HadCM3 and ECHAM4 global circulation models (GCMs) adjusted to The Gambia's climate. Every effort is made to ensure that socioeconomic scenarios prescribed are consistent with SRES A2.

Climate scenarios

The climate change scenarios used in the analysis are derived from projections of the Max Planck Institute's ECHAM4 general circulation model and the Hadley Centre's HadCM3 model for the A2 emission scenario of the Intergovernmental Panel on Climate Change (IPCC). The A2 scenario of

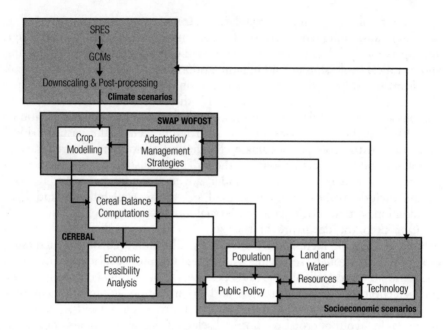

Figure 7.1 *Analytical framework for assessing economic feasibility of climate change adaptation*

greenhouse gas emissions is characterized by high population growth, slow and regionally oriented economic growth, and slow technological change (Nakicenovic and Swart, 2000). Details of the procedure for downscaling the climate projections of the global scale models to the study region can be found in Gomez et al (2005).

Both the ECHAM4 and HadCM3 models project an average temperature rise of 3 to 4°C by 2100 but differ significantly in their projections of precipitation changes. Whereas ECHAM4 shows no significant change in mean rainfall, and some increase in extreme values, HadCM3 shows a drastic drop of 400mm in annual rainfall in the distant future (2070–2099). This situation presents us with two scenarios: (1) global warming only and (2) global warming and increasing aridity. From recent changes in Sahel rainfall, we take a neutral position and assume both are plausible scenarios. To feed the climate model-derived information into the environmental and biophysical models used in this study, monthly data are transformed into daily values by interpolation and statistical modelling (Richardson, 1981; Racsko et al, 1991).

Socioeconomic scenarios

The upper envelope of population projections from different growth models (Njie, 2002b) is used in this study. This corresponds to the cohort survival method that assumes unchanging fertility rates. Under this scenario, total population of The Gambia grows from under 2 million persons to over 11 million by the year 2100. Rural to urban population ratio, currently 50:50, is assumed to evolve linearly over time to 20:80 by the end of the century. In this scenario, however, absolute decline in rural population will occur late in this century. Bolstered by food security and poverty alleviation policies, agricultural production is, therefore, expected to be a dominant factor in the economy.

Land availability, a crucial factor in cereal production, is assessed in light of other competing uses – economic, nature conservation and residential. Priorities of use, regeneration and degradation rates, together with suitability for agriculture, are also incorporated in the land availability calculus. In like manner, the feasibility of putting 20 per cent of millet production under irrigation is assessed by comparing projected water demand with renewable water resources. The reader may note that 20 per cent irrigation is a benchmark only so far achieved in developed countries.

For the adaptation/management strategies analysed in this study, the issue of significant technological change and innovation is only considered for research and development outcomes of regional crop breeding programmes. No major revolution is expected in already mature irrigation technology, but there is room for improvement of water delivery efficiencies. Costs are also likely to change, but projections are not attempted because of large uncertainties and use of constant prices for other variables in the study. Little sophistication is required for fertilizer application.

Public policy in food security and poverty alleviation seeks to 'stay ahead of the curve', so to speak. Policy variables used in this study include per capita cereal consumption of 250kg per year, based on the upper limit of local production and imports from 1995 to the present (Department of Planning, 2005). Strategic food reserves are defined as food reserves sufficient for one month up to a maximum of two months but not more. As already mentioned, we ensure at the problem specification stage that conflicts between land and water management and other policies are eliminated. At the analysis stage, changes in policy variables and food preferences are made to see what impact they might have on the economic performance of the adaptation strategies and food security.

Crop modelling

Crop yields are simulated with the linked SWAP-WOFOST models (Feddes et al, 1978; Van Dam et al, 1997; Kroes et al, 1999). The SWAP (soil water atmosphere plant) model simulates one-dimensional water, solute and heat transport in saturated and unsaturated soils (Feddes et al, 1978; Droogers, 2000). WOFOST (the world food studies model) simulates the phenological development of a crop from emergence to maturity on the basis of the crop's genetic attributes, and environmental conditions (Spitters et al, 1989; Supit et al, 1994).

In the SWAP model, rainfall, irrigation water and solar radiation reaching the soil surface are related to the leaf area index (LAI) of the crop. Solute, heat and water transport within the soil, governed by laws of mass and energy conservation, is modulated by heat and moisture transmission and storage properties, concentration, temperature and pressure gradients, and fluxes at the boundaries of the study domain. While precipitation amounts are input from the downscaled climate scenarios, irrigation amount and scheduling is specified by the SWAP model user. Irrigation is triggered by soil in the root zone drying beyond a critically low value.

The Penman-Monteith equation is used to compute evapotranspiration, or the sum of evaporation and transpiration. For crops with closed canopies, or densely planted crops, soil evaporation decreases but transpiration increases as crop development progresses. Density of foliage, characterized by the LAI, decreases the amount of rainfall and radiation that directly reaches the soil surface.

Water not retained within the unsaturated zone, or taken up by the crop, flows to adjacent drains or groundwater or drains freely according to the boundary conditions specified by the model user. Runoff is generated when surface infiltration or storage capacity is exceeded. Crop water uptake is directly related to soil wetness, potential evapotranspiration and root length. Soil temperature exercises some influence on the bioavailability of nutrients, but less on water dynamics, especially when the crop in place has a well-developed root system. Essentially isothermal at depths below 100cm, the soil temperature regime depends on surface heating, soil thermal properties and wetness. Estimates of soil properties used in the study are obtained from the literature (Williams, 1979; Campbell, 1985; FAO, 2002).

The SWAP model is linked to the WOFOST model through water and nutrient uptake by crop roots and LAI. The main processes simulated by WOFOST are the partitioning of assimilates from photosynthetic activity into root, stem, leaves and storage organs.

In photosynthesis, CO_2 from the air is transformed into glucose. The energy for this transformation originates from sunlight, or, more precisely, from the photosynthetically active radiation (PAR). Part of the glucose produced is used to provide energy for respiration and crop maintenance, depending on the amount of dry matter in the various living plant organs, the relative maintenance rate per organ and the temperature.

The remaining assimilates are partitioned among roots, leaves, stems and storage organs in fractions depending on the phenological development stage of the crop. Time-dependent partitioning coefficients change with crop development stage (Van Diepen et al, 1989). For a grain crop such as millet, the dry weight of storage organs per hectare at the end of the crop cycle, equivalent to crop yield, is an important model output. The net increase in leaf structural dry matter and the specific leaf area determine leaf area development and, hence, the dynamics of light interception, except for the initial stage when the rate of leaf appearance and final leaf size are constrained by temperature, rather than by the supply of assimilates. Leaf senescence occurs because of water stress and shading, and also because of lifespan exceedance. The death rate of stems and roots is related to the development stage and crop genotype. Crop parameters in this study were taken from the literature (de Willingen and Noordwijk, 1987; Van Diepen et al, 1989). Some of these were changed in one of the adaptation strategies to see the impact on yields.

Economic feasibility analysis

CEREBAL, a simple spreadsheet model, is used to update and compute running totals of cereal stocks in The Gambia. Population size and per capita consumption constitute key variables on the demand side of the model. Cereal production in any year is obtained by summing up production from rain-fed/irrigated rice with that from upland cereals, the latter derived from SWAP-WOFOST crop yield and cultivated area. Variations in cultivated area are handled through a land-use submodel that incorporates competing land uses within the socioeconomic context of The Gambia. At the end of every year in the time window studied, CEREBAL compares demand for cereals with production and computes commercial grain imports and food aid requirement in line with national food security policy.

The economic feasibility of different adaptation options is evaluated from net adaptation benefits and imposed climate change damages associated with a particular adaptation strategy. Climate change damages are measured as losses in the economic value of cereal production, as simulated with the SWAP-WOFOST and CEREBAL models, for climate change scenarios of the near and distant future relative to a baseline reference case for current climate and current crop yields. The future climate change simulations include business-as-usual management cases for low-intensity rain-fed cereal production and cases that incorporate changes in management to adapt to changes in climate. Net adaptation benefits are measured as the reduction in climate change damages, comparing the business-as-usual case with an adaptative management strategy, less the cost to implement the adaptation strategy. The residual damage after taking into account the net benefits of adaptation are referred to as the imposed climate change damages.

Results and Discussion

Crop yields

Statistics of simulated yields are presented in Table 7.1 for business-as-usual and adaptive management strategies combined with observed and future climates corresponding to the historical reference period 1961–1990 and projections of ECHAM4 and HadCM3 for the periods 2010–2039 and 2070–2099. The simulated yields reveal a systematic element of dependence on climate scenario. This is hardly surprising, especially in the case of distant future simulations with HadCM3, which we recall prescribes a 400mm decrease in rainfall relative to the period 1961–1990.

The effects of climate change on yields under business-as-usual management vary depending on the time period and the climate model. For the near future period, ECHAM4 and HadCM3 alike project an increase in average yields of 13 per cent and 2 per cent respectively. Increasing yields could be explained by carbon dioxide fertilization and a shift in the climate toward optimum temperature for C4 crops (Wand et al, 1999). The advantages of the higher average yield estimated for the HadCM3 climate projection are counteracted by an increase in the variability of yields due to an 8 per cent decrease in rainfall, which, as we will see later, results in economic losses despite the marginally higher average yield. For the more distant future of 2070–2099, results for the ECHAM4 climate projection show a nearly 40 per cent increase in cereal yield for business-as-usual management. But for the severe rainfall decrease projected by the HadCM3 model, estimated yields collapse by almost 80 per cent.

Table 7.1 *Average millet yields (kg/ha) and variability (CV) for current and future climates with business-as-usual and adaptive management strategies*

Adaptation strategy	ECHAM4		HadCM3	
	Yield	CV	Yield	CV
Reference period: 1961–1990				
Business-as-usual	923	23	1115	30
Near future: 2010–2039				
Business-as-usual	1046	24	1141	33
High-yielding cultivar	1186	22	1294	22
N-fertilizer (100kg/ha)	1450	20	1517	25
Irrigation (150mm)	1496	13	1563	11
Distant future: 2070–2099				
Business-as-usual	1274	29	354	167
High-yielding cultivar 1	500	30	583	135
N-fertilizer (200kg/ha)	1733	20	610	125
Irrigation (500mm)	1110	32	1811	13

Table 7.1 also presents results for yields under three different adaptation strategies: substitution of an improved millet cultivar, fertilization with nitrogen and irrigation. As previously mentioned, there is nothing revolutionary about such an approach. The technology to implement some strategies already exists but has not been fully harnessed. In the future, one may also expect improved cultivars from crop breeding programmes. Mimicking the outcome of selective breeding and genetic engineering programmes, we make changes to the following attributes of *P. typhoides*: (1) increased drought tolerance, (2) increased yield and (3) shorter growing cycle. Nitrogen fertilization is introduced at rates of 100kg and 200kg per hectare per year and irrigation is introduced at rates of 150mm and 500mm per year for the time periods 2010–2039 and 2070–2099 respectively.

For the period 2010–2039, average yields increase with adaptation relative to business-as-usual yields by 13 to 43 per cent and 13 to 37 per cent for the ECHAM4 and HadCM3 projected climates respectively. The substitution of a higher yielding cultivar produces the smallest gain in average yield of the three adaptation strategies, 13 per cent for both climate model simulations. In comparison, nitrogen fertilization and irrigation would increase yields by 33 to

43 per cent. All the adaptation strategies would also reduce the variability of yields, ranging from 8 to 67 per cent. The greatest reduction in variability comes from irrigation. For smallholder farming households, stability in yields is one important aspect of poverty and survival. The consequences of a poor harvest year are not so devastating if the following year gives a normal harvest. Periods of two or more successive years of poor harvest, however, are rather difficult to overcome without external assistance.

For the climate projection of the ECHAM4 model for the distant future (2070–2099), the adaptation strategies analysed result in average yield changes of between –12 and +36 per cent. Adoption of the improved cultivar and fertilization with nitrogen would amplify yield gains projected for the ECHAM4 climate with no adaptation. In these cases, adaptation and climate change would bring net benefits for cereal yields. In contrast, irrigation at the rate of 500mm would reduce yields by 12 per cent relative to business-as-usual management, although yields would still be higher with climate change than for the reference climate of 1961–1990. Waterlogged conditions caused by overirrigation explain the drop in yield relative to business-as-usual management under the ECHAM4 climate.

Simulations based on the HadCM3 climate projection for the distant future show an increase of 64 to 411 per cent relative to the business-as-usual case, depending on the adaptation strategy simulated. Yet, except for the irrigation case, the gains are not enough to overcome the severely negative effects of climate change for the highly water stressed climate of the HadCM3 model and yields are less than for the reference case with no climate change. Implementation of irrigation at a rate of 500mm is projected to increase average yields by a factor of more than 5 and decrease variability by an order of magnitude. This more than compensates for the yield effects of the HadCM3 drop in mean annual rainfall relative to the reference period. But these benefits need to be weighed against the substantial costs of irrigation.

In general, as crop yields increase, interannual variability decreases under all adaptation options. Using the criteria of reduced variability and increased average yield, irrigation, except when it is overdone, outranks other adaptation options. Crop fertilization ranks second best by these criteria. In the remainder of this paper, a pairwise comparison of the economic performance is made between irrigation and crop fertilization.

Economic performance of selected adaptation practices

Economic analysis presented in this paper uses costs and benefits of adaptation strategies within a national cereal self-sufficiency/import substitution framework. The economic analysis focuses on the HadCM3 climate scenario as it is this scenario that poses the greatest threat to cereal production in The Gambia. However, it should be kept in mind that a wetter climate, as projected by the ECHAM4 model, would generate benefits which we have not estimated.

Key variables in the analysis include the cost of inputs, market price of cereals, consumer preferences and food security policy. Constant cereal prices of US$150/ton are assumed. Other general assumptions include (1) cultivated area of millet increases proportionally with population growth, subject to a 0.15ha/capita limit in order to avoid conflict with other competing land uses; (2) food imports and aid are triggered by buffer stocks falling below a critical stock-to-utilization ratio (STU); and (3) food import and aid costs are expressed in constant dollar values. We further assume that economic agents with a central role in climate change adaptation, that is traders and *dabadas*, respond to market signals in a rational way that is influenced by government regulation, taxation, incentives and other economic instruments.

Crop fertilization

For analysis of crop fertilization, it is assumed that this strategy is applied to the entire area under millet cultivation. Two different assumptions are used for the stock utilization ratio that triggers food imports and aid: 10 and 20 per cent. Per capita cereal consumption is set at 250kg per person per year, closely matching current levels (Department of Planning, 2005). Results of the economic analysis of fertilization are shown in Table 7.2. Visual inspection of net adaptation benefits indicates the economic potential of fertilization as an adaptation option in the near future. Positive net benefits into the distant future also reinforce the evidence of economic viability under a changing climate. Note that while average crop yields are projected to increase in the near term for some of the scenarios (see Table 7.1), climate change damages are estimated to result due to increases in interannual variability that result in more frequent production shortfalls under a food security policy scenario. In essence, climate change damages are equivalent to commercial import of cereals and/or food aid required to maintain national food security. Average annual damages for the 2010–2039 period are roughly US$150 million for both stock utilization ratios and in excess of US$1 billion for the 2070–2099 period.

Fertilization generates average annual benefits of US$29 million to 38 million in the near term period at an annual cost of US$6.3 million, for a net adaptation benefit of US$22 million to US$32 million. With adaptation, residual damages, hereafter referred to as imposed climate change damages, are the difference between climate change damages without adaptation and net adaptation benefits. Positive imposed climate change damages, such as the ones that appear in Table 7.2, indicate that fertilization alone is not sufficient to make up for cereal production shortfalls, under the combined effect of climate and demographic changes. The imposed climate change damages in the near term are US$123 million to US$130 million per year. For the more distant 2070–2099 period, results are more sensitive to assumptions about the stock utilization ratio. Estimated net adaptation benefits range from US$17 million to US$95 million and imposed climate change damages range from US$955 million to US$1.0 billion. Sensitivity of results in Table 7.2 to the percentage area treated with fertilizer is analysed by changing this fraction to several values between 5 and 100 per cent, without qualitative changes to the results.

Table 7.2 *Climate change damages and costs and benefits of fertilization – Average annual values in millions of US dollars*

Period	Economic indicator	Damages, costs and benefits (million US$)	
		10% stock utilization ratio	20% stock utilization ratio
2010–2039	Climate change damages	155.1	151.9
	Adaptation benefits	37.9	28.6
	Adaptation costs	6.3	6.3
	Net adaptation benefits	31.6	22.3
	Imposed climate change damages	123.5	129.5
2070–2099	Climate change damages	1049.8	1049.8
	Adaptation benefits	28.0	105.4
	Adaptation costs	10.7	10.7
	Net adaptation benefits	17.2	94.6
	Imposed climate change damages	1032.6	955.2

The stock utilization ratio is an important food security variable, especially when natural hazards or disruption of supplies are anticipated. Results of increasing the stock utilization ratio shown in Table 7.3 are, however, ambivalent. An increase in imposed climate change damages in response to an increase in the ratio in the near future could be seen as a misallocation of resources. With higher interannual variability, however, an increased stock utilization ratio seems to have a positive payoff by stimulating an increase in net adaptation benefits from 17 million to nearly 95 million US dollars.

In the long run, especially when individuals' economic situations improve, it is reasonable to assume change in peoples' food preferences. For the distant future, therefore, we posit and analyse the impact of a shift in food choices, marked by a reduction in cereal consumption from 250kg to 175kg per person per year. Observe that reducing consumption to 175 kg does not imply food rationing but simply reflects a change in dietary habits, born out of improved economic status. This results in a reduction in imposed climate change damages of from 65 to almost 80 per cent in the near term and 30 to 40 per cent in the more distant future, depending on the stock utilization ratio assumed to trigger food imports.

Irrigation
Assumptions for the analysis of irrigation are as follows:

1 irrigated area of coarse cereals increases linearly with time, from its current value of 2 per cent to 20 per cent by the end of the 21st century;
2 rice irrigation from surface water is accelerated after commissioning of Sambangalou dam;
3 rice yields increase to 4 metric tons/ha under controlled irrigation; and
4 irrigated millet and rice are harvested twice a year.

Table 7.3 shows results for economic performance indicators under irrigated conditions. A major point of observation in this table concerning the near future (2010–2039) is the negative net adaptation benefits, which range from 81 million to 88 million US dollars per year. While irrigation would increase yields rather substantially, the even more substantial costs of irrigation outweigh the value of the increased production. This clearly indicates that resources could be more efficiently allocated to procurement of food supplies on the world grain markets than to investment in irrigation of cereals, at least for the 2010–2039 time period.

This situation may change in the more distant future when water becomes the major limiting factor for crop production, depending on the stock utilization ratio that triggers food imports. As shown in Table 7.3, the economic efficiency of irrigation is related to the policy stock utilization ratio variable, which governs cereal imports. Indeed, increasing the trigger level reverses the outcome of the net benefit calculus from a net benefit from irrigation of 52 million US dollars per year to an annual loss of 44 million.

Sensitivity analysis shows that, for the latter case, net adaptation benefits from irrigation only become positive when water costs drop below US$0.09 per cubic metre. Considering, however, that this is 25 per cent less than the unit cost of water in the lowest tariff block, it is extremely unlikely that small-holder irrigation schemes can achieve economies of scale sufficient to bring down water costs to a profitable level. It suffices to point out that operation and maintenance account for 80–90 per cent of pump irrigation costs and how this fraction evolves would depend on future world energy markets, technological innovation and the state of The Gambian economy.

An oblique approach to the problem of cost reduction is how to increase the market value of crops harvested. The conundrum essentially reduces the

Table 7.3 *Climate change damages and costs and benefits of irrigation –*
Average annual values in millions of US dollars

Period	Economic indicator	Damages, costs and benefits (million US$)	
		10% stock utilization ratio	20% stock utilization ratio
2010–2039	Climate change damages	155.1	151.9
	Adaptation benefits	43.3	36.4
	Adaptation costs	124.5	124.5
	Net adaptation benefits	−81.2	−88.0
	Imposed climate change damages	236.3	239.9
2070–2099	Climate change damages	1049.8	1049.8
	Adaptation benefits	303.2	207.5
	Adaptation costs	252.2	251.2
	Net adaptation benefits	52.0	−43.7
	Imposed climate change damages	997.8	1064.8

number of choices of crops. High value, non-food crops (for example, flowers) and non-staple crops (for example, vegetables) currently fetch higher prices than cereals on the market, but on the scale of production envisaged, this may no longer be the case. The problem, however, does not have to be articulated in such dichotomous terms. What one probably needs is to find an optimal mix of crops fetching the highest economic returns, subject to land, water, labour and other constraints.

Analyses were also performed for a decrease in cereal intake from 250kg per person to 175kg. This reduction in per capita cereal consumption would reduce the imposed climate change damages by 85 per cent in the 2010–2039 time period and 30 to 35 per cent in the more distant time period. One way of interpreting such a reduction is foreign exchange savings if the bill for foreign goods and services relating to the production and/or importation of alternative foods is incorporated into the calculus.

Conclusions

The estimated impacts of climate change on cereal crop yields range from increases to decreases, depending strongly on projected changes in rainfall. For a climate that is warmer but not drier than the present climate in The Gambia, simulated cereal yields would increase and generate economic benefits. But for warmer and drier scenarios, yields decrease and become more variable, resulting in economic losses, or climate change damages. For the severely dry climate derived from the HadCM3 model projection for the end of the century, cereal yields would collapse by an estimated 80 per cent.

Crop selection and fertilization, used as insurance against climate variability by smallholder farming households, are shown to be effective in offsetting adverse climate change impacts in the near future, or amplifying beneficial impacts. The yet untested practice of irrigation gives the highest increase in average productivity and the greatest reduction in variability under the different climate projections. But the economic efficiency of adaptation options is strongly influenced by unit costs of implementation. High units costs of irrigation development result in negative net benefits from irrigation in the near term, making irrigation an unattractive option, at least in the near term. In contrast, adoption of fertilization practice would generate net benefits of roughly 20 million to 30 million US dollars per year.

For the very dry climate projected by the HadCM3 for the end of the century, the economic performance of irrigation changes. Depending on government policies that determine food imports and aid, irrigation could generate substantial net benefits. Indeed, irrigation may become an imperative in the distant future if precipitation declines as sharply as indicated by the HadCM3 projections, or if world cereal markets become seriously affected by conditions in countries with historically surplus production. Whether expanded irrigation would be feasible in a much drier climate, however, has not been investigated.

Essentially, there is no single best adaptation strategy, and instead of import substitution, one should be looking at complementing business-as-usual (in other words food imports) with fertilization and irrigation of locally grown cereals. In the short run, expanding crop fertilization, in particular, has significant advantages. It requires no technological sophistication and promises high returns. These results call for an immediate response from government, the service sector and the farming community. Notwithstanding this, some challenges still remain. Food policy development and analysis require more sophisticated projections of commodity prices, costing of research and development in crop science, and technology trends.

A commitment to food security, rooted in developing The Gambia's agricultural potential, is indispensable to reducing the country's sensitivity and vulnerability to climate risks. In this regard, it may be quite important to discuss the role of key stakeholders, as well as to examine the conditions under which adaptation options are most likely to be taken up by *dabadas*. Without any doubt, the government of The Gambia, responsible for social, economic and related policies, should take the first step to ensure that valid research findings get translated into tangible benefits. The government's role is to pick up research results, demonstrate their validity and create incentives for integration of the options into current agricultural practice.

Considering, however, the low level of returns on investments in cereal production compared to other crops, it is fair to say that farmers, when given the choice, will opt for irrigation of non-cereal crops in the dry season. There is already ample evidence of this in community and women's gardens across the country. Even in some lowland environments, non-cereal, high-value horticultural crops, such as pepper (*Capsicum annum*), okra (*Hibiscus esculentus*) and tomatoes (*Lycopersicon esculentum*), are grown instead of rice under irrigation in the dry season. Two reasons may explain this practice. First, part of the incomes generated by households from horticultural production is used to purchase imported rice when their cereal stock gets depleted. Second, and perhaps overlooked, planting decisions also make sense in light of the agronomic practice of crop rotation.

A future government outreach/extension programme therefore stands the best chance of success when *dabadas* pursue multiple objectives, including food security. In this scenario, discussed above as an optimization problem, a fraction of the area under full water control could be allocated to non-cereals, and part of the extra revenue generated used to expand cereal production in subsequent years. *Dabadas* have a good knowledge of these issues, experience of climate extremes and a strong stake in harnessing adaptation options to ensure their food security.

From a broader perspective, The Gambia government's national disaster reduction strategic framework, in the making, provides an attractive opportunity to build partnerships for food security. Community empowerment, efficient marketing structures, competitive prices and price stability are some of the factors that hold the key to the transformation of agricultural production in The Gambia.

References

Campbell, G. S. (1985) 'A simple method for determining unsaturated conductivity from moisture retention data', *Soil Science*, vol 117, pp311–314

Carney, E. (1986) 'The social history of Gambian rice production: An analysis of food security strategies', PhD thesis, Michigan State University, Ann Arbor, MI

Cole, A., K. Sanyang, A. J. Marong and F. Jadama (2005) 'Vulnerability and adaptation assessment of the agricultural sector of the Gambia to climate change', consultancy report prepared for NAPA Project LDL 2328 2724 4699, Banjul

de Willingen, P. and M. van Noordwijk (1987) 'Roots, plant production, and nutrient efficiency', PhD thesis, Wagenigen Agricultural University, Wagenigen, The Netherlands

Department of Planning (2001) 'Statistical yearbook on Gambian agriculture for the year 2000', annual report, Department of State for Agriculture, Banjul

Department of Planning (2003) 'Statistical yearbook on Gambian agriculture for the year 2002', annual report, Department of State for Agriculture, Banjul

Department of Planning (2005) 'Statistical yearbook on Gambian agriculture for the year 2004', annual report, Department of State for Agriculture, Banjul

Droogers, P. (2000) 'Estimating actual evapotranspiration using a detailed agro-hydrological model', *Journal of Hydrology*, vol 29, pp50–58

FAO (2002) 'Digital soil map of the world and derived soil properties', on CD-ROM, www.fao.org/ag/agl/agll/dsmw.htm

Feddes, R. A., P. J. Kowalik and H. Zarandy (1978) *Simulation of Field Water Use and Crop Yield*, Simulation Monographs, Center for Agricultural Publishing and Documentation (PUDOC), Wageningen, The Netherlands

Gomez, B. E., M. Njie, B. P. Jallow, M. E. Hellmuth, J. M. Callaway and P. Droogers (2005) 'Adaptation to climate change for agriculture in The Gambia: An explorative study on adaptation strategies for millet', AIACC Working Paper No 37, International START Secretariat, Washington, DC

GOTG (Government of The Gambia) (2003) 'First national communication of the Republic of The Gambia to the United Nations Framework Convention on Climate Change', Government of The Gambia, Department of State for Fisheries, Natural Resources and the Environment, Banjul

Jallow, S. S. (1995) 'Identification of the response to drought by local communities in Fulladu West District of The Gambia', *Singapore Journal of Tropical Geography*, vol 6, pp22–41

Kroes, J. G., J. C. van Dam, J. Huygen and R. W. Vervoort (1999) 'User's guide of SWAP version 2.0. Simulation of water flow, solute transport, and plant growth in the Soil-Water-Atmosphere-Plant environment', Technical Document 48, Alterra Green World Research, Wageningen, Report 81, Department of Water Resources, Wageningen University and Research, Wageningen, The Netherlands

Nakicenovic, N. and R. Swart (eds) (2001) *Emissions Scenarios. A Special Report of Working Group III of the Intergovernmental Panel on Climate Change*, Cambridge University Press, Cambridge, UK, and New York

National Environment Agency (1997) 'State of the environment report – The Gambia', National Environment Agency, Banjul

Njie, M. (2002a) 'National water security in the first half of the 21st century', report prepared under GAM/93/003 as a contribution towards The Gambia's National Water Resources Management Strategy, UNDP/DWR, Banjul

Njie, M. (2002b) 'Second assessment report of climate change induced vulnerability of Gambian water resources sector, and adaptation strategies' report prepared for The Gambia's National Climate Committee, Banjul

Racsko, P., L. Szeidl and M. A. Semenov (1991) 'A serial approach to local stochastic weather models', *Ecological Modelling*, vol 57, pp27–41

Richardson, C. W. (1981) 'Stochastic simulation of daily precipitation, temperature and solar radiation.' *Water Resources Research*, vol 17, pp182–190

Spitters, C. J. T., H. Van Keulen and D. W. G. van Kraalingen (1989) 'A simple and universal crop growth simulator: SUCROS87', in R. Rabbinge, S. A. Ward and H. H. van Laar (eds) *Simulation and Systems Management in Crop Protection*, Simulation Monographs, Center for Agricultural Publishing and Documentation (PUDOC), Wageningen, The Netherlands

Supit, I., A. A. Hooijer and C. A. Van Diepen (eds) (1994) *System Description of the WOFOST 6.0 Crop Growth Simulation Model Implemented in CGMS. Vol 1: Theory and Algorithms*, European Commission, Luxembourg

Van Dam, J. C., J. Huygen, J. G. Wesseling, R .A. Feddes, P. Kabat, P. E. V. Van Walsum, P. Groenendijk and C. A. van Diepen (1997) 'Theory of SWAP version 2.0', Technical Document 45, Wageningen Agricultural University and DLO Winand Staring Centre, Wageningen, The Netherlands

Van Diepen, C. A., J. Wolf, H. Van Keulen and C. Rappoldt (1989) 'WOFOST: A simulation model of crop production', *Soil Use and Management.*, vol 5, pp16–25

Verkerk, M. O. and C. P. M. van Rens (2005) 'Saline intrusion in Gambia river after dam construction: Solutions to control saline intrusion while accounting for irrigation development and climate change', research report, University of Twente, The Netherlands

Wand, S. J. E., G. F. Midgley, M. H. Jones and P. S. Curtis (1999) 'Responses of wild C4 and C3 grass (Poaceae) species to elevated atmospheric CO_2 concentration: A meta-analytic test of current theories and perceptions', *Global Change Biology*, vol 5, pp723-741

Williams, J. B. (1979) 'Soil water investigation in The Gambia' Technical Bulletin 3, Land Resources Development Centre, Ministry of Overseas Development, Surbiton, Surrey, England

8

Past, Present and Future Adaptation by Rural Households of Northern Nigeria

*Daniel D. Dabi, Anthony O. Nyong, Adebowale A. Adepetu
and Vincent I. Ihemegbulem*

Introduction

There is sufficient information now to indicate that the climate is changing with far reaching effects for the well-being and livelihoods of rural people in the northern, semi-arid areas of Nigeria. This part of the country has characteristics similar to the rest of the arid and semiarid regions of West Africa commonly known as the Sahel. The Sahel is characterized by scant rainfall, with an annual average of between 150 and 600mm. Annual rainfall levels have been decreasing in the region over the course of the last century, with an increase in inter-annual and spatial variability (Glantz, 1987; Tarhule and Woo, 1998; Ozer, 2003). This region has experienced fluctuations in rainfall on all time-scales – from decadal (ten-day cycle) to monthly, seasonal, annual and longer term. The longer-term fluctuations are caused by the oscillations of the climatic borders of the Sahara Desert. The arid and semiarid part of Nigeria experiences similar characteristics with droughts occurring due to rainfall variability (Tarhule and Woo, 1998).

Drought is the main climatic hazard affecting the socioeconomic activities and livelihoods of rural households of our study region (Mortimore, 1989). For the purpose of this chapter, we have adopted the National Drought Mitigation Center's conception of drought as a normal, recurrent feature of climate that originates from a deficiency of precipitation over an extended period of time relative to long-term average or normal conditions and results in a water shortage for some activity, group or environmental sector (National Drought Mitigation Center, 2005). Drought is also related to the timing of rainfall (principal season of occurrence, delays in the start of the rainy season, occurrence of rains in relation to principal crop growth stages) and the effectiveness of rain (rainfall intensity, number of rainfall events) and can be aggravated by other climatic factors such as high temperature, high wind and low relative humidity.

In Nigeria, food shortages are attributable to poor socioeconomic conditions and government policies, increases in population and declining crop production following the oil boom of the 1970s, when the agricultural sector was abandoned for oil revenue. Problems of food shortages and insecurity are aggravated by drought, as occurred in 1973 to 1974. Since then, the agricultural sector has not lived up to expectations (Dabi, 2006).

In its most recent assessment of regional climate change projections, the Intergovernmental Panel on Climate Change (IPCC) concludes that warming in Africa is very likely to be greater than the global average and that warming will be greater in the drier subtropical regions, which include northern Nigeria, than in the moister tropics (Christensen et al, 2007). Changes in rainfall are uncertain for the Sahel, with the 25–75 percentiles of model projections ranging from decreases to increases in the region. But despite the uncertainty in rainfall changes, governments and policy makers should plan for the possibility of more frequent and more severe dry spells and droughts and should develop strategies for coping and adapting accordingly. Preparing for a potentially more droughty future is prudent because drought will continue to be a major threat to food security, livelihoods and rural development in the Sahel, even if the climate becomes moister than at present, and because the current capacity to cope with drought is poorly developed. These efforts should emphasize enhancing the adaptive capacity of poor rural households, who are highly vulnerable to drought and other climate pressures (Nyong et al, 2003).

Several definitions of climate change vulnerability, sensitivity and adaptation are found in the literature, for example, Smithers and Smit (1997), Pielke (1998), Smit et al (1999, 2000 and 2001) and Adger (2001), as summarized in Huq et al (2003). For the purposes of this chapter, we have adopted the definitions of Working Group II of IPCC (McCarthy et al, 2001): Vulnerability is 'the degree to which a system is susceptible to, or unable to cope with, adverse effects of climate change, including climate variability and extremes. It is a function of the character, magnitude and rate of climate change and variation to which a system is exposed, its sensitivity, and its adaptive capacity'. Sensitivity is defined as 'the degree to which a system is affected, either adversely or beneficially, by climate-related stimuli', while the adaptive capacity is 'the ability of a system to adjust to climate change, including climate variability and extremes, to moderate potential damages, to take advantage of opportunities, or to cope with the consequences' (McCarthy et al, 2001). Increasing the adaptive capacity of a system represents a way of coping with changes and uncertainties in climate, reducing vulnerabilities and promoting sustainable development (Huq et al 2003).

Objectives and Methods

Field observations have indicated that many poor rural households are already adapting to drought and water scarcity. Because there is general agreement that climate change adaptation can usefully begin with strategies to reduce present vulnerabilities (Downing et al, 1997; Adger, 2001), we seek to learn from the experiences of these households. The aims of this chapter are to examine how rural households of northern Nigeria have adapted to drought and water scarcity and to consider what lessons past and present strategies may hold for future adaptation. The specific questions we investigate are as follows:

1   Which members of the community are most vulnerable and what aspects of their livelihood are threatened?
2   How have they coped with the problem of drought and water scarcity?
3   How might they adapt in the future?
4   What factors influence their capacity to adapt?
5   What opportunities are available to them to enhance their adaptive capacity?

Our analysis is largely based on primary data collected directly in the field from households across the Sahelian belt of northern Nigeria. The field activities included a rapid rural appraisal, questionnaire surveys and focus group discussions. The rapid rural appraisal was carried out during the reconnaissance survey in October 2002. Observations were made, photographs taken, communities selected for data collection and poor rural households identified based on interactions with key informants. Twenty-seven communities from nine states were selected for administration of household surveys (see Figure 8.1). The selected communities include three settlement types: hamlets, villages and small towns, with populations of roughly 1000, 5000 and 10,000 respectively. The communities were selected in a ratio of 1:2:3 for small towns, villages and hamlets in that order to represent the first stratum in the sampling process. Thirty households were randomly selected from each of the communities for participation in the survey, giving a total of 810 households surveyed. A household is defined for our study as a group of people living in the same compound, consisting of an enclosed set of buildings or huts, eating from the same pot and recognizing one person as head, usually a husband and father or guardian. The household is chosen as our unit of analysis because livelihood decisions are usually made at the household level and vulnerability also resides at that level.

The questionnaire is made up of a combination of open- and closed-ended questions related to aspects of drought, activities undertaken by households, problems they encounter, strategies they have adopted to cope with droughts and their opinions regarding future adaptation. Information generated by the questionnaires provided the basis for identifying vulnerable households, establishing how they have been coping (adaptation measures) and revealing factors that influence their capacity to adapt to droughts.

The household survey was followed up with a series of focus group discussions. The focus group discussions used a bottom–up approach to adaptation strategizing and planning in which 10 key informants were invited from each of the 27 communities for in-depth discussions of present coping strategies identified by the survey, difficulties faced and possible strategies that might be useful for future adaptation. The informants include community and religious leaders, farmers' representatives, and leaders and representatives of associations, women's groups and local government.

The bottom–up approach emphasizes initiatives developed locally at a micro or local community level that are based on indigenous knowledge and local perceptions, opinions and preferences. Aspects of indigenous knowledge

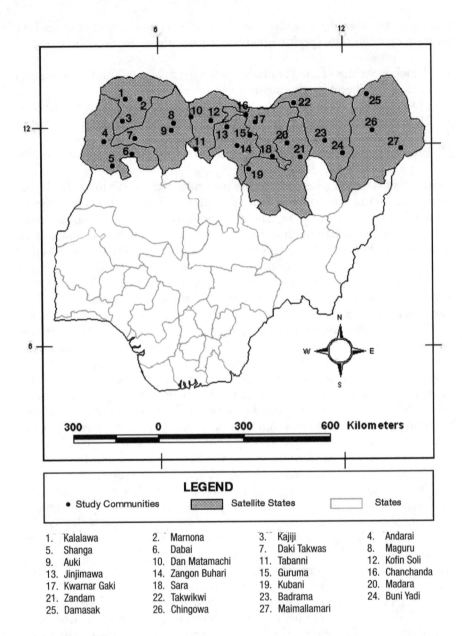

Figure 8.1 *Communities surveyed in northern Nigeria*

systems that are important for planning climate change adaptation include what the households did in the past to survive earlier droughts, what they are doing now to cope with current droughts, and the factors that influence their choices and activities. A bottom–up approach helps to ensure that adopted strategies are responsive to, and consistent with, local priorities, preferences and capacities and that the strategies are likely to benefit local people. This contrasts with a top–down approach in which measures are externally planned and imposed on a locality without the local people's input or participation. Such projects run a high risk of failure.

Data collected from the household survey and focus group discussions were analysed using qualitative methods and descriptive statistics in the form of frequencies and percentages presented in tables and graphs. We assessed the vulnerability and adaptive capacity of rural households using the livelihood systems approach (Carney et al, 1999; Davies, 1996). The livelihood systems approach examines five different categories of household resources or capital used to support a household's livelihood strategy and provide it with resilience and capacity for coping with, and adapting to, shocks. These are natural, human, financial, physical and social capital. Davies (1996) distinguishes differential vulnerability and livelihood system vulnerability and argues that vulnerability research should focus more on livelihood system vulnerability because people in different livelihood systems are vulnerable in different situations and seasons. Different livelihood systems experience different trigger events that can cause food and livelihood stress. Following her approach, we identify three major livelihood strategies in the study region – crop farming, herding livestock and fishing – and distinguish the different ways in which these livelihood groups are vulnerable to drought.

Results

Vulnerability of rural households

Vulnerability is a relative term differentiating between socioeconomic groups or regions, rather than an absolute measurement of deprivation. The analyst or decision maker must assign the thresholds of vulnerability that warrant specific responses. For our study areas, we developed a methodology for classifying households based on their levels of current vulnerability using a vulnerability index constructed from 13 indicators that measure different aspects of vulnerability (Table 8.1).

The indicators and their weights were determined based on factors identified by households and stakeholders as important determinants of vulnerability, as well as information from the published literature. Indicator values were normalized such that the highest value for an indicator was set at 100 per cent. Adding up the household scores on the 13 indices resulted in an overall vulnerability score for each household in the sample. Households were grouped into three levels of vulnerability: very vulnerable, vulnerable and less vulnerable. In the questionnaire, we had asked the respondents to place themselves in one of these classes of vulnerability. This allowed us to factor in their perceptions and self-reported assessments of vulnerability. We took all those who put themselves in each group, found the average scores for each group and used these as the midpoints of the various vulnerability classes and then built class intervals about them.

Table 8.1 *Indices and weights for vulnerability assessment in northern Nigeria*

Index	Index Value	Measured/Calculated as	Range	Average
Acreage under cultivation	1	Hectares/consumer units	0.1–2.8	0.6
Dependency ratio	1	Labour units/consumer units (inverted)	0.3–0.8	0.5
Livestock ownership	1	Tropical livestock units/consumer units	0.0–8.2	3.7
Gender of household head	1	Value given to sex of household head	1.0–2.0	1.8
Livelihood diversification	1	Weighted number of non-agricultural income generating activities/consumer units	0.0–2.4	0.7
Annual cash income	1	In 1000 naira/consumer units	2.5–9.7	4.2
Drought preparedness	1	Value given to use of drought resistant crops and livestock, and receiving drought-related information and advice	0.0–2.0	1.1
Educational background of the household head	0.5	Value given to highest school level attained by the head of the household	0.0–4.0	1.8
Land tenure situation	0.5	Value given to land tenure situation	1.0–3.0	2.5
Type of house	0.5	Value given to type of house lived in	1.0–3.0	1.8
Self-sufficiency in food production	0.5	Number of years surplus foodstuffs were sold minus number of years foodstuffs were bought in the past 10 years	0.0–20.0	11.2
Family and social networks	0.5	Value given to strength of family and social networks.	1.0–4.0	2.3
Quality of household	0.5	Number of able persons/number of disabled and or sick persons in the household (inverted)	1.5–12.0	7.6
Overall vulnerability	10	Sum of (index scores × index value)	235.1–833.9	472.1

All three livelihood groups of the Sahelian zone of northern Nigeria (crop farmers, herders and fishers) are vulnerable to drought and water scarcity. But within these groups, the poorer rural households are identified as the most vulnerable. The rural poor are highly vulnerable because their low asset base exposes them to high risk of impacts from climate stresses and limits their resilience and capacity to adapt. The landholdings and other natural resource assets of the poor are small in amount and degraded in quality, leaving a slim margin between meeting basic needs in good years and suffering severe depravation in poor years in this water scarce region. Their stock of human capital includes knowledge of traditional, indigenous practices as well as some new innovations that are used to cope with drought and scarcity. But educational attainment is low and constrains further innovation that requires new knowledge and skills. Food stores, financial savings, marketable assets and access to credit are too little to assure recovery from losses of crops and livestock that may be experienced due to drought or other shocks. Farm equipment and other physical assets of poor households are meagre and can act as a barrier to the adoption of some adaptive practices, while physical infrastructure in poor rural communities are also lacking and constrain options. Social networks exist and provide some security, but often these networks are weakest for the poorest households of a community

Comparing across the three livelihood groups, we find that the proportion of households that is very vulnerable is greatest for crop farmers and lowest for fishing. We illustrate this for three hamlets, each dominated by a different livelihood strategy. Figure 8.2 shows the number of households classified as very vulnerable, vulnerable and less vulnerable for the hamlets of Zangon Buhari, Dabai and Takwikwi. The number of very vulnerable households is greatest in Zangon Buhari, a predominantly crop farming community located in Kano State in the north central region of Nigeria, followed by Dabai, a community of mostly livestock herders in Kebbi State in the northwest. The fewest number of very vulnerable households is found in Takwikwi, a fishing community located in Yobe State in the northeast. The lesser vulnerability of households in Takwiki may be attributed to the fact that the fishing community has access to the Nguru wetlands, a fairly reliable resource that supports fishing livelihoods. In Dabai, livestock farmers may have an advantage over crop farmers because of their mobility, moving cattle, sheep and goats to areas with greener pasture and greater availability of water, while the crop farmers depend on highly variable rainfall for their rainfed crops, which can easily be affected by droughts. The results for these hamlets are representative of the general pattern of vulnerability of the different livelihood groups.

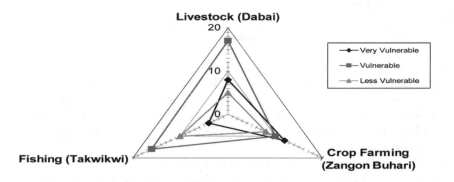

Figure 8.2 *Number of households classified as very vulnerable, vulnerable and less vulnerable in the hamlets of Zangon Buhari, Dabai and Takwikwi*

Strategies for coping with drought

From the analysis of the survey data, a number of coping strategies have been identified, most of which are autonomous (implemented without external intervention) and responsive (implemented in reaction to climatic events and impacts) (Smit et al, 1999). We subdivide coping strategies into two categories, past and present. Past coping strategies are strategies adopted by the respondents during droughts that occurred in 1994 or earlier, while present coping strategies are more recent strategies adopted by the respondents after 1994. Table 8.2 identifies coping strategies reported by households and summarizes the number of respondents who adopted each strategy in the past and present. Past and present coping strategies are not mutually exclusive since a respondent might have adopted a strategy in the past and still use it in the present.

Table 8.2 *Respondents' coping strategies*

Serial No	Strategy	Number of Respondents		
		Past	Present	Total
1	Drought resistant variety	206	168	374
2	Crop diversification	201	235	436
3	Livestock diversification	164	150	314
4	Early maturing crop varieties	183	325	508
5	High yield varieties	128	312	440
6	Replanting	198	128	326
7	Herd movement	89	104	193
8	Herd supplementation	54	127	181
9	Culling animals	57	56	113
10	Labour migration	105	95	200
11	Selling assets	142	150	292
12	Herd sedentarization	70	75	145
13	Farm relocation	230	96	326
14	Herd/farm sizes	132	82	214
15	Water exploitation methods	120	93	213
16	Water use	204	115	319
17	Water storage methods	161	166	327
18	Food storage	361	203	564

Strategies that have been widely used in the past or present include food storage; cultivation of drought resistant, early maturing or high yield varieties of crops; diversification of crops or livestock; replanting when crops are lost; selling assets; relocating farms; and a variety of strategies for water use and storage. Food storage was the coping strategy adopted by the largest number of the respondents in the past. More recently, however, the number of people using food storage as a coping strategy has declined by 40 per cent. The most probable reason is that most households have had insufficient food even in good years to enable them to store food as a hedge against drought. Other strategies that have declined in use are cultivation of drought resistant varieties, replanting of crops, changes in farm locations, and changes in water exploitation methods and water use. The reasons for these decisions may be that benefits from the strategies are less than what was expected or that the costs of inputs needed to implement them, such as chemical fertilizers, water pumps and fuel to operate pumps, are not affordable to poor households. A number of coping strategies have increased in use, most notably the adoption of early maturing and high yielding crop varieties. Other strategies that have increased in use include diversification of crops and herd movement and supplementation.

Respondents gave a series of reasons for the adoption of food storage and early maturing and high yield crop varieties. Figure 8.3 gives a summary of the reasons provided by the respondents for the adoption of food storage as a coping strategy. Food security was cited by the largest number of households as the reason for storing food (44 per cent), followed by lack of new information (20 per cent) and new method (19 per cent). The lack of new information may be attributed to the fact that most of the respondents had never been visited by government agricultural extension workers.

The next most widely used past and present coping strategies are the cultivation of early maturing and high yield crop varieties. Figures 8.4 and 8.5 show the reasons for their adoption. The major reason for planting early maturing crops is to ensure early harvest (41 per cent), while the reason for planting high yield crops is, of course, to attain higher yields, as indicated by more than half (55 per cent) of the respondents who use this strategy. For both strategies, the second most important reason for the adoption of these strategies is lack of new information. The availability of early maturing and high yield crop varieties are recent developments, which explains why more respondents are adopting the strategies at present than in the past. This shift is also evidence that respondents are willing to change and adopt new coping strategies.

Three groups of factors affect the adoption of the coping strategies discussed earlier: (1) the resource base for sustainable livelihoods, (2) government policy and (3) availability of information and warning signals. Households' capabilities for adopting coping strategies are determined by their resource base, composed of their endowments of five types of livelihood capital: financial, human, natural, social and physical. We measure households' endowments of these capitals using household income, level of education of household head and members, water availability for household use, community groups and social networks, and distance to road or market. Using the survey data and information from the focus group discussions, threshold values are established for these measures to classify the level of vulnerability of households along the dimensions of the five livelihood capitals. Figure 8.6 shows the number of households classified as very vulnerable, vulnerable and less vulnerable with respect to the five capital types.

From Figure 8.6 it is evident that most households (80 per cent) are very vulnerable financially. Most households of the study sites have very low incomes, financial savings and saleable assets. They lack opportunities for off-farm labour income and the collateral necessary to access financial credit. This limits their ability to purchase farm inputs, equipment and other technology to improve agricultural production or adapt it to climate variability and extremes.

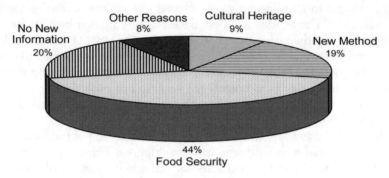

Figure 8.3 *Reasons for food storage*

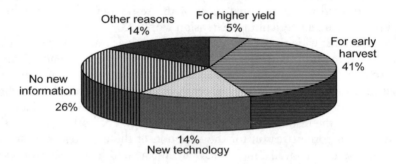

Figure 8.4 *Reasons for planting early maturing crop varieties*

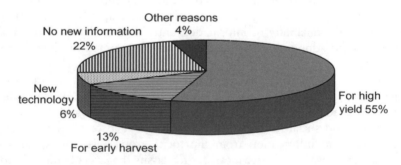

Figure 8.5 *Reasons for planting high yield crop varieties*

Consequently, most farmers are restricted to their old traditions of low input farming practices, which bring about poor yields and high risk crop losses and failure. Unable to invest in improvements to their farms or earn off-farm labour income, poor rural households are trapped in poverty, placing them at risk of hunger and other adverse impacts of drought.

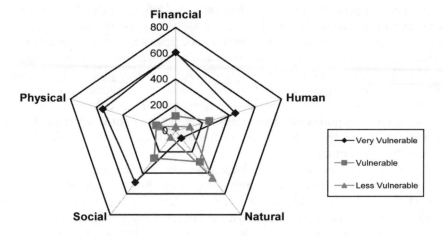

Figure 8.6 *Household vulnerability levels for the five livelihood capitals*

A large proportion of households (68 per cent) are also very vulnerable with respect to physical capital. This is due to the absence of infrastructural facilities such as roads, which affects accessibility to sources of inputs and markets for products; water supply and sanitation, which affects human health and well-being; and energy in the form of electricity, which affects food storage and food processing. Many households (61 per cent) are highly vulnerable in terms of social capital, or family and community networks. Drought will often impact many or even all members of a social network simultaneously, weakening the ability of the network to assist and support its members during times of drought. This condition will affect self-help programmes such as food sharing, labour assistance and remittances, which serve as safety nets that can reduce vulnerability; religious beliefs and prayers, which give hope and relief to households during and after droughts but can also have negative effects (for example, the purdah system which restricts women may affect the search for work and food for the household); and the possibilities of conflict (for example, conflicts between arable farmers and herdsmen over grazing land and watering points, especially during and immediately after droughts).

Many households (55 per cent) are also very vulnerable with respect to their endowment of human capital. This is attributed to the fact that most households have had no formal education, which may constrain access and use of information about climate risks, risk management strategies and new technologies, which affects in turn the adoption of coping strategies. The majority of households (55 per cent) are less vulnerable with respect to natural capital, for example, water availability. The most probable reason for this perception is because no major drought events, such as the 1973/74 and 1982/83 drought periods, have been recorded recently.

There have been many policies on agriculture and rural development in Nigeria since the pre-colonial through the independence and post independence periods up to the end of the last century. Unfortunately, these policies have not been favourable to the rural poor (Dabi, 2006). More recently, however, with the advent of the current democratic dispensation, government policies have been targeted at rural areas and for poverty alleviation, although most of the rural poor are yet to see the dividends of such policies. For example, agricultural inputs are not readily available to farmers. Whenever fertilizers are provided or subsidized to farmers, farmers either do not have access to the commodity or the quantity available does not reach them at the appropriate time. About 90 per cent of households did not receive fertilizers during the last planting season.

Extension workers are based at local government authorities to assist farming households in rural areas, and more than 90 per cent of surveyed households indicate awareness of extension services. But many extension workers are not adequately trained to assist farmers to improve their farm practices. In most cases, there is little or no aid coming from government during and after droughts. The extension service and workers represent a potentially valuable resource that could help to create the right environment for appropriate adaptation to climate change by raising farmers' knowledge and skills for improved management and use of new methods and technologies.

Climatic information for rural areas of Nigeria is greatly limited by the sparse distribution of weather stations. The number of stations is greater now than 20 years ago, but the number is still insufficient to meet the needs for information about recent climate trends or to prepare forecasts at scales useful for adaptation. Furthermore, many of the stations are not adequately maintained due to lack of commitment from government. Farmers generally have poor access to what little climatic information is collected. Occasionally, some information may be disseminated via the electronic media, but few if any rural farmers possess television sets. The radio is the only dependable source of information for most of the rural poor, who own small transistor radios that use dry cell batteries. However, the language of transmission (English) is not usually understood by the majority of the rural poor. Issues translated into local languages are normally far from climatic; rather they are political. From the standpoint of the poor rural households, weather and climatic issues may not necessarily interest them due to ignorance and lack of awareness regarding the use of this kind of information.

Enhancing Future Opportunities

Although the coping capacity of poor rural households is limited by a variety of factors as discussed in the preceding section, opportunities for future adaptation can be enhanced. Three strategies for enhancing opportunities are explored below: (1) development of past and present coping strategies; (2) introduction of alternative strategies; and (3) improvements in government policies and assistance.

A wide range of strategies are in use for coping with drought and water scarcity that have helped to limit vulnerability, as demonstrated by our fieldwork. Still, these practices are not as widely or effectively used as they might be. Consequently, there exists a significant adaptation deficit (Burton, 2004) among rural households of the dry northern areas of Nigeria. Support and encouragement of selected current strategies would help to close this deficit by expanding their use to more households and improving the effectiveness with which they are applied. Efforts should focus in particular on four coping strategies that have been used by more than half of the surveyed households, an indication of their wide acceptance and effectiveness. These include crop diversification, cultivation of early maturing and high yield crop varieties and food storage. Other strategies that have been less widely used may nonetheless be effective and should also be encouraged as part of a diversified portfolio of coping mechanisms (see Table 8.2).

Extension services could support coping strategies by increasing awareness of them and their utility for reducing climate related risks and helping farm households to develop the knowledge and skills needed to adopt and apply them effectively. Better distribution of early maturing and high yielding seed varieties and other farm inputs is needed, as is access to credit to enable farmers to purchase them. Assistance is also needed for marketing to support diversification of farm products.

In addition to supporting the use and expansion of existing strategies, the introduction of alternative or new strategies should also be promoted. There exist alternate strategies that are not used presently by farmers in the study area but that have been used elsewhere (Dabi and Anderson, 1999) and which can be implemented within the household livelihood structures of crop farming, livestock herding and fishing.

Coping strategies for crop farming can be expanded through introduction of new water sourcing and water use practices. Examples include sinking of boreholes and tube wells, cultivating Fadama or floodplain areas where water can be accessed via shallow wells or groundwater, and water management techniques such as conjunctive use. Farming practices to conserve water resources and reduce environmental degradation can also be introduced. These changes include reducing irrigation practice in water scarce areas, changing the timing of farm operations such as planting, irrigation and harvesting, and adopting new crop varieties and types.

Livestock farming can also be enhanced with changes in grazing methods or reduction of herd sizes and animal diversification, herd sedentarization and supplementation. These changes will help to reduce the risk of animal malnutrition or even death. In our earlier analysis of vulnerability, fishing households were found to be the least vulnerable. Improvements in fishing techniques and the introduction of aquaculture can help them to cope even better and to serve as an adaptation strategy in the future.

Success in introducing new strategies will depend on the willingness of farmers to change their practices, which has been explored in our survey. Farmers were asked their views on the adoption of selected options. As shown in Figure 8.7, most households are willing to adopt the selected options. Households expressed the greatest willingness to change the type of crop they cultivate and the timing of cultivation practices. They also indicated a willingness to change grazing practices and to relocate their farms. Farmers are less willing to reduce the size of their herd, as indicated by a high number of respondents who 'disagree' and 'strongly disagree' with this option. Livestock often are a family's major asset and serve as security during droughts and famines. Consequently, many are reluctant to reduce their herds, even if the animals are vulnerable to lack of food and water during droughts.

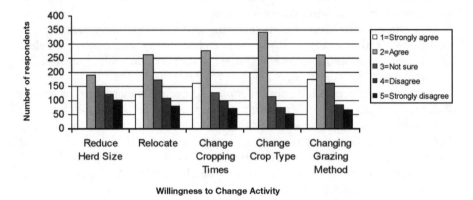

Figure 8.7 *Farmers' willingness to change practices to reduce vulnerability to drought*

Besides the use of existing and new coping strategies, supporting government policies are also critical for creating an enabling environment for adaptation. Important policy areas include education, research and development, institutional change and political will, religious and traditional institutions, and provision of infrastructural facilities. In the area of education, government should provide formal and informal educational institutions in order to improve literacy levels. Education will enable households to be informed and benefit from any strategy that may be introduced. Education will also enable households to access and utilize weather information. However, government must improve on information dissemination, provision of extension services and capacity building. Research and development are needed to better understand the risks to rural livelihoods from climate variability and change, as well as to develop and test risk management strategies and technologies. Important areas for research and development include climate change, water resources management, agricultural practice and development, land use and land use changes, farm inputs, and harvest and post harvest activities. These efforts will facilitate the monitoring of weather elements and the provision of warning signs as well as the development of good practice.

Institutional change and political will are needed to ensure that policy and decision makers develop measures that will assist households in their adaptation process. Examples of such measures include credit facilities for households, favourable pricing and market policies, storage facilities, and the distribution of food commodities. Religious and traditional institutions can help to ensure community cooperation, assistance and participation through self-help. Finally, there must be an improvement in the provision of infrastructure facilities in rural areas for roads, electricity and communication networks, and development of industries such as food processing and marketing.

References

Adger, N. W. (2001) 'Scales of governance and environmental justice for adaptation and mitigation of climate change', *Journal of International Development*, vol 13, pp921–931

Burton, I. (2004) 'Climate change and the adaptation deficit', Occasional Paper 1, Adaptation and Impacts Research Group (AIRG), Meteorological Services of Canada, Environment Canada, Toronto, Ontario, Canada

Carney, D., M. Drinkwater, T. Rusinow, K. Neefjes, S. Wanmali and N. Singh (1999) 'Livelihoods approaches compared: A brief comparison of the livelihoods approaches of the UK Department for International Development (DFID), CARE, Oxfam and the United Nations Development Programme (UNDP)', November, Department for International Development (DFID), London, UK

Christensen, J. H., B. Hewitson, A. Busuioc, A. Chen, X. Gao, I. Held, R. Jones, R. Koli, W. Kwon, R. Laprise, V. Rueda, L. Mearns, C. Menendez, J. Raisanen, A. Rinke, A. Sarr and P. Whetton (2007) 'Regional climate projections', in S. Solomon, D. Qin, M. Manning, Z. Chen, M.C. Marquis, K. Averyt, M. Tignor and H. L. Miller (eds) *Climate Change 2007: The Physical Science Basis*, contribution of Working Group I to the Fourth Assessment Report of the Intergovernmental Panel on Climate Change, Cambridge University Press, Cambridge, UK, and New York

Dabi, D. D. (2006) 'Agricultural and rural development policies in the third world: Lessons for Nigeria in the new millennium', *Abuja Journal of Geography and Development Sciences*, vol 1, no 3

Dabi, D. D. and W. P. Anderson (1999) 'Water use for commodity production in Katarko village, northern Nigeria', *Applied Geography*, vol 19, no 1, pp105–122

Davies, S. (1996) *Adaptable Livelihoods: Coping with Food Insecurity in the Malian Sahel*, Macmillan Press, UK

Downing, T. E., L. Ringius, M. Hulme and D. Waughray (1997) 'Adapting to climate change in Africa', *Mitigation and Adaptation Strategies for Global Change*, vol 2, pp19–44

Huq, S., A. Rahman, M. Konate, Y. Sokona and H. Reid (2003) *Mainstreaming Adaptation to Climate Change in Least Developed Countries(LDCs)*, The International Institute for Environment and Development, Russell Press, Nottingham, UK

Mortimore, M. (1989) *Adapting to Drought*, Cambridge University Press, Cambridge, UK

McCarthy, J., O. Canziani, N. Leary, D. Dokken and K. White (eds) (2001) *Climate Change 2001: Impacts, Adaptation and Vulnerability*, contribution of Working Group II to the Third Assessment Report of the Intergovernmental Panel on Climate Change, Cambridge University Press, Cambridge, UK, and New York

National Drought Mitigation Center (2005) 'What is Drought? Understanding and defining drought', *Drought Impact Reporter*, University of Nebraska–Lincoln, Lincoln, NE, US, www.drought.unl.edu/whatis/concept.htm

Nyong, A., A. Adepetu, V. Ihemegbulem and D. Dabi (2003) 'Vulnerability of rural households to drought in northern Nigeria', *AIACC Notes*, November 2003, vol 2, issue 2, pp6–7

Pielke, R. A. (1998) 'Rethinking the role of adaptation in climate policy', *Global Environmental Change*, vol 8, pp159–170

Smit, B., I. Burton, R. J. T. Klein and R. Street (1999) 'The science of adaptation: A framework for assessment', *Mitigation and Adaptation Strategies*, vol 4, pp199–213

Smit, B., I. Burton, R. J. T. Klein and J. Wandel (2000) 'An anatomy of adaptation to climate change and viability', *Climatic Change*, vol 45, pp223–251

Smit, B., O. Pilifosova, I. Burton, B. Challenger, S. Huq, R. Klein and G. Yohe (2001) 'Adaptation to climate change in the context of sustainable development and equity', in J. McCarthy, O. Canziani, N. Leary, D. Dokken and K. White (eds) *Climate Change 2001: Impacts, Adaptation and Vulnerability*, contribution of Working Group II to the Third Assessment Report of the Intergovernmental Panel on Climate Change, Cambridge University Press, Cambridge, UK, and New York

Smithers, J. and B. Smit (1997) 'Human adaptation to climatic variability and change', *Global Environmental Change*, vol 7, no 2, pp129–146

Ozer, P., P. Erpicum, G. Demaree and M. Vandiepenbeeck (2003) 'The Sahelian drought may have ended during the 1990s', *Hydrological Sciences*, vol 48, no 3, pp489–496

Tarhule, A. and M. Woo (1998) 'Changes in rainfall characteristics in northern Nigeria', *International Journal of Climatology*, vol 18, no 11, pp1261–1271

9

Using Seasonal Weather Forecasts for Adapting Food Production to Climate Variability and Climate Change in Nigeria

James Oladipo Adejuwon, Theophilus Odeyemi Odekunle and Mary Omoluke Omotayo

Introduction

An overarching anxiety among peasant farmers concerns the unpredictable onset and cessation of the rainy season. A foreknowledge of the weather for an upcoming growing season can enable farmers to plan with greater confidence to forestall negative consequences of poor or late rains and exploit beneficial opportunities when more favourable weather is in the offing. Extended-range or seasonal weather forecasts are made for West Africa and are a potential tool that could be used to great advantage for adapting farm decisions to climate variability. Successful application of seasonal forecasts in farming would also increase resilience to climate change. But seasonal forecasts are little used by Nigerian farmers. The reasons for this include inadequacies of the forecasts themselves and also failures in the communication of forecasts to farmers.

In this chapter we describe the effects of climate variability on peasant farmers, strategies that are used by farmers to cope with the variability that do not depend on seasonal weather forecasts, and decisions that could benefit from the use of forecast information. Current efforts and capacity for extended-range forecasting in West Africa and Nigeria are examined and the skill and utility of the forecasts are evaluated. Based on our analysis, recommendations are made for improving forecasts and the communication of forecasts so that the potential benefits offered by this tool might be realized.

The situation in Nigeria is relevant more broadly as a case study for sub-Saharan West Africa. The country encompasses climatic zones of the subcontinent from very wet to semiarid and all the indicator vegetation types of these zones are present, including evergreen rainforests, Southern Guinea Savannah, Northern Guinea Savannah, Sudan Savannah and Sahel Savannah.

Of all the countries in West Africa, Nigeria alone is able to produce the complete range of foodstuffs characteristic of the subcontinent. Apart from the presence of the major tropical climate and vegetation types, the range of latitude, the varied relief and soils, together with differing peoples with their contrasted methods and crops, make this possible (Harrison-Church, 1956).

This contribution is a synthesis of several investigations of climate variability, climate change and food security in sub-Saharan West Africa, the details of which have been provided in the technical report to the Assessment of Impacts and Adaptation to Climate Change (AIACC) Project (Adejuwon, 2006). Some of the work, including Adejuwon (2002, 2005 and 2006), Adejuwon and Odekunle (2004 and 2006), Odekunle (2003 and 2004) and Odekunle et al (2005), has been published in academic journals, in which details of the methods used can be found. The review of the farm operations and possible choices for responding to forecasts is based mainly on data collected during field surveys. The field surveys were conducted in five major ecological zones in Nigeria, including: rainforest (Atakumosa Local Government Area of Osun State), Southern Guinea savannah (Irepodun Local Government Area of Oyo State), Northern Guinea savannah (Oorelope Local Government Area of Oyo State), Sudan savannah (Askira Local Government Area of Bornu State) and Sahel savannah (Konduga Local Government Area of Bornu State). The assessment of the capacity represented by the major forecasting organizations (see Existing Capacity and Practice for Extended-range Weather Forecasting, on page 168) is based on a variety of sources, including Folland et al (1986 and 1991), Ward et al (1990), Philippon and Fontaine (2000), Colman and Richardson (1996), Colman et al (1997, 2000), websites of relevant organizations and climate records, and other information collected from the offices of the Nigerian Meteorological Agency at Oshodi, Lagos State. The methods used for the assessment of forecasting skills are described in detail in Adejuwon and Odekunle (2004).

Farm Operations and Variable Weather

Farmers, whether in the humid or semiarid zones, realize the need to time farm operations to correspond with specific weather patterns. The dry season is used to prepare the land for cultivation. If the rains come too early, preparation operations could be adversely affected. The main anxiety about the onset of the rainy season concerns the planting date. Farmers like to sow their crops as early as possible. The earlier the crops are sown, the earlier the food products will be made available to end the annually occurring period of food deficiency. Moreover, crop yields are typically higher when planted early in the rainy season (Adejuwon, 2002; Fakorede, 1985) and farmers who get their crops early to the market are likely to enjoy better prices. The higher yields are explained by the early season nitrogen flush from farm residues of the previous season and higher incident solar radiation levels in the period before the heavy rains come. All of these factors explain why farmers are anxious about when the rains will come.

Sometimes there are false starts of the rainy season and farmers rush to plant their crops. This can result in disaster, as the seedlings may be completely lost. Replanting could be expensive if the seeds have to be purchased. If replanting seeds are sourced from the farmers' stores, they could deplete food needed by the household during the period of low food supply. The amount of resources wasted in this way could be considerable when the crop concerned is yam. The yam seed is cut from the same tuber used as food, and could be up to 25 per cent of production. The farmers would, therefore, benefit considerably from a foreknowledge of the onset of the rainy season.

Crop yield response to rainfall variability was investigated in the semiarid zone of Nigeria (Adejuwon, 2005). June rainfall turned out to be a more powerful predictor of crop yield than any of the other monthly rainfall variables. This is explained by the fact that June is the month of the onset of the rainy season. Low or insufficient June rainfall implies a delayed onset and a rainy season not long enough for the needs of most crops. September rainfall also served as a powerful predictor of yield. September is the month of cessation of the rainy season. Low or inadequate rain in September results in inadequate moisture for crops during the critical phases of grain filling, truncating the growing season prematurely.

Crops such as cowpeas or late maize are planted so that they could be ready for harvesting after the rains have ceased. A late cessation of the rains means that the crops would be harvested under wet conditions, and much of the crop could be lost to mould and other pests and diseases. On the other hand, if the rains cease too early, the entire crop could fail due to inadequate moisture.

There is a system of yam production practised at the drier margins of the rainforest zone which requires the seeds to be planted at the end of the rainy season, just before the rains cease. The seed remains dormant for the whole period of the dry season. As soon as the rains come during the following year, yam vines shoot up and harvestable tubers are produced two or three months ahead of the normal yam harvest season. New yam produced in this way commands very high prices, as it is preferred to the old yam, which would by then be losing its taste. The critical weather requirement of this system is that at least one heavy downpour must fall on the planted seeds before the dormancy period. In the absence of this, as much as 50 per cent of the crop could be lost. Thus the farmers can also benefit considerably from a foreknowledge of when the rains would cease.

The length of the rainy season and the amount of rain that falls during the peak rainfall period are also watched with anxiety by farmers. Palm fruits are harvested by climbing the tree to cut the fruit. This is a very hazardous operation, especially during the rainy season when the trunks are slippery. Fewer climbers are available during the rainy season, and for this reason, palm products such as palm oil are in short supply and expensive. The fruits are left to waste and the longer the rainy season the greater the loss. Furthermore, during the rainy season rural roads become impassable and crops like cassava are unable to reach the market.

Heavy tropical rainstorms can make all of the difference between a good harvest and crop failure. Farmers complain that such storms could cause a heavy loss to flowers before they become fruits. Cowpeas are an example of crops that could be damaged in this way. Heavy rainfall also reduces the number of pods per stand of cocoa, while it increases the degree of infestation by black pod disease (Thoroid, 1952; Adejuwon, 1962). Years with heavy rainfall, therefore, usually correspond to years with low yields of the crop. Moreover, cocoa harvested at the height of the rainy season has low grades and may not be able to make the export market. This is due to the prevailing heavy clouds and the little sunlight available for drying the produce. A prolonged little dry spell midway into the rainy season is a blessing to cocoa farmers.

A minimum amount of rainfall during the dry season is needed for the establishment phase of tree crops. Cocoa is usually planted as seeds or seedlings during the rainy season. The new crop plants will die during the first dry season if no rain falls or if there is a spell of desiccating harmattan winds from the Sahara. Thus in the forest zone farmers meet with varying degrees of success in developing a new plot for the crop, depending on how much rain falls during the first dry season after planting (Adejuwon, 1962). Farmers' needs from the weather forecaster thus include statements on whether or not the dry season will be completely dry.

Coping Without the Benefits of Weather Forecasts

It needs to be noted that the peasant farmers are not altogether helpless in the absence of weather forecast information. Traditional agricultural practices include a number of options designed to mitigate the negative consequences of unfavourable weather which are applicable each year, whether or not the weather turns out to be unfavourable. One good example of such practices is the storage of water and the use of shallow wells to extend the period of adequate water supply at the end of the rainy season. Such practices will help to cushion the impacts of abrupt termination of the rainy season. Mulching is another practice which helps to protect seedlings against dry spells during the earlier parts of the growing season. The use of wetlands, whether extensive floodplains or local valley bottoms, is also part of traditional agricultural practice. Wherever such lands are available, they help to reduce the impacts of milder droughts on peasant communities.

As of now, farmers use a number of hedging strategies while expecting the worst of weather and hoping for the best. Such strategies include multiple cropping, relay cropping and intercropping. These are recognizable features of traditional farming practices evolved over time all over the country and are designed to make one crop serve as an insurance against the failure of another. In the case of multiple cropping, crops that occupy different ecological niches are planted together on the same plot. For example, maize may be planted with melon. While maize is a standing crop, melon is a creeper. In the case of relay cropping, several crops are planted in succession on the same plot to make use of different parts of the growing season. For example, in the Guinea Savannah Zone of south-western Nigeria, maize is planted in April, harvested in July; sorghum is planted in June and harvested in November. In the case of intercropping, some crops are planted at low density among the major crop, which is planted on every heap. For example, maize could be planted at low density on a farm plot primarily meant for yam production.

In the forest zone, tree food crops are maintained as insurance against the failure of field crops. In normal years, the fruits of the tree crops are not harvested. However, during years of inadequate rainfall, when the yields of the field crops are not able to sustain the peasants' livelihood, they fall back on the tree crops whose food products are considered to be of lower quality. Another example: cocoyam, planted under cocoa, is treated as a weed during normal years. However, whenever the major food crops fail, it becomes a dependable source of food.

There are also measures for damage control as soon as it is realized that an abnormally dry season is in the offing. One such measure is replanting with crops with shorter growing seasons or that are more tolerant of arid conditions. Water yam with a maturity period of 3 months could be planted to replace white yam, whose maturity period lasts seven months; millet, with a maturity period of two months, could be planted to replace sorghum, which takes six months to mature.

Decisions that Might Benefit from Forecast Information

In a typical tropical region like West Africa, rainfall is the principal controlling element of crop productivity (Nieuwolt, 1982; Stern and Coe, 1982). The crop plants are sensitive to the moisture situation both during their growth and development, and especially as they reach maturity. This is reflected in a definite soil and atmospheric moisture range in which field preparations are expected to commence and also in which such farm operations as sowing, thinning; transplanting, weeding, irrigation, insecticide and fertilizer application, as well as harvesting are scheduled to take place. Thus, as seasonal weather changes from one year to another, the most suitable crops, cropping systems and operations schedules also change. The need thus arises to choose from a range of options.

As an adaptation strategy, extended-range weather forecasting represents an early warning system that could be used at the farm level for decision making. A foreknowledge of seasonal weather affords the farmer the opportunity to make decisions that could enhance the productivity of his farm and maximize returns on his inputs of land, labour and capital. Ex ante decisions that might benefit from a foreknowledge of seasonal weather include the timing of farm operations such as land preparation, tillage, planting, transplanting, thinning, weeding, irrigation and harvesting. Whether, when and how much insecticide, herbicide, fungicide and fertilizer to apply can also be optimized with forecast information. Choices of crops, crop varieties, tillage method and depth, and density of planting can be adapted to the anticipated weather. Whether or not to adopt water conserving practices, which type to adopt, and whether to store water for irrigation and the mode of irrigation to use are other decisions that can benefit from forecasts. If drought conditions are expected, poorly drained *fadama* soils can be productively cultivated, or deep, loamy soils can be cultivated if abundant rainfall is anticipated. The best choice of farming system, for example, single, multi-cropping or intercropping, depends on the weather. How much credit to secure and how much of the harvest to store or sell are also decisions that can profit from forecast information.

Existing Capacity and Practice for Extended-range Weather Forecasting

West Africa depends to a large extent on organizations based in Europe and North America for its operational weather forecasting. The UK Meteorological Office (UKMO or Met Office), the French Centre de Recherche de Climatologie of the Centre National de la Recherche Scientifique (CNRS) and the National Oceanic and Atmospheric Administration (NOAA) in the US, routinely make forecasts directed at the West African subcontinent. Within Africa, the Africa Centre for Meteorological Applications for Development (ACMAD) is responsible for gathering, collating and disseminating forecast information, while the national meteorological services are responsible for issuing weather forecasts for their own countries and sub-regions.

The statistical forecasting methods of the Met Office, CNRS and NOAA are described in Adejuwon (2007). The Met Office has been making experimental forecasts of seasonal rainfall in the Sahel since 1986. Forecasts are made for four regions for the months of June, July, August and September using ocean and atmospheric information that is available in early May (Colman et al, 1996 and 1997; Graham and Clark, 2000).The predictions use seasonally averaged sea surface temperature anomalies (SSTA) with a spatial resolution of $10° \times 10°$. The CNRS forecasts cumulative rainfall for June to September in West Africa, also using sea surface temperature as well as other indices, but aggregated to a spatial resolution of $5° \times 5°$ (Philippon and Fontaine, 2000). In order to make the forecasts available before the beginning of the growing season, only information available by the end of April is employed. NOAA produces experimental forecasts for rainfall anomalies for July, August and September in the Sahel using predictors with a spatial resolution of $2° \times 2°$ (Thiaw and Barnston, 1996, 1997 and 1999). So far, the experiments confirm that the global SSTA field is the best predictor, a view shared by the European and African forecasting teams. Additional fields such as upper air geo-potential heat, tropical low level wind and outgoing long-wave radiation could enhance forecast skill. However, the data do not extend far enough into the past (minimum of 25 years for an adequate control period).

ACMAD (www.acmad.ne), based in Niamey, Niger Republic, was created in 1987 by the Conference of Ministers of the United Nations Economic Commission for Africa and the World Meteorological Organisation to meet the challenge of weather prediction for the continent. The strategy is for the centre to lead both in training personnel for capacity building within the continent and implementing operational activities, such as the issuance of weather information products. In the execution of its programmes, ACMAD operates within a network with international partners, regional partner institutions and national focal points. The international partners include the CNRS, the Met Office and NOAA, among others. These organizations provide the primary weather information to be collated and transmitted to the end users within the various African countries. The regional partner institutions have specific sector and regional responsibilities that require weather and climate information. Among such institutions are ICRISAT and AGRHYMET. The focus of ICRISAT is on agricultural development in semiarid tropics while AGRHYMET deals with agriculture and hydrology in the Sahel region of West Africa. These organizations are expected to help in broadcasting information to sector end users. To reciprocate, they also use the website of ACMAD to advertise their products for the benefit of their stakeholders. The primary focal points are the national meteorological services of 53 African countries. Focal points for ACMAD have also been established within the operational structure of sub-regional economic groupings, such as the Economic Community of West African States and the South Africa Development Community. The focal points are the primary recipients of the products emanating from ACMAD, meant for end users in agriculture, energy, water resources and other sectors within the various countries.

The operational products of the Numerical Weather Prediction Unit within ACMAD include the daily Meteorological Bulletin for national meteorological services and the daily continent-wide 24-hour public significant weather forecasts. These products give relatively accurate forecasts but do not go far enough into the future for our purposes. The numerical weather production unit of ACMAD also issues 5-day guidance bulletins on specific days of the week (Mondays, Wednesdays and Fridays). On other days (Tuesdays and Thursdays) the unit issues 3-day guidance bulletins to monitor and follow up the previous day's 5-day guidance bulletin. The terms 'guidance' and 'forecast' are understood in the context that ACMAD issues weather guidance to African National Meteorological Services, which have the responsibility to issue the weather forecasts for the specific countries or sub-regions. The conventional network of surface and upper air observations within ACMAD's area of responsibility is generally too poor to make longer-term relevant numerical forecasting feasible.

However, ACMAD, through its African Climate Watch page (www.acmad.ne), advertises two relevant products for short-term, medium-term and seasonal weather prediction: 'El Niño/La Niña Update on Impact over Africa' and 'Rainfall Onset over West Africa'. The former gives early-warning type of forecasts extending over three to nine months. However, this is a very new product that has yet to be tested. The other product predicts the onset of the rainy season within three to six weeks (Omotosho, 1990). However, the predictions are confirmed only within three weeks.

The Cotonou workshop on 'Climate variability prediction, water resource and agricultural productivity: Food security in tropical sub-Saharan Africa' led to ACMAD being charged with organizing the Seasonal Climate Prediction Forum for West Africa, PRESAO, with assistance from START to arrange funding (Fleming et al, 1997). PRESAO pools expertise in the subcontinent for the purpose of weather prediction and assembles producers and users of predictions for two-way interactions to improve the forecasts and the ability of the agricultural and water resource sectors to use forecast information. The PRESAO forecast map appears on the webpage of ACMAD and is meant to be accessed and downscaled by the national meteorological services.

The Nigerian Central Climate Forecasting Office (CFO) of the Nigerian Meteorological Agency (NIMET) is responsible for short-term and extended-range weather forecasts in Nigeria. Ahead of each cropping season, the CFO issues a bulletin on the weather outlook for the season. The CFO makes forecasts based on analogue, statistical and dynamic methods as well as forecasts downscaled from European and American forecasting organizations. While making its local forecasts, the CFO makes use of SSTA data of $2° \times 2°$ resolution (Toure, 2000), compared with SSTA data on a resolution of $5° \times 5°$ employed by the CNRS (Philippon and Fontaine, 2000) and the $10° \times 10°$ grid SSTA data used by the Met Office (Colman et al, 2000). For its own local forecasts, the CFO makes use of the current trend of the weather, the pressure systems, the position of the Intertropical Convergence Zone (ITCZ) and global SSTA. The CFO also makes forecasts of the dates of the first rains (which is not the same thing as the date of onset of the rainy season as defined by Ilesanmi (1972a). Very often, the first rains represent only a false start.

Assessment of Seasonal Forecast Skill

The forecasts assessed have been prepared by various meteorological organizations, including: the Met Office, the CNRS, NOAA and the CFO. The forecasts of the international organizations were obtained from articles published in *Experimental Long-Lead Forecasts Bulletin*. The forecasts of the Nigerian CFO were obtained directly from their offices in Lagos. The stations selected to test the forecasting skills of the various tools include Benin City, Lagos, Ibadan, Ilorin, Enugu, Minna, Jos, Kaduna, Lokoja, Maiduguri and Kano, all located within Nigeria. The stations have been selected to represent the various climatic and ecological zones between the Gulf of Guinea in the south and Sahara Desert in the north. Their selection is also based on the availability of rainfall data from 1961 to 2000. The forecasts assessed were for the five years from 1996 to 2000.

The scheme used for the assessment of forecast skill is described in detail in Adejuwon and Odekunle (2004). Skilful forecasts are those that are subsequently confirmed by observations. High skills are demonstrated when forecasts are very close to observations, whereas low skills are recorded when the two are substantially different. One practical problem in assessing the skills is the fact that observations and forecasts are not presented in the same units of measurement. Observations are usually presented on an interval scale with the amounts of rainfall given in millimetres. On the other hand, forecasts are stated using ordinal categories. The most common are quint categories varying from very wet to wet, average, dry and very dry. Determination of what is very wet, wet and so on in this exercise was based on the records from 1961 to 1995. Rainfall values for each year, whether annual, seasonal or monthly, were arranged in descending order of magnitude and divided into five groups. The resulting highest quint consists of the values for the seven wettest years and the lowest those of the seven driest years. The years with rainfall values falling within the range in the highest quint are classified as very wet, while years with values falling within the range of rainfall in the lowest quint are classified as very dry. Other years are similarly classified as wet, average or dry.

Sometimes tercile categories are used by simply forecasting near normal, above normal or below normal. Determination of what is above normal, near normal or below normal was also based on the records from 1961 to 1995. Rainfall values for each year, whether annual, seasonal or monthly, were arranged in descending order of magnitude and divided into three groups. The resulting highest tercile (above normal) consists of the 12 wettest years and the lowest (below normal) those of the 12 driest years. The 11 middle years define the near-normal range. The quint and the tercile limits provide the framework for converting both observations and forecasts to the same units of measurement (Adejuwon and Odekunle, 2004). In assessing the skills of forecasts, the same criteria were used to classify observations as were used for the forecast categories. Where observations and forecasts fell within the same category, skill was assessed as high. Where there was a one-category difference – for example, forecast was average but observation was very wet – skill was assessed as moderate. In situations of more than one category disparity between observation and forecasts, the skill was assessed as low. Tables 9.1 and 9.2 provide the framework for the assessment of the skills of the forecasts.

Table 9.1 *Skill assessment of quint forecast categories*

Forecast Observations	Very Wet	Wet	Average	Dry	Very Dry
Very Wet	High	Moderate	Low	Low	Low
Wet	Moderate	High	Moderate	Low	Low
Average	Low	Moderate	High	Moderate	Low
Dry	Low	Low	Moderate	High	Moderate
Very Dry	Low	Low	Low	Moderate	High

Table 9.2 *Skill assessment of tercile forecast categories*

Forecast Observations	Above Normal	Near Normal	Below Normal
Above Normal	High	Moderate	Low
Normal	Moderate	High	Moderate
Below Normal	Low	Moderate	High

The skill of the forecasts for June, July, August and September annual rainfall totals, as determined in the analysis, is 26 per cent rated as high, 45 per cent moderate and 29 per cent low. There is thus considerable room for improvement. The results ranked NOAA and the CFO ahead of the CNRS and the Met Office in weather forecasting skill over West Africa (Table 9.3). A careful look at the background of the methods of data collection and analysis appears to explain the relative successes of the forecasting organizations. Virtually all of the forecasting organizations made use of the SSTA as a predictor among others. However, a noticeable difference in the nature of the sea surface temperature data used by various organizations is with respect to the spatial resolution. Although the CFO and NOAA made use of SSTA data of 2° × 2° resolution (Tourre, 2000), the CNRS used seasonally averaged 5° × 5° grid SSTA data (Philippon and Fontaine, 2000) and the Met Office used seasonally averaged 10° × 10° square SSTA data (Colman et al, 2000). It thus appears that the finer the SSTA resolution, the better the forecasting skill.

Table 9.3 *Organizational skill performance assessment of the June, July, August and September annual rainfall totals*

Forecasting Organization	Percentage Contribution of Each Skill Category		
	High	Moderate	Low
UKMO (all)	21%	44%	35%
NOAA	56%	22%	22%
CNRS	18%	55%	27%
CFO	20%	70%	10%

The number of predictor variables used in the forecast models also seems to have played a role in determining the level of skill. Although the Met Office, whose tools seem to be less skilful than the others, made use of SSTA data alone in the construction of their prediction models, other forecasting centres made use of additional rainfall formation-related factors. For instance, the CFO, which came first on the basis of having the smallest low-skill score, made additional use of synoptic data, including current weather, pressure systems, equivalent potential temperature and the position of the ITCZ. NOAA, which ranked first on the basis of the highest high-skill score made additional use of upper air geo-potential heat, tropical low level wind and outgoing long-wave radiation (Thiaw and Barnston, 1999). The CNRS, which was third, used additional factors, such as geo-potential indexes, describing near surface humidity and moist static energy values (Philippon and Fontaine, 2000). Note that the ITCZ is one major factor not used by the CNRS in the construction of their prediction model. The comparison indicates that the inclusion of these additional predictor variables in the construction of a forecast model improves the skill of the forecasts.

The study clearly demonstrates regional disparities in the skills of the forecasting tools (Table 9.4). The prediction skill is generally higher for southern coastal locations than for the northern continental locations. It is well known that the Atlantic Ocean is the major, if not the only, source of moisture into the West Africa subcontinent. The moisture is brought to the land areas by the southwesterly winds moving in after the northward migrating ITCZ. The characteristics of the southwesterly winds, which bring the moisture to the continent, are, in turn, determined by the nature of the sea surface temperature of the Gulf of Guinea (Adedokun, 1978). It is thus logical that the conditions of the southwesterly winds, as determined by the SSTA and its associated ITCZ, would be least modified near the coast. As the ITCZ advances and the southerly winds progress further inland, their thermodynamic transformations become more pronounced. The changes in the nature of southwesterly winds and other rainfall-associated factors are thus a function of space and time. The space connection between rainfall over the land and the activities over the Atlantic Ocean weakens with distance between a location over the land and the sea. It is, therefore, not surprising that a prediction model based on SSTA is more skilful in the south, near the ocean, than in the interior of the continent.

Table 9.4 *Regional disparities in forecasting skill*

Stations		Percentage Contribution of Each Skill Category		
		High	Moderate	Low
Maiduguri	Sahel	17%	33%	50%
Kano	Sudan	17%	17%	66%
Kaduna	Northern Guinea	20%	60%	20%
Minna	Northern Guinea	25%	25%	50%
Jos	Plateau	17%	50%	33%
Ilorin	Southern Guinea	17%	83%	0%
Lokoja	Southern Guinea	50%	25%	25%
Ibadan	Dry Forest	29%	71%	0%
Enugu	Dry Forest	33%	33%	34%
Ikeja	Rainforest	29%	57%	14%
Benin	Rainforest	33%	67%	0%

Source: Adejuwon and Odekunle (2004)

Inadequacies in the Existing Forecast Capacity

A number of inadequacies are associated with the forecasts made between 1996 and 2000. First, the skill was not sufficiently high, and there was no evidence to the effect that it was improving (Adejuwon and Odekunle, 2004). Second, the forecasts were directed at the total rainfall amount, whereas the more relevant determinants of crop yield were neglected. As noted by Odekunle and Gbuyiro (2003), all of the recent studies on rainfall predictability in West Africa, including those from the Department of Meteorology, University of Oklahoma (Berte and Ward, 1998), Nigerian Central Forecasting Office, (Omotosho et al 2000), Centre de Recherche de Climatologie

(Philippon and Fontaine, 2000) and UK Meteorological Office (Colman et al, 2000), Gbuyiro et al (2002) and Wilson (2002), were directed at rainfall amounts only. The number of rain days, which is a measure of rainfall effectiveness, was hardly ever included. The same total amount of rainfall is expected to benefit crops more coming as drizzle over several days than when it comes in one heavy downpour (Odekunle and Gbuyiuro, 2003). Other important parameters of rainfall that were not considered as predictands are the rainfalls of onset and retreat periods. These are of paramount importance in the subregion because they affect regional economies (Walter, 1967; Olaniran, 1983; Adejuwon et al, 1990). A failure in early establishment of rainfall onset, for example, usually indicates a drought in the early part of the rainy season, as noted earlier.

A third inadequacy, as noted by Odekunle et al (2005), is that the forecasting tools currently deployed are regional in approach and general in perspective. Models developed at such coarse spatial scales often fail at individual farm-level sites. Within the same zone, in respect of which the forecasts are usually made, a wet season at one station may be a dry season at another (Adejuwon et al, 1990). The disparities in forecasting skills between stations lying in the same ecological zones, as demonstrated in Table 9.4, are probably a result of such intrazonal variability.

The fourth and most serious inadequacy of the existing capacity is the absence of an effective means of communicating the forecasts with any end user in the agricultural sector. During stakeholder field surveys conducted in all ecological zones, from the rainforest belt in the south to the Sahel savannah zone in the north, in 50 village communities not a single respondent affirmed receiving either directly or indirectly weather-related information from the meteorological services (Adejuwon, 2005). Extension personnel routinely advise the farmers in their charge on the timing of farming activities. It turned out that such advice is based not on weather information received from the meteorological services, but on the extension agents' knowledge of the climate of the area.

Measures for Upgrading Forecasting Effectiveness

New attempts are being made to generate extended-range weather forecasting models in Nigeria that address the inadequacies noted above (Odekunle and Gbuyiro, 2003; Odekunle, 2004, 2006; Odekunle et al 2005; Adejuwon and Odekunle, 2006). The models, which are yet to be incorporated into the seasonal weather forecasting programmes of the country, are designed to improve the skills, credibility and applicability of the existing capacity at the local level. For example, Odekunle et al (2005) examined the rainfall onset and retreat dates between 1962 and 1996 in Nigeria, and generated models for their prediction. The study adopted a composite of rainfall-promoting factors: the sea surface temperature of the tropical Atlantic Ocean, land-sea thermal contrast between selected locations in Nigeria and the tropical Atlantic Ocean, surface location of the Intertropical Convergence Zone, and the land surface temperature in selected locations. Rainfall and temperature data were collected from Ikeja, Benin, Ibadan, Ilorin, Kaduna and Kano, in Nigeria. Cumulative percentage mean rainfall was employed to generate the rainfall onset and retreat dates series, while the method of stepwise multiple regression analysis was used to construct the required prediction models.

The results obtained showed that the hypothesized rainfall-promoting factors are efficient in predicting rainfall onset and retreat dates. Both the statistical goodness-of-fit (using both the R^2 and a-values) and the actual goodness-of-fit (comparing the observed rainfall onset and retreat dates with the predicted values using 1962–1969 and 1995–1996 data) of the models obtained in this study support the reliability of the models for predicting rainfall onset and retreat dates in Nigeria.

Measures for Communicating Forecasts to End Users

More than forecasting skill is needed if forecasts are to be useful to, and used by, farmers. Much depends on the communication of forecasts, how they are perceived and how they are understood.

First, the forecasts must be credible (Patt and Gwato, 2002). Where previous forecasts were perceived as incorrect, the tendency would be for subsequent ones to be ignored. Credibility is determined primarily by the level of skill, but it is also a function of the difference between what was promised by the forecasts and what was realized. At their best, forecasts are probabilistic. By couching forecasts in deterministic language, skill will invariably be perceived as low whenever there is an incorrect forecast. On the other hand, a probabilistic forecast is accorded the benefit of the doubt if there are infrequent, incorrect forecasts. Considerable efforts, therefore, need to be put into how the forecasts are interpreted before they are transmitted. It is not always easy to find a simple expression to convey the concept of probability, especially in a non-mathematical context such as that in which peasant farmers operate. Another requirement for a credible forecast is that it must be presented at a spatial resolution or scale with which individuals can identify. As indicated above, forecasts are made for extensive zones, within which there could be considerable differences in the actual weather observed. The CFO divides Nigeria into five zones and makes its forecasts for each zone. In order to achieve a high skill level for each zone, the skill levels for the localities within the zone, with which the farmers are familiar, have been compromised.

Second, the forecasts must be presented in simple, understandable language. Forecasts need to be presented in the local languages. Where forecasts are rendered in strange and confusing language, users will either not incorporate them or they will do so in a way that is counterproductive (Patt and Gwato, 2002). However, forecasters cannot be expected to be experts in all the local languages, and leaving the translation entirely in the hands of local agricultural extension personnel may introduce considerable distortions. Sometimes, to increase farmers' acceptance of the forecasts, doses of exaggeration are introduced. Apart from this, technical terminologies in English may not have local language equivalents. Hence, there is the need for the forecasters to work repeatedly with users or the intermediary conveyers of forecasts to be able to arrive at an appropriate forecasting language for each locality.

Third, the meteorological department should also bear the responsibility of issuing forecasts early enough to be useful in planning the following season's operations. Planting normally starts as soon as the rainy season is established. The decision as to what farm operations to adopt for the season, therefore, must have been made before then. While the rainy season may come as late as June in the semiarid zone, it could come as early as February in the humid coastal zone. Thus, weather forecasts made in April or May are practically useless to half of the farming community in Nigeria – the expected end users will simply not be in a position to make use of them.

Fourth, the forecasts should include information about the consequences of forecasted weather, for example, effects on yields of different crops, that can enable farmers to understand potential outcomes of decisions based on tradition or observations and compare them to results that might be attained if the forecasts are acted on. If forecasts do not contain enough new information to alter specific decisions, the intended users will not organize their activities in response to them (Patt and Gwato, 2002). As indicated previously, the peasant farmers are not altogether helpless in the absence of weather forecast information. Traditional agricultural practices include a number of no-regret options designed to mitigate the negative consequences of unfavourable weather, applicable each year, whether or not the weather turns out to be unfavourable. These cover a considerable proportion of the risks posed by interannual climate variability. One way of providing the type of information that could offer credible choices is to indicate the changes in crop yield that could be expected given the forecast in question. The Agricultural Meteorology Department of NIMET must be equipped in terms of personnel to issue such forecasts.

Even though there is an Agricultural Meteorology Department within NIMET, it appears that the organization is not explicitly charged with activities that could expedite the transmission of the forecasts. Its extended-range weather forecasts are consigned to the archives as soon as they are made. Either by amendment to the Act Establishing the Agency or amendment to the relevant statute or regulations, the responsibility of NIMET in this regard should be clearly stated. The Agricultural Meteorology Department should be strengthened and charged with the responsibility of translating the forecasts into forms that could be understood and are relevant to the needs of farmers. The government at the federal level should also accommodate budgetary provisions for communicating the forecasts to the agricultural extension services.

A related problem concerns the channel of communicating the forecasts with the farmers. Our suggestion is that the existing administrative structure for local government be used. Just as ACMAD recognizes the national meteorological services as the focal points within each member country for the dissemination of its products, NIMET should identify focal points close to the farmers for purposes of passing on weather forecast information to them. Because NIMET is a federal agency, the logical location of such focal points is within the administrative structure of the states. However, this would tend to prolong the line of communication and reduce its efficiency. Therefore, we suggest the Local Government Administration (LGA) as the location of NIMET focal points.

The precedent for this is the direct transfer of the LGA component of federally generated revenue to the local governments without passing through the states. There are more than 700 LGAs in the country. Each LGA is structured to have a department of agriculture, whose political head is the Supervisory Councillor for Agriculture. There are also technical staff within such departments with qualifications comparable to those of holders of the Higher National Diploma, obtained after four years' post-secondary education and one year of industrial training. Focal points established within such departments will be close enough to the farmers for effective and efficient transmission of forecasts. The LGA represents government at the grass roots and their technical staff in the agricultural departments are expected to operate among the rural populace, especially among farmers. They speak the same language as the farmers and are well placed to ensure that vital weather information gets to them within a few days.

As part of the bottom–up approach, we suggest that the forecasts be based on the existing 28 synoptic weather stations in the country. On average, each station will be expected to represent the weather of an area of some 30,000 km². This, it is hoped, will enhance the skill of the forecasts, as they could be accessed from each farm site. This means that 28 weather forecast pamphlets will be produced to cover the country and that it is the forecasts for the weather station closest to each focal point that will be sent to the farmers who need them.

However, communication should not just be one-way. There should be opportunities for feedback from end users about the skill of the forecasts. This could be in the form of specific requests for forecasts or in the form of complaints about the level of skills demonstrated.

Conclusions

Extended-range weather forecasting is a basic step for creating an early warning system for farmers in West Africa. Skilful, timely and effectively communicated forecasts of climate variables that matter to farmers can be applied in comprehensive adaptation strategies to forestall the negative consequences of climate variability and climate change. However, the existing forecast system in Nigeria, and West Africa generally, has a number of inadequacies in both the prediction skill and the usefulness of the forecasts to the end users. A high percentage of the forecasts are found to have moderate or low skill, while a low percentage of forecasts exhibit high skill. In addition to problems with forecast skill, the usefulness of forecasts to the end users is limited by lack of forecasts of rainfall characteristics that are important to farmers such as onset, cessation, length of the growing season and number of rain days; lack of forecasts for specific localities instead of extensive zones; and lack of forecasts for the coastal and middle belts of West Africa by some of the forecasting organizations.

Seasonal forecasts of rainfall characteristics of West Africa can be made more skilful and useful. As described in this chapter, we have made skilful predictions of the onset and cessation dates of the rainy season using SST and land/sea thermal contrast alone. In other published works, we have also generated prediction models for the length of the rainy season, the length of the 'safe' period and the number of rain days (Odekunle and Gbuyiro, 2003; Odekunle, 2004). The models generated represent an improvement on the existing weather forecasting tools, as they could be used to make forecasts for specific locations (in other words the areas around the weather stations) and for all the zones from the coast to the Sahel.

As well as improving the skills of the forecasts, constraints that prevent the use of forecasts by peasant farmers need to be addressed. These include inadequate or lack of credibility, late publication of forecasts, cognitive deficiencies, and absence of clear choices associated with specific forecasts. The existing agricultural extension system has an extensive local presence throughout the farming communities of Nigeria that could be used as channels for communicating forecasts. To be effective, close cooperation between forecasters and extension personnel is needed to ensure that forecasts are timely, are rendered in forms that are relevant to farmers' decisions, are presented in ways that make their probabilistic nature understood, are in languages used by the farmers and have credibility.

References

Adedokun, J. A. (1978) 'West African precipitation and dominant atmospheric mechanisms', *Archives for Meteorology Geophysics and Bioclimatology*, Series A, vol 27, pp289–310

Adejuwon, J. O. (1962) 'Crop-climate relationship: The example of cocoa in Western Nigeria', *Nigerian Geographical Journal*, vol 5, pp21–31

Adejuwon, J. O. (2002) 'Extended weather forecasts as management tool for the enhancement of crop productivity in sub-Saharan West Africa', *Nigerian Meteorological Society Journal*, vol 3, pp25–38

Adejuwon, J. O. (2005) 'Food crop production in Nigeria: I. Present effects of climate variability', *Climate Research*, vol 30, pp53–60

Adejuwon, J. O. (2006) 'Climate variability, climate change and food security in sub-Saharan West Africa', technical report of AIACC Project No AF23, International START Secretariat, Washington, DC (www.aiaccproject.org/Final%20Reports/final_reports.html)

Adejuwon, J. O., T. O. Odekunle and M. O. Omotayo (2007) 'Extended-range weather forecasting in Sub-Saharan West Africa: Assessing a potential tool for adapting food production to climate variability and climate change', AIACC Working Paper No 46, International START Secretariat, Washington, DC (www.aiaccproject.org)

Adejuwon, J. O., E. E. Balogun and S. A. Adejuwon (1990) 'On the annual and seasonal patterns of rainfall fluctuations in sub-Saharan West Africa', *International Journal of Climatology*, vol 10, pp839–848

Adejuwon, J. O., and T. O. Odekunle (2004) 'Skill assessment of the existing capacity for extended-range weather forecasting in Nigeria', *International Journal of*

Climatology, vol 24, pp1249–1265

Adejuwon, J. O. and T. O. Odekunle (2006) 'Variability and severity of the Little Dry Season in south-western Nigeria', *Journal of Climate*, vol 19, pp1–8

Berte, Y. and M. N. Ward (1998) 'Experimental forecast of seasonal rainfall and crop index for July–September 1998 in Cote D'Ivoire', *SODEXAM, Meteorologie Nationale de Cote D'ivoire*, Abidjan, Department of Meteorology, University of Oklahoma, Norman, OK

Colman, A.W. and D. Richardson (1996) *Multiple Regression and Discriminant Analysis Predictions of July–Aug–Sept 1996: Rainfall in the Sahel and other Tropical North Africa Regions*, Ocean Application Branch NWP (National Weather Prediction) Division, UK Met Office, Bracknell, UK

Colman, A. W., D. M. Harrison, T. Evans and R. Evans (1997) *Multiple Regression and Discriminant Analysis Predictions of July–Aug–Sept 1999 Rainfall in the Sahel and other Tropical North Africa Regions*, Ocean Application Branch NWP Division, UK Met Office, Bracknell, UK

Colman, A. W., M. K. Davey, R. Graham and R. Clark (2000) *Experimental Forecast of 2000 Seasonal Rainfall in the Sahel and other Regions of Tropical North Africa, Using Dynamical and Statistical Methods*, Met Office, Bracknell, UK

Fakorede, M. A. B. (1985) 'Response of maize to planting dates in a tropical rainforest location', *Experimental Agriculture*, vol 21, pp19–30

Fleming, C., M. Hoepftner, A. Freise and M. Sobtj (1997) *Climate Variability, Climate Prediction, Water Resource and Agricultural Productivity: Food Security in Tropical Sub-Saharan Africa*, proceedings of the Cotonou-Benin workshop, July, Cotonou, Benin

Folland, C. K., T. N. Palmer and D. E. Parker (1986) 'Sahel rainfall and worldwide sea temperature', *International Journal of Climatology*, vol 13, pp271–285

Folland, C. J., J. Owen, M. N. Ward and A. Colman (1991) 'Prediction of seasonal rainfall in the Sahel region using empirical and dynamical methods', *Journal of Forecasting*, vol 10, pp21–56

Gbuyiro, S. O., M. T. Lamin and O. Ojo (2002) 'Observed characteristics of rainfall over Nigeria during ENSO years', *Nigerian Meteorological Society Journal*, vol 3, pp1–17

Graham, R. and R. Clark (2000) *Experimental Forecast of 2000 Seasonal Rainfall in the Sahel and Other Regions of Tropical North Africa Using Dynamical and Statistical Methods*, Meteorological Office, Bracknell, UK

Harrison-Church, R. J. (1956) *West Africa*, Longman Green and Co, London

Nieuwolt, S. (1982) *Tropical Climatology: An Introduction to the Climates of the Low Latitudes*, John Wiley and Sons, New York

Odekunle, T. O. (2004) 'Rainfall and the length of the growing season in Nigeria', *International Journal of Climatology*, vol 24, pp467–477

Odekunle, T. O. (2006) 'Determining rainy season onset and retreat over Nigeria from precipitation amount and number of rainy days', *Theoretical and Applied Climatology*, vol 83, pp193–201

Odekunle, T. O. and S. O. Gbuyiro (2003) 'Rain days predictability in south-western Nigeria', *Nigerian Meteorological Society Journal*, vol 4, pp1–17

Odekunle, T. O., E. E. Balogun and O. O. Ogunkoya (2005) 'On the prediction of rainfall onset and retreat dates in Nigeria', *Theoretical and Applied Climatology*, vol 81, pp101–112

Olaniran, O. J. (1983) 'The onset of the rains and the start of the growing season in Nigeria', *Nigerian Geographical Journal*, vol 26, pp81–88

Omotosho, J. B. (1990) 'Theory of rainfall onset prediction', *The African Climate Watch*, ACMAD (www.acmad.ne/uk)

Omotosho, J. B., A. A. Balogun and K. Ogunjobi (2000) 'Predicting monthly and seasonal rainfall, onset and cessation of the rainy season, in West Africa, using only surface data', *International Journal of Climatology*, vol 20, pp865–88

Patt, A. and C. Gwato (2000) 'Effective seasonal climate applications: Examining constraints for subsistence farmers in Zimbabwe', *Global Environmental Change*, vol 12, pp185–195

Philippon, N. and B. Fontaine (2000) *West African June to September Rainfall: Experimental Statistical Forecasts Based on April Values of Regional Predictors*, Centre de Recherche de Climatologie, 21000 Dijon, France

Stern, R. D. and R. Coe (1982) 'The use of rainfall models in agricultural planning' *Agricultural Meteorology*, vol 26, pp35–50

Thiaw, W. and A. Barnston (1996) *CCA Forecast for Sahel Rainfall in Jul–Aug–Sept 1996*, Climate Prediction Centre, NOAA, Camp Springs, MD

Thiaw, W. and A. Barnston (1997) *CCA Forecast for Sahel Rainfall in Jul–Aug–Sept 1997*, Climate Prediction Centre, NOAA, Camp Springs, MD

Thiaw, W. and A. Barnston (1999) *CCA Forecast for Sahel Rainfall in Jul–Aug–Sept 1999*, Climate Prediction Centre, NOAA, Camp Springs, MD

Thoroid, C. A. (1952) 'The epiphytes of Theobroma cacao in Nigeria in relation to the incidence of the black pod disease', *Journal of Ecology*, vol 40, pp126–142

Toure, Y. M. (2000) *CLIMAB: 2000. Version 1.0.*, International Research Institute for Climate Prediction, Columbia University, New York

Walter, M. W. (1967) 'The length of the rainy season in Nigeria', *Nigerian Geographical Journal*, vol 10, pp123–128

Ward, M. N., J. A. Owen, C. K. Folland and G. Farmer (1990) 'The relationship between sea surface temperature anomalies and summer rainfall in Africa 4–20°N', *Long-Range Forecasting and Climate Memorandum*, no 32, available from the National Meteorological Library, Met Office, Bracknell, UK

Wilson, S. M. (2002) 'Empirical prediction of seasonal rainfall in Nigeria', *Nigerian Meteorological Society Journal*, vol 3, pp25–29

10

Adapting Dryland and Irrigated Cereal Farming to Climate Change in Tunisia and Egypt

Raoudha Mougou, Ayman Abou-Hadid, Ana Iglesias,
Mahmoud Medany, Amel Nafti, Riadh Chetali,
Mohsen Mansour and Helmy Eid

Introduction

Agriculture in North Africa is both the main use of the land in terms of area and the principal water-consuming sector, accounting for over 70 per cent of total water consumption (Iglesias, 2003; Iglesias et al, 2003). Rainfall is low and characterized by large year-to-year variation. Water is scarce and the countries of North Africa are considered to be water stressed, withdrawing large proportions of available surface waters for agricultural, domestic and other uses. Climate change is projected to bring warmer temperatures and changes in rainfall that, together, have the potential to reduce water availability. The impacts that climate change will have on agriculture and water supplies, and the potential for conflicts over water, are critical concerns in the region.

Under current conditions, North African countries import significant amounts of grain and it seems likely that climate change will lead to an expansion of this import requirement. The effects of sea level rise in North Africa, especially on the coast of the Delta Region of Egypt, would reduce the area under cultivation and probably reduce agricultural production (Desanker and Magadza, 2001; El-Raey, 1999; Abd-El Wahab, 2005). In addition to the effects of climate change, agricultural production in the region is expected to change rapidly due to technological advancements, economic changes, such as potential trade agreements with the European Union, and projections of high population increase (Iglesias et al, 2004).

Over 80 per cent of the cropland in the region is used for rain-fed production, which produces very low and variable yields (FAOSTAT, 2005). In Tunisia, an even larger percentage of cropland, 93 per cent, is used for rain-fed agriculture, yet irrigation is the largest consumptive use of water (Mougou and Salem, 2003). Egypt is an exception in the region, with almost 90 per cent of cropland being irrigated (FAOSTAT, 2005). Water demand for irrigation is expected to increase in all countries of North Africa and it is important to define adaptation strategies that take into account the possible deficit of water for irrigation in the future (Strzepek et al, 1995; Iglesias et al, 2003).

Northern Africa's adaptation capacity to climate change is challenged as it is compounded by high development pressure, increasing population, water management that is already using most of the available water resources, and agricultural systems that are often not well adapted to current conditions (Abdel et al, 2003; Eid, 1994; Iglesias and Moneo, 2005). Evidence of limits to adaptation of socioeconomic and agricultural systems in the North African region has been documented in recent history. For example, water reserves were not able to cope with sustained droughts in the late 1990s in Morocco and Tunisia, causing many irrigation dependent agricultural systems to cease production (Iglesias and Moneo, 2005).

Adaptation to climate change in North Africa is a major issue from the perspectives of food production, rural population stabilization and distribution of water resources. Previous studies have addressed adaptation in a top–down approach, evaluating theoretical options with little relation to current agricultural management. But there is also a need to incorporate management knowledge for formulating adaptation measures in a bottom–up approach (Abdel et al, 2003; Eid, 1994; Eid et al, 1993, 1995 and 1996).

The aim of our study is to evaluate adaptation measures for rain-fed and irrigated agriculture in Tunisia and Egypt. Using surveys of farmers and interviews with technical resource managers, information is collected about the sensitivity of farm operations to climate and available adaptation measures. Information was also provided to farmers and other stakeholders to increase their understanding of the interactions between climate and agriculture. Selected measures are evaluated for Tunisia using the expert judgements of agricultural managers and farmers, while in Egypt, adaptation measures are evaluated quantitatively using agricultural simulation models.

Case Study of Rain-fed Cereal Farming in Kairouan, Tunisia

Tunisia belongs to the group of water scarce countries. Rainfall is characterized by its scarcity and spatial and temporal variability (Mougou et al, 2002). The average of total annual rainfall varies from 1500mm in the north to 100mm in the south. High temperatures and solar radiation result in high rates of evapotranspiration and an environment of water stress. Variability and scarcity of water resources and high temperatures negatively affect agricultural production, most particularly rain-fed agriculture.

The Tunisian case study focuses on cereal farming Kairouan in central Tunisia. The central region is a climatic transition zone between the Mediterranean zone and the Sahara region. Severely dry years with a deficit of evapotranspiration of over 50 per cent are frequent in this region and represent a recurrent risk for rain-fed agriculture, which is practiced on nearly 90 per cent of farmlands in Kairouan. High temperatures during the grain filling period can negatively affect crop growth and development. Another major risk to crops is the very dry wind blowing from the Sahara known as the Sirocco, which occurs between May and September. If the Sirocco occurs during the grain filling stage, an irreversible loss of water from the plant can take place (Mougou and Henia, 1998).

Tunisian farmers apply a variety of strategies to lessen the negative effects of high temperatures, high rates of evaporation and periods of below normal rainfall. Some of these practices may be useful as strategies for adapting to future climate change. To identify and evaluate adaptation measures in use, a survey was conducted of farmers in the Kairouan region. One hundred farmers of rainfed cereals in the region were selected at random and surveyed (Mougou et al, 2003). The sample represents 5 per cent of the 2000 cereal farmers in the study area. The survey sought to identify farmers' behaviours for coping with climate variability to characterize the capability of the farmers to adapt to climate variability, define the 'know-how' of the farmers in relation to predicted climate change, list the adaptation methods already in use, and identify factors that prevent farmers from adapting to current and future climate variability.

Survey results

Cereal yields fluctuate in the surveyed farms, especially for rain-fed cereals. There is also variation in yields across farms due to differences in soils, farm inputs and management practices. For example, yields of durum wheat and barley range between 24 and 47 quintals (100kg) per hectare and 38 and 54 quintals per hectare for irrigated durum wheat and barley respectively, whereas for rain-fed wheat and barley the respective ranges are 0–33 and 0–40 quintals. However, scarcity of water and financial considerations prevent irrigation on 88 per cent of farms in the study area.

On rain-fed farms, variability in cereal production is explained by the variability of rainfall, with values of 78 per cent for the north, 50 per cent for the centre and 40 per cent for the south regions (see Figure 10.1 and Mougou et al, 2003). For the study region of Kairouan, variability of rainfall explains 56 per cent of the cereal yield variability. Differences in average cereal yields for wet and dry years over the period 1995–2003 are shown in Table 10.1 for sites at Kairouan, Sbikha, Haffouz and Ouslatia. The average yield in dry years at the four sites is 6.9 quintals per hectare, while in wet years the average is 15.6 quintals, or 132 per cent more.

The survey highlighted the fact that agriculture in the region is mainly practised by old farmers with an average age of 58 years and with low levels of schooling. Illiterate farmers represent 55 per cent of the sample; 36 per cent have primary education, 5 per cent have secondary education and 3 per cent have had schooling at higher levels. Average farm size is 29.8ha, which is large by Tunisian standards. Approximately one third of farmers have farms smaller than 10ha, while 54 per cent have farms between 10 ha and 50ha. Only 12 per cent of farmers have farms larger than 50ha. The participation of family members in agricultural activities is high; we found that in 62 per cent of the farms family members participate in farming.

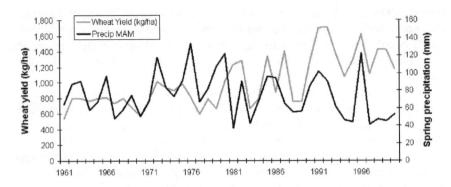

Figure 10.1 *Wheat yield and spring precipitation in Tunisia, 1961–2000*

Source: Mougou et al (2003).

Table 10.1 *Average cereal yields during wet and dry years at four sites in the Kairouan region of Tunisia*

	Rainy years (1995/1996 and 2002/2003)		Dry years (1996/1997 to 2001/2002)		Yield increase during rainy years compared to dry years (%)
	Mean rainfall (mm)	Mean yields (qt)	Mean rainfall (mm)	Mean yields (qt)	
Kairouan	231.55	15.5	114.2	6.5	138
Sbikha	275.2	17.5	112.6	6.3	178
Haffouz	269.15	14	99.	1 6	133
Ouslatia	317.7	15.5	130.9	8.6	80
Average	**273.4**	**15.6**	**114.2**	**6.9**	**132.3**

Surface ploughing is used by almost all farmers since it is relatively cheap. Only 29 per cent of the farmers practise crop rotation, fertilizers are used by only 17 per cent, and this is done only during years with favourable weather. While a majority of farmers use commercial seeds, nearly 40 per cent use seeds that they produce themselves or buy from other farmers.

Supplemental irrigation, defined as the application of a limited amount of water to the crop when rainfall fails to provide sufficient water for plant growth to increase and stabilize yields (Oweis et al, 1999), is used by roughly one quarter of the farmers. Supplemental irrigation is applied mainly to fodder crops for livestock. The irrigated area represents less than 4 per cent of the total cultivated area because of the small amount of water available and the financial constraints on purchasing irrigation equipment and materials. All farmers who use supplemental irrigation are conscious of the advantages of fertilization and its management. It is important to note that rain-fed cereal production is the principal activity even for the farmers who have access to water.

Nearly three quarters of surveyed farmers supplement their cereal production by raising sheep. This serves to diversify farm households' income and food sources and acts as a hedge against crop losses. In poor years, livestock are sold to provide income, while in good years the number of livestock is increased. Nearly 30 per cent of farmers reported liquidating their herds because of the drought experienced in 1997–2002.

All farmers expressed their suffering from difficult climatic conditions. Indeed they suffered from long years of drought. Even during the rainy years rainfall distribution can be inadequate for the crops. Dry conditions in March decrease cereal production and result in a loss of income for the farmer. According to farmers' responses, over 95 per cent of agricultural output is determined by climate conditions and by water scarcity. This perception results in a feeling of frustration, and it is obvious that this is the result of the drought during recent years (1997 to 2002). Concerning the farmers' understanding of climate change, only 52 per cent know about climate change phenomena and only 12 per cent are aware of the necessity to adapt to climate change in a different way than to climate variability.

Although only a small number of farmers understand the concept of adapting to climate change, over 90 per cent employ measures to adapt to current climate variability. Practices currently used include digging a well (48 per cent), shifting sowing date if the autumn is drier than normal (48 per cent), storing fodder to ensure supply of food for livestock (57 per cent), changing cultivation techniques to decrease management costs (75 per cent) and applying supplemental irrigation when rainfall is not enough to ensure a minimum water requirement (44 per cent). Other practices that are used yearly regardless of climate are breeding sheep, which are resilient to adverse climate conditions and consume a range of fodder resources, and growing cactus and other crops for use as fodder.

Despite the diversity of adaptation methods used by Kairouan farmers, yields are low and variable and adaptation is incomplete and inefficient. Factors that limit adaptation include the low level of education of farmers, difficulties for the extension services to change the farmers' behaviours and difficulties for farmers to adopt new techniques even if they agree with them. Survey results indicate that only 20 per cent of the sampled farmers adopt the advice of extension services. Others use some adaptation methods but do them in their own way and do not follow exact prescriptions for implementing recommended technologies.

Implications for adapting to climate change

Farmers in the survey area appear to overestimate the effect of climate variability on crop yields and so underestimate the potential effectiveness of adaptive strategies for improving yields and reducing the variability of yields. While nearly all farmers use adaptation measures to reduce risks from climate variability, many confirm that they do not use them consistently with the recommendations of extension service agents. This may reduce their effectiveness. In addition, some investments by farmers may not be implemented because low prices for agricultural products result in low economic returns on the investments.

Our results show that only 20 per cent of farmers adopt the advice of extension services. Indeed, it arises also from the survey data that the extension services are effective only in medium and large farms. Only one third of farmers that follow extension service advice are smallholder farmers with farms less than 10ha in size. This is confirmed by Chennoufi and Nefzaoui (1996), who find that 'technologies generated appear to be more readily adopted by large-scale farmers, agricultural development agencies, rural development societies and cooperative farms, rather than by the majority of medium-and small-scale farmers'. This behaviour goes contrary to the strategies of rural development in Tunisia. In fact the strategies of rural development are directed towards a participative approach and providing technical aid mainly for small farmers, who represent about 80 per cent of farmers in Tunisia. In the Tunisian private sector the primary agricultural areas, 24 per cent of farm land, are cultivated by 1 per cent of farmers with farms of at least 100ha (Souki 1994). Their large incomes allow large-scale farmers to use new techniques and to be advised by qualified people.

New agricultural techniques need to be developed and introduced to farmers, with a focus on smallholder farmers, to build their capacity for coping with and adapting to climate change. The involvement of the rural population is essential to ensure that appropriate technologies are being developed and that farmers are provided with the knowledge and skills necessary for effective adoption of the technologies.

Research and development on improvement of cereal varieties started in Tunisia in the 1970s. Packages of agronomic varieties and techniques have been developed mainly for wet regions. But attempts to diffuse new technologies to farmers have not been entirely successful. New approaches that engage farmers are being used to improve dissemination of new technologies and to transfer to farmers knowledge and skills that are necessary for their effective adoption. Cooperation with development agencies has been improving through short-term training, field days, joint research programmes, research contacts, utilization of libraries and databases, use of laboratories and research stations, and joint publications and reports. The most innovative move has been the establishment of seven 'regional development poles' for research, one per large agro-ecological zone, which offer a good framework for bringing together all partners in research at the regional level, including development agencies and farmers' representatives (Lasram and Mekni, 1999).

During the last ten years, the agronomic research programmes in Tunisia have mainly been focused on the assessment of drought impacts on agricultural production. Because of the potential for the climate to become drier and more drought-prone, results of this research are expected to provide more options for adapting to climate change. Research on water resources management, the improvement and the genetic selection of cultivars, the improvement of techniques already used, and the use of new techniques such as fertilization and irrigation methods are particularly promising.

Case Study of Irrigated and Rain-fed Wheat Farming in Egypt

Precipitation in Egypt is only significant in the northern Mediterranean coast, where average annual rainfall is roughly 180mm, and is extremely low in the rest of the country's desert territory. Consequently, agriculture is restricted to the fertile lands of the narrow Nile Valley from Aswan to Cairo and the flat Nile Delta north of Cairo. Together this comprises only 3 per cent of the country's land area.

The study in Egypt focuses on the Nile Delta, represented by Sakha in the Kafr El-Shikh Governorate and Mersa Matrouh on the northwest coast of Egypt. The objective of the Egyptian case study is to detect important adaptation measures of the wheat crop production system under irrigated and rain-fed agriculture systems. The study was conducted in three steps: stakeholders engagement, evaluation of adaptation measures and simulation of effectiveness of selected adaptation measures using crop models.

The Nile Delta region is characterized by high-production irrigated smallholder agriculture, high urban water demands and rapidly growing population. Water for irrigation, which comes entirely from the River Nile, varies due to changes in freshwater availability and to competition among water users. The total area cultivated for wheat in Egypt reached nearly one million hectares in 2004, just over half of which are located in the Delta region. The region also accounts for just over half of the wheat production of Egypt, 7.2 million tons, and has an average yield of 6.6 tons of wheat per hectare (MALR, 2004).

The Mersa Matrouh region is one of the few regions of rain-fed agriculture in Egypt. The region is characterized by low population and low agricultural productivity. The total cropped area in the governorate is about 95,000ha, of which approximately 20 per cent is cultivated for wheat in the winter season (MALR, 2004). Wheat yields average only two tons per ha in Mersa Matrouh, or about 30 per cent of the national average. The low yields are due to scarce and highly variable water availability for rain-fed agriculture in this arid region and to low levels of farm inputs and management.

The stakeholders engaged in the study represent the smallholder farmers, commercial farmers and strategic resource managers. Training programmes, field demonstration and workshops were organized by the study to orient stakeholders with the issues of climate change and adaptation, and to build an information exchange between stakeholders and the study team. The impacts of climate change on agriculture and possible adaptation measures were discussed with stakeholders. Adaptation measures proposed by stakeholders include changing cropping patterns, adopting new drought-tolerant cultivars, reducing or constraining the cultivation of high water consuming crops, changing sowing dates, changing fertilization practices, and improving the efficiency of water application and water use.

Simulation of Adaptation Measures

Simulation models were used to quantify some of the adaptation measures proposed by the stakeholders. The modelling study considered on-farm adaptation techniques such as the use of alternative varieties of wheat, optimizing the time of planting and optimizing the application of irrigation water. Simulations were performed for irrigated wheat production in the Sakha district of the Kafer El-Sheikh governorate in the Delta region and for rain-fed wheat production at the Sidy Baraney district in the Marsa Matrouh governorate on the northern coast. The CERES-Wheat model developed by the International Benchmark Sites Network for Agrotechnology Transfer and included in the DSSAT 3.5 package was used for the simulations (Tsuji et al, 1995).

The simulations were performed for current climate conditions and for two scenarios of future climate change that were constructed using the MAGICC/SCENGEN software of the University of East Anglia (UK) and input data from climate projections of the HadCM3 general circulation model for the A1 and B2 SRES emissions scenarios (Nakicenovic and Swart, 2001; Eid et al, 2001). The regional temperature increases corresponding to the A1 and B2 scenarios are 3.6°C and 1.5°C respectively. Precipitation changes in most of the territory are about ±10 to 20 per cent, depending on the season. Data for daily maximum and minimum temperatures, precipitation, solar radiation, soil properties and water quality were collected for the period 1975–1999 for Sakha and 1993–1995 for Sidy Baraney.

Irrigated wheat in the Nile Delta region of Egypt

For irrigated wheat, the simulation experiments are designed to investigate variations in crop yield and crop water demand in response to higher temperatures for different choices of wheat cultivars (Sim 1), water conservation measures (Sim 2), and sowing dates (Sim 3). The irrigation requirements of the crop, applied by flood irrigation, and application of fertilizer are set according to recommendations of the agricultural extension services. Other assumptions incorporated into each of the simulation cases are summarized in Table 10.2. Two outputs of the DSSAT model are used as indicators of the effects of climate and adaptation measures: the percentage change in wheat yield and the percentage change in evapotranspiration (Et_{crop}).

For Sim 1, three cultivars common to Egypt and the Delta region are selected. The selected cultivars are Giza-168, which is planted on 24 and 22 per cent of land cultivated for wheat in Egypt and the Delta respectively, Sakha-8, which is planted on 2.0 and 1.6 per cent of land cultivated for wheat in Egypt and the Delta respectively, and Sakha-69, which is planted on 10 and 13 per cent of land cultivated for wheat in Egypt and the Delta respectively. Simulations were performed for each cultivar for the current climate, a climate that is 1.5°C warmer on average and a climate that is 3.6°C warmer on average. In each of these cases irrigation water is applied at a level of 600mm and the crop is planted on 20 November, which reflect the current recommendations for these management variables.

Table 10.2 *Assumptions for simulations of irrigated wheat production in the Nile Delta region*

	Sim 1	Sim 2	Sim 3
Adaptation measure	Change cultivars	Change water amount	Change sowing dates
Climate scenario	Current +1.5°C +3.6°C	Current +1.5°C +3.6°C	Current +1.5°C +3.6°C
Wheat cultivar	Giza-168 Sakha-8 Sakha-69	Sakha-8	Sakha-8
Sowing date	20 November (recommended)	20 November (recommended)	20 November (0 shifting days) 1 December (11 shifting days) 10 December (20 shifting days) 20 December (30 shifting days)
Irrigation amount	600mm/season (recommended)	300mm/season (50% water saving) 400mm/season (33% water saving) 500mm/season (17% water saving) 600 mm/season (0% water saving)	600mm/season (recommended)

Sim 2 varies the amount of irrigation water over a range of 300 to 600mm for the Sakha-8 cultivar. For Sim 3, the effects of delaying planting by 11, 20 and 30 days are investigated for the Sakha-8 cultivar and an irrigation application of 600mm. Results for the three irrigated wheat simulation cases are presented in Table 10.3.

In Sim 1, the productivity of all three wheat cultivars is reduced by higher temperatures relative to current yields, while Et_{crop} is increased. Under current climate conditions, Giza-168 provides the highest yields of the three and is less water demanding than Sakha-8. However, the effects of higher temperatures are more pronounced on Giza-168 than on the two other cultivars, with the exception that Sakha-69 is not viable for the 3.6°C warming case. Yields of Giza-168 are reduced by 30 and 37 per cent for warming of 1.5 and 3.6°C respectively. In comparison, yields of Sakha-8 are reduced by 4 and 26 per cent for 1.5 and 3.6°C warming. These results suggest that the Sakha-8 cultivar is more stable under climate change conditions than Giza-168 and Sakha-69. For this reason, Sakha-8 was selected for use in the two other simulation cases.

Water conservation is an important objective for irrigated agriculture in the Nile Delta because of high national agricultural water demands. The joint effects of climate change and water conservation measures are simulated in the Sim 2 cases, the results of which are illustrated in Figure 10.2(a). Under the

Table 10.3 *Results of simulations for irrigated wheat in the Nile Delta region (percentage changes are relative to current climate)*

Indicator	Climate scenario	Sim 1 (cultivars)			Sim 2 (water amounts)				Sim 3 (sowing dates)			
		Giza-168	Sakha-8	Saka-69	300 mm	400 mm	500 mm	600 mm	20 Nov	01 Dec	10 Dec	20 Dec
Yield (t/ha)	Current	6.8	4.7	6.1	3.6	4.5	4.7	4.7	4.8	4.7	4.7	4.7
	+1.5°C	4.7	4.5	4.7	3.0	4.2	4.5	4.5	4.5	4.5	4.5	4.5
	+3.6°C	4.2	3.5	na	2.3	3	3.6	3.9	3.1	3.1	3.4	3.4
Yield change (%)	+1.5°C	−30	−4	−28	−16	−6	−4	−5	−6	−5	−4	−5
	+3.6°C	−37	−26	na	−35	−32	−25	−18	−35	−34	−28	−27
Et$_{crop}$ (mm)	Current	354	464	352	313	403	464	476	469	462	467	476
	+1.5°C	357	489	358	316	409	489	513	498	490	490	491
	+3.6°C	503	498	na	310	405	498	579	507	501	498	494
Et$_{crop}$ change (%)	+1.5°C	1	5	2	1	1	5	8	6	6	5	3
	+3.6°C	42	7	na	−1	0	7	22	8	8	7	4

current climate, reducing flood irrigation from the recommended rate of 600mm to 500mm or even 400mm have small effects on yields while generating significant water conservation benefits. At 300mm, a 50 per cent reduction in water use, yields would be reduced 25 per cent relative to what can be attained with 600mm in the current climate. But this is considered an acceptable trade-off in this water scarce region.

In comparison, under climates that are 1.5° and 3.6°C warmer, a 50 per cent reduction in water use would result in unacceptably high losses of wheat yield of 33 per cent and more. In the warmer climate scenarios, water conservation measures would probably be limited to less aggressive targets due to economic considerations. With 3.6°C warming, a reduction to even 500mm of water, which has negligible effects on yields in the current climate or in a climate that is 1.5°C warmer, would cut yields by nearly 10 per cent.

Another adaptation option proposed by farmers is to shift sowing dates. This strategy is represented by the Sim 3 cases, for which the date of sowing is delayed by 10, 20 and 30 days compared to the current norm. The results are illustrated in Figure 10.2(b). Delaying planting dates results in negligibly changed yields in the current climate and under a climate 1.5°C warmer. But for 3.6°C warming, delaying sowing by 20 to 30 days would increase yields by 10 to 11 per cent. Changes in sowing dates had only small effects on evapotranspiration.

The simulation results indicate that changes in wheat cultivars can be an effective strategy for adapting to a warmer climate. Of the cultivars tested for the selected climate change scenarios, the Sakha-8 cultivar appears to have advantages over the Giza-168 cultivar, which is more widely used at present. But the performance of different cultivars is controlled by local environmental conditions such as the local climate, soil condition and water availability, and

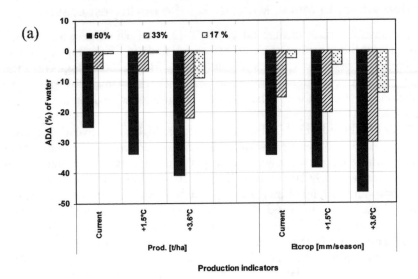

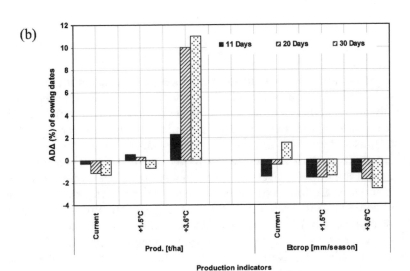

Figure 10.2 *Simulated changes in irrigated wheat yields and evapotranspiration under different climate conditions for (a) water conservation measures and (b) variations in sowing dates*

the best choice of cultivar will vary by location and will differ according to the change in climate. The results also show that climate change may constrain water conservation efforts, as yields are more sensitive to reduced flood irrigation at higher temperatures. Changes in planting dates are found to have little benefit except for the warmest climate scenario studied.

Rain-fed wheat in Mersa Matrouh on the northwest coast

The impacts of climate change on yields of rain-fed wheat grown in Mersa Matrouh on the northwest coast are simulated for warming of 1.5°C combined with assumed changes in rainfall that include no change from the current average of 240mm per year, ±10 per cent and ±20 per cent. Simulations are also made for changes in sowing dates as an adaptation to the changes in climate.

Yields of rain-fed wheat in this dry climate are highly sensitive to variations in rainfall. At current temperatures, a 10 per cent reduction in rainfall would reduce yields by 26 per cent, while a 20 per cent reduction in rainfall would reduce yields by nearly 50 per cent (Figure 10.3a). Yields respond positively but less strongly to rainfall increases. Warming of the climate would amplify the yield losses for rainfall reductions and would erode the yield gains from increases in rainfall. Yields can be improved 10 to 35 per cent by delaying planting dates, both in the current and warmer climates, with the best results being achieved with a delay of 40 days (Figure 10.3b). These results indicate that a change in sowing dates could be utilized to reduce the negative effects of water shortage from reduced rainfall.

Conclusions

Climate is perceived by farmers in the study areas of Tunisia and Egypt as a major risk to agricultural production. A variety of practices are in use to help manage climate-related risks, but their adoption is incomplete and their use is often not consistent with the advice of agricultural extension services. The survey of Tunisian farmers found that they tend to over-estimate the impact of climate variability and consequently fail to recognize the potential of farm management techniques to offset the effects of adverse climate conditions. The extension services are found to be effectively used primarily by large-scale farms and some medium-scale farms. Therefore, new agricultural techniques and extension programmes need to be developed and introduced to small- and medium-scale farmers to build up their capacity to manage climate risks.

Changes in wheat cultivars, irrigation practices and sowing dates were suggested by farmers and other stakeholders in Egypt as options for adapting to climate risks and were evaluated using crop simulation models. The results demonstrate that it is possible to adapt and partly offset the effects of climate change on wheat yields using these strategies. The simulations also pointed to the potential of climate change to constrain water conservation efforts as wheat yields are more sensitive to reductions in irrigation water at higher temperatures. Further work is needed to explore synergistic effects of different combinations of farm management strategies to identify promising options, both through model simulations and in field tests with farmers. But a key challenge will be to improve the communication and diffusion of agricultural methods for managing climate risks.

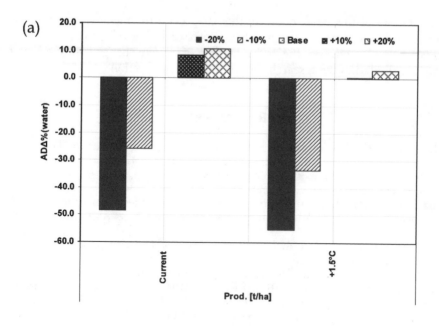

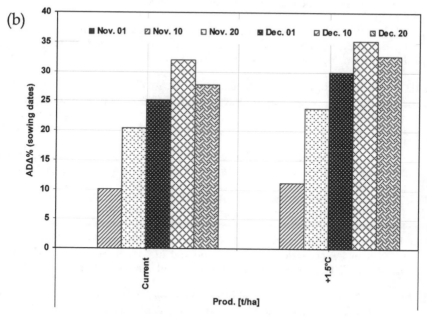

Figure 10.3 *Simulated changes in rain-fed wheat yields for (a) different rainfall scenarios and (b) variations in sowing daes*

Dedication

The authors of this work dedicate this chapter to the memory of Professor Helmy Eid, who passed away in 2005. Professor Eid, a valued colleague and friend, helped to advance the understanding of climate change issues in Egypt and contributed to the research reported in this publication.

References

Abdel, H., N. G. Ainer and H. M. Eid (2003) 'Climate change impacts on Delta crop productivity, water and agricultural land', *Journal of Agricultural Science*, special issue: 'Scientific symposium on problems of soils and water in Dakahlia and Damietta governorates', 18 March, pp15–26

Abd-El Wahab, H. M. (2005) 'The impact of geographical information system on environmental development', MSc thesis, Faculty of Agriculture, Al-Azhar University, Cairo

Chennoufi, A. and A. Nefzaoui (1996) *The Role of Universities in the National Agriculture Research Systems of Egypt, Jordan, Morocco, Sudan and Tunisia. Case Study: Tunisia*, Food and Agriculture Organization, Rome

Desanker, P. and C. Magadza (2001) 'Africa', in J. McCarthy, O. Canziani, N. Leary, D. Dokken and K. White (eds) *Climate Change 2001: Impacts, Adaptation and Vulnerability*, contribution of Working Group II to the Third Assessment Report of the Intergovernmental Panel on Climate Change, Cambridge University Press, Cambridge, UK, and New York

Eid, H. M. (1994) 'Impact of climate change on simulated wheat and maize yields in Egypt', in C. Rosenzweig and A. Iglesias (eds) *Implications of Climate Change for International Agriculture: Crop Modeling Study*, US Environmental Protection Agency, Washington, DC

Eid, H. M., A .A. Rayan, K. A. Mohamed, M. M. A. El-Refaie, M. M. Attia, H. A. Awad, K. M. R. Yousef and M. M. A. El-Koliey (1995) 'Impact of climate change on yield and water requirements of some major crops', in *2nd Conference of On-Farm Irrigation and Agroclimatology, 2–4 January*, Agricultural Research Center, Egypt, pp492–507

Eid, H. M., N. M. Bashir, N. G. Ainer and M. A. Rady (1993) 'Climate change crop modeling study on sorghum', *Annals Agricultural Science*, Special Issue 1, pp219–234

Eid, H. M., S. M. El-Marsafawy, N. G. Ainer, M. A. Ali, M. M. Shahin and N. M. El-Raey (1999) 'Impact of climate change on Egypt', special report, EEMA, Cairo, Egypt

Eid, H. M., S. M. El- Marsafawy and N. G. Ainer (2001) 'Using MAGICC and SCEN-GEN climate scenarios generator models in vulnerability and adaptation assessments', Conference on Meteorology and Environmental Issues, March, Cairo, Egypt

El-Mowelhi, N. M. and O. El-Kholi (1996) *Vulnerability and Adaptation to Climate Change in Egyptian Agriculture*, Country Study Report, Country Studies Program, Washington, DC

Iglesias, A. (2003) 'Climate, drought and prediction in the Mediterranean: Opportunities for agricultural adaptation', *Revista de Ingenieria Civil*, vol 131, pp25–31

Iglesias, A. and M. Moneo (2005) *Drought Preparedness and Mitigation in the Mediterranean: Analysis of the Organizations and Institutions*, Options Mediterranean, Series B, No 51,Universidad Politécnica de Madrid, Madrid, Spain

Iglesias, A., M. N. Ward, M. Menendez and C. Rosenzweig (2003) 'Water availability for agriculture under climate change: Understanding adaptation strategies in the Mediterranean', in C. Giupponi and M. Shechter (eds) *Climate Change and the Mediterranean: Socioeconomic Perspectives of Impacts, Vulnerability and Adaptation*, Edward Elgar Publishers, Northampton, MA, US

Iglesias, A., N. X. Tsiourtis, D. A. Wilhite, A. Garrido, L. Garrote, M. Moneo, A. Gomez-Ramos, M. J. Hayes and C. Knutson (2004) 'Terms of reference for drought risk management: Drought identification studies, drought risk analysis and best practices', MEDROPLAN working paper, Mediterranean Agronomic Institute of Zaragoza, Zaragoza, Spain

Lasram, M. and M. Mekni (1999) *The National Agricultural Research System of Tunisia*, WANA NARS study, International Center for Agricultural Research in the Dry Areas (ICARDA), Aleppo, Syria

MALR (Ministry of Agriculture and Land Reclamation) (2004) *Agricultural Statistics*, vol 1, Economic Affairs Sector, Ministry of Agriculture and Land Reclamation, Cairo, Egypt

Mougou, R. and M. Ben Salem (2003) 'Meteorological conditions in arid regions and effects of climate change in dryland crops', Proceedings of the Training on Agricultural Techniques for Rain-fed Agriculture and Communication to Farmers, Arab Center for Studies in Dry land Agriculture (ACSAD), Tunis, Tunisia

Mougou, R. and L. Henia (1998) 'Contribution à l'étude des phénomènes à risques en Tunisie cas du sirocco', *Les Publications de l'Association Internationale de Climatologie*, vol 9

Mougou. R., S. Rejeb and F. Lebdi (2002) 'The role of Tunisian gender issues in water resources management and irrigated agriculture in Tunisia', in 'The First Regional Conference on Perspectives of Arab Water Cooperation: Challenges, Constraints and Opportunities', workshop on gender and water management in the Mediterranean countries, Cairo

Nakicenovic, N. and R. Swart (eds) (2001) *Emissions Scenarios*, special report of Working Group III of the Intergovernmental Panel on Climate Change, Cambridge University Press, Cambridge, UK, and New York

Oweis, T., A. Hachum and J. Kijne (1999) 'Water harvesting and supplementary irrigation for improved water use efficiency in dry areas', SWIM Paper 7, International Water Management Institute, Colombo

Souki, K. (1994) 'Analysis of the effects of water and nitrogen supply on the yield and growth of durum wheat under semiarid conditions in Tunisia', doctoral thesis, University of Reading, Reading, UK

Strzepek, K., D. Yates, G. Yohe, R. Tol and N. Mander (2001) 'Constructing "not implausible" climate and economic scenarios for Egypt', *Integrated Assessment*, vol 2, pp139–157

Tsuji, G. Y., J. W. Jones, G. Uhera and S. Balas (1995) *Decision Support System for Agrotechnology Transfer, V 3.0* (three volumes), International Benchmark Sites Network for Agrotechnology Transfer (IBSNAT), Department of Agronomy and Soil Science, University of Hawaii, Honolulu, HI

11

Adapting to Drought, *Zud* and Climate Change in Mongolia's Rangelands

*Punsalmaa Batima, Bat Bold, Tserendash Sainkhuu
and Myagmarjav Bavuu*

Introduction

When climatic events adversely impact Mongolia's grasslands and livestock herds, the effects reverberate throughout the country. The people and economy are highly dependent on pastoral livestock herding, a livelihood that is extremely sensitive to climate variations and extremes. Roughly 80 per cent of Mongolia's 127 million hectares is used as open pasture for year-round grazing of sheep, goats, cattle (including yaks), horses and camels. Herding these animals and processing livestock products engages nearly half of the Mongolian population, provides food and fiber to the other half, and generates about one-third of the country's foreign exchange earnings (Mongolian Statistical Yearbook, 2004).

The arid to semiarid climate of Mongolia supports extensive grasslands that, while fragile, have sustained pastoral herding for centuries. In recent decades the climate has become warmer and drier. Partly due to this trend, the productivity of Mongolia's pastures has declined by 20 to 30 per cent. Another observed trend is an increase in the frequency and intensity of climatic extremes such as drought and severe winters, or *zud*. Drought and *zud* events have caused livestock deaths, hardship for herders and, in some instances, large rural to urban migrations, unemployment, deep poverty and economy-wide losses of income.

The pastoral system of Mongolia has also been strongly impacted by the change from a socialist system to a market economy. The change is not complete in the livestock sector as pastureland remains state owned and, with both traditional and socialist systems for livestock management disintegrated in the wake of the transition, is treated as an open access commons. The changes have increased the vulnerability of the pastoral system to climate and other stresses and led to practices that have placed heavy grazing pressures on many pasture areas. These pressures, combined with the trend toward a drier climate, have contributed to the degradation of some of Mongolia's pasturelands.

Projections of future climate change indicate that Mongolia will become warmer still, potentially drier in summer and wetter in winter. There is also a threat of even more frequent and intense droughts and *zud* in the future. Our study of projected climate change finds that the rangelands, livestock herds and pastoral livelihoods of Mongolia would be strongly impacted (Batima, 2006; Batima et al, 2008). Finding strategies to adapt to the changes will be critical if pastoral livelihoods are to be sustained and adverse impacts on Mongolia's economy and future development avoided.

We briefly review below the effects of social and economic changes on the livestock sector and the vulnerability of the sector to climate extremes and change. We then describe the stakeholder process used to identify and evaluate options for adapting to current and future climatic risks, present some of the options that emerged as high priorities, and explore barriers, opportunities and responsibilities for adaptation.

Social and Economic Changes

Up until the 1960s, Mongolia's pastures and livestock herds were managed according to traditional pastoral practices that included four seasonal migrations each year, maintenance of emergency pasture reserves, and grazing schedules that consider vegetation growth phases and the recovery periods needed by previously grazed pasture. This system provided for efficient and sustained use of pasture in the arid and semiarid climates of Mongolia's grasslands, limitation of grazing pressures, and management of risks from drought and *zud*.

In the 1960s, the livestock industry was collectivized. The land, animals and tools of production became the property of collectives called *negdels* that managed livestock production. Inputs and services such as fodder, transport, veterinary services, marketing, maintenance of wells, improved breeds and fencing were subsidized and provided by the central government through the *negdels*. With these subsidies, herders became highly dependent on the state. Seasonal migration of herds was still practised but it become more restricted with centralized management and provision of services.

In 1990–1992, with the transformation of the country from a socialist system to a market economy, the collectives were disbanded and ownership of livestock was privatized. Land, however, was not privatized and remains the property of the state. State-subsidized inputs and services to the livestock sector were cut sharply and infrastructure such as water supply, electricity, schools and hospitals collapsed in the rural areas. The state-owned land effectively became an open access commons in which herders lacked incentives to conserve pastures or make investments in the land such as maintenance of water supplies. Herders responded by increasing the numbers of animals stocked on pastures. Many reduced their seasonal migrations to twice a year, migrated shorter distances because of lack of infrastructure and services near distant pastures, moved their herds closer to urban centres with services that were increasingly unavailable in rural areas, or even stayed in one place to graze the same pasture year-round (Chuluun and Enkh-Amgalan, 2003; Batima et al, 2005).

The higher animal stocking rates and more limited movement of herds raised grazing pressures on many pastures and disrupted ecological balances. This has contributed to land degradation and reduced pasture productivity. Some areas have been abandoned because of lack of water supply or overuse (Tserendash et al, 2005). In the current situation, traditional and socialist systems for managing livestock activities are no longer operative, while incentives in the market system are distorted and imperfect with privatized livestock but open access land. The resulting practices and behaviours have heightened vulnerability in the livestock sector to climate and other stresses.

Vulnerability to Climate Extremes and Change

Changes in the climate of Mongolia observed in recent decades have impacted the ecosystems and livestock sector of the country (Batima, 2006; Batima et al, 2005; Tserendash et al, 2005). From 1940 to 2003, annual mean temperature rose by 1.8°C, precipitation increased slightly in autumn and winter and decreased slightly in spring and summer, potential evapotranspiration rose by 7 to 12 per cent, the duration of ice cover on lakes and rivers decreased by 10 to 30 days, maximum ice thickness decreased by 40 to 100cm, and the date of snow cover has shifted earlier by 3 to 10 days (this last since 1981). The warming has been most pronounced in winter, during which average temperatures have risen by 3.6°C.

Peak above ground biomass on Mongolia's pastures, which is usually attained in August, has declined by 20 to 30 per cent over the past 40 years. In the forest-steppe and steppe zones, the emergence of pasture plants comes earlier but above ground biomass has decreased by 10 to 20 per cent in April and by 30 per cent in May. The warmer and drier spring and early summer have contributed to the decline in biomass. Also observed is a shift in the species composition of pastures towards low nutrient plants such as *Carex duriuscula Artemisia sp*, which has become dominant in pasture communities.

These changes have reduced forage production for livestock animals. Livestock grazing is also directly impacted by rising summer temperatures. Mongolia's livestock animals are accustomed to relatively low temperatures. When mid-day summer temperatures exceed thresholds, animals cease grazing. One impact of the warming of past decades is a reduction in the time spent grazing by livestock. Over the past 20 years, the time spent grazing during the months of June and July has reduced by an average of 0.8 hours per day and the number of days with more than 3 hours less grazing time increased by 7 days. The reduced pasture productivity and grazing time show their effects in reduced animal weights observed since 1980 (Bayarbaatar et al, 2005).

Changes in extremes have also been observed. Drought has increased significantly in the last 60 years. The worst period was 1999–2002, during which droughts affected 50 to 70 per cent of the land area for four consecutive summers. In normal years, animals start to gain weight in early summer, attaining their maximum weight by the end of autumn and having built up the capacity to withstand winter and spring weather. But in years of summer drought, reduced forage limits animal weight gain and leaves them more vulnerable to the stresses of winter and spring.

Mongolian winters are harsh and livestock losses occur every winter. But unusually severe winters, called *zud*, can cause conditions of livestock famine that result in abnormally high rates of animal deaths. There are several forms of *zud. Tsagaan*, or white *zud*, the most common and disastrous form if it affects a large area, results from high snowfall that covers the grass and prevents animals from grazing. *Har*, or black *zud*, occurs when a lack of snow causes water supply shortages. An impenetrable ice cover that forms when snowmelt refreezes and prevents access to grass is called *Tomor*, or iron *zud*. Extreme cold temperatures and strong winds that prevent animals from grazing and cause them to expend most of their energy to maintain body heat is called *Huiten*, or cold *zud. Havarsanzud* results when two or more of the above phenomena occur together.

Very severe *zud* in the winters of 1944–45, 1967–68, 1978–79 and 1999–2002 each resulted in millions of animal deaths. The harsh winters during the 1999–2002 period, which also coincided with severe and extensive summer droughts, were particularly devastating. An estimated 12 million animals died, representing almost one-third of Mongolia's livestock. More than 12,000 families lost all their animals, while thousands more lost substantial portions of their herds and were forced into poverty. Many migrated from rural areas to urban centres, raising the urban population in Mongolia from just under 50 to over 57 per cent and increasing unemployment in the urban areas. Agricultural output declined by 40 per cent in 2003 relative to 1999 and the contribution of agriculture to national gross domestic product decreased from 38 per cent to 20 per cent (Mongolian Statistical Yearbook, 2004).

Projections of future climate from five different general circulation models, interpolated to 2.5×2.5 degree grid cells in Mongolia, indicate that the climate will become warmer, in both summer and winter, and that precipitation is likely to increase in winter but change only slightly in summer (Natsagdorj et al, 2005; Batima, 2006). Winter temperatures are projected to warm by 2.8 to 8.7°C by the 2080s relative to 1961–1990, while summer temperatures are projected to warm by 3.2 to 6.9°C. Projected changes in precipitation range from +6.9 to +67.0 per cent in winter and –2.3 to +10.9 per cent in summer by 2080. The relatively small precipitation changes projected for summer imply that, with higher temperatures, evapotranspiration will increase and soils will become drier in summer. It is also estimated that drought incidence will increase (Natsagdorj and Sanjid, 2005).

Estimates of net primary productivity for scenarios of future climate change project decreases in pasture biomass in the forest-steppe and steppe zones of up to 37 and 20 per cent respectively by 2080 and increases in the high mountains and desert steppe of up to 28 and 52 per cent respectively (Tserendash et al, 2005; Batima, 2006). Carbon:nitrogen ratios are projected to decrease for all zones, indicating a decrease in nutritional quality of forage. Weight gain of livestock would be negatively affected in all but one zone. For example, weight gain by ewes would decrease by up to 58 per cent in the forest-steppe, 39 per cent in the steppe and 9 per cent in the high mountains, while the positive change in the dessert steppe would be less than 1 per cent.

With warming, winter temperatures will be milder in the future, which may bring some benefits to herders. The effect of climate change on the occurrence of *zud*, however, is not clear. With greater winter precipitation, more snow will fall, which can cause white *zud* events. Short periods of high temperatures followed by a return to sub-freezing temperatures can cause iron *zud*. Wind storms in the coldest months of December and January, which were rare in the past but have become more frequent in recent years, can bring cold *zud*. Despite the warming observed in past decades, the frequency of *zud* that covered more than 25 per cent of the land rose between the 1960s and 1990–2000 (Batima, 2006). Thus it cannot be assumed that *zud* will become less common with climate change, and increases in *zud* frequency are possible. As demonstrated by past events, the livestock sector and Mongolian society as a whole are highly vulnerable to *zud*. Continued increase in their frequency, particularly if in combination with summer droughts, would have detrimental impacts on the sector and on development in Mongolia.

The impacts of climate variability and climate change on the pastoral live-stock system of Mongolia are thus both near term and long term. In the near term, extreme events such as droughts and *zud* hold greater significance. Over the longer term, climate change will bring changes in temperatures, precipita-tion, snowfall and the duration of snow cover, as well as changes in the frequency of droughts and *zud*. These long-term trends in combination with socioeconomic changes in the country could potentially degrade pasturelands and significantly impact on human livelihoods.

Identifying and Screening Adaptation Options

Recognizing the near- and long-term risks due to climate variability and change, adaptation planning should take into account both the need to increase the current ability of pastoral communities to lessen and cope with the impacts of extremes as well as the need to conserve and improve the resilience of pastureland. The near-term productivity and longer-term resilience of the pastoral system of Mongolia depends on three primary components of the system: the stock and condition of natural resources, primarily pasture, which are strongly affected by climatic conditions; animals' biocapacity to cope with environmental stresses; and the human element that manages and depends on livestock and pasture lands. Our investigation of adaptation strategies, performed with the participation of stakeholders in Mongolia's pastoral system, focuses on measures to bolster these three components.

The identification and evaluation of options for adapting the Mongolian pastoral livestock system to climate change followed a two-step process. In the first step, a team of technical experts identified a large number of adaptation options, screened the options against a small number of broad criteria, and selected a subset for further evaluation with stakeholders. In the second step, workshops and consultations were held with three different groups of stakeholders.

Technical experts from our case study team prepared a preliminary list of 89 adaptation options. The options were drawn first from responses to a household survey that was carried out during 2002–2004. The survey involved more than 700 herders' households in 16 of Mongolia's 21 provinces, called *aimags*, eliciting information about their perceptions of major risks to their livelihoods and strategies used to cope with problems caused by climatic phenomena. Additional options were identified from the findings and recommendations of previous studies of climate change in Mongolia (see, for example, Batima et al, 2000) and expert judgements of the team.

The preliminary list of options was then screened to select options that warrant further consideration. A number of factors were judged to be important in the context of Mongolia's livestock sector for screening options. As in many developing countries, the people and government of Mongolia give priority to immediate and pressing domestic problems such as economic development, poverty, public health, education and environmental degradation. Consequently, emphasis is given to adaptation measures that are consistent with and therefore might be more easily integrated into existing policies, plans and programmes in these areas.

More specifically, the preliminary list of options was screened to identify those that satisfy three criteria: (1) Would the option advance climate change adaptation as well as existing objectives for development, poverty reduction, public health and education? (2) Is the option consistent with government policy, plans and programmes for agriculture? and (3) Would the option cause any adverse impacts on the environment? Options that were judged to satisfy at least two of these criteria were passed for further evaluation. This shortened the list of adaptations to be considered from 89 to 56.

The shortened list of adaptation options that passed the initial screening was evaluated by three different levels or groups of stakeholders in a series of workshops and consultations. These included workshops with local community stakeholders, with scientists, and with policy and decision makers from national ministries. Participants in the workshops applied six additional criteria to evaluate the potential of each option:

1   *Current adaptive capacity*: What is the existing capacity to implement the option successfully?
2   *Importance of climate as driver of outcomes*: How important is climate relative to other exogenous factors as a driver of the risk that is targeted by the adaptation option?
3   *Near-term effectiveness*: How effective is the option expected to be for reducing negative near-term impacts of drought, *zud* and other extremes that are important sources of current climatic risks?
4   *Long-term benefits*: Will the option produce long-term benefits for reducing vulnerability to climate change by, for example, improving the condition and sustainable use of pastureland?

5   *Cost*: What are the expected investment, operation and maintenance costs of the option?
6   *Barriers*: Are there significant technical, social, financial or institutional obstacles that could impede the implementation or performance of the option?

The workshop participants qualitatively rated the adaptation options for each of the six criteria, giving rankings of high (H), medium (M) and low (L). The aggregated results are shown in Table 11.1 for those adaptation measures that emerged as high priorities. The participation of local stakeholders, scientists, and policy and decision makers in the evaluation of adaptation measures is described below.

Workshops with local stakeholders

Much of the actual implementation of adaptation measures will be carried out at the level of the household or community and key factors for consideration include (1) local community and herders as the primary beneficiaries of successful adaptation, (2) the substantial indigenous knowledge and experience regarding pasture and animal management accumulated within local communities, and (3) the devolution of responsibility for the management of the livestock sector to local authorities, including the governors' offices of the *aimags*, or provinces, and *sums*, or subregions of the *aimags*. The incorporation of the perspectives and priorities of local actors in the implementation of climate change adaptation is considered critical.

Four local community workshops were held, one each in the eco-regions of the Gobi-steppe, steppe, high-mountain and forest-steppe. More than 200 participants attended the workshops, including local governors, herders, environmentalists, climatologists and animal experts such as veterinarians.

Almost all of the adaptation options prepared by the expert team were accepted as feasible by the local stakeholders, including the herders. An important exception are measures that would promote private ownership of pasture and water supply, which were rejected by more than 98 per cent of participants. Seasonal migration to access pasture, water and shelter are considered essential for optimal and sustainable management of Mongolia's pastoral system. Private land ownership, in the opinion of herders and other community stakeholders, is not conducive to maintaining the traditional or even current pastoral livestock system of Mongolia. Almost all the participants expressed the view that individualized, private ownership of pastureland, under Mongolian conditions, is likely to increase conflict and jeopardize environmental stability. But despite the opposition to private ownership of land generally, herders expressed openness to the possibility of private ownership limited to small pastures for emergency use near winter and spring camps.

Some of the major constraints to the implementation of adaptation measures identified in the local workshops include financial and material shortage, inadequate information, and remoteness from markets. A shortage of educated people in the rural areas to take care of rural education, grazing management,

Table 11.1 *Evaluation of adaptation options*

Objective	Adaptation measures	Current adaptive capacity	Importance of climate drivers	Near-term effectiveness	Long-term benefits	Cost	Barriers
Conserve natural resources	Improve grazing management	M	H	H	H	L	S& I
	Introduce cultivated pasture	L	M	H	H	H	F
	Improve pasture yield	L	M	M	H	H	T&F
	Improve pasture water supply	L	H	H	H	H	T&F
	Legislate possession of pasture	M	M	M	M	L	S& I
	Introduce taxation of pasture	L	M	M	M	L	S
	Livestock population control according to the pasture capacity	M	M	M	M	H	S
Strengthen animal bio-capacity	Improve shelter for animals	M	H	H	M	M	F
	Increased supplementary feed	M	H	H	H	M	R&F
	Improve per animal productivity	L	H	H	H	H	T&F
	Introduce genetic engineering	L	H	H	H	M	T&F
	Improve veterinary services	M	H	H	H	H	T&F
	Introduce high productive cross breeds	L	H	H	H	H	T&F
Enhance rural livelihoods	Promote collective communities	M	H	M	H	M	S&I
	Develop/transfer new technologies	M	H	H	H	H	T&F
	Expand access to credit and generate alternative income	L	H	M	H	H	I&F
	Expand the supply of renewable energy applications to herders	M	H	H	H	M	T&F
	Promote and support the establishment of different kinds of enterprises	L	M	M	H	H	T&F
	Establish insurance system for animals	L	M	M	M	H	T&F
	Establish risk fund	L	H	M	H	H	F&I
	Prepare educated herders	L	M	M	H	M	F
	Training of young herders	H	H	M	H	M	F
Increase food security and supply	Expand dairy and meat farms close to big cities to meet the demand of milk and other dairy products	M	H	H	H	H	T&F
	Promote and expand other food supply farms (e.g. eggs and vegetables)	M	H	H	H	H	T&F
Improve understanding	Establish climate change monitoring stations	H	H	H	H	H	T&F
	Improve forecasting system of extreme events	H	H	H	H	H	T&F

Note: H-high, M-medium, L-low, F-financial; T-technical, S-social, I-institutional, R-resource and social and economic issues was also raised as a serious concern. Participants of the local workshops emphasized the importance of education and training, the return of students after graduation to their home provinces, and financial and management support in the implementation of adaptation measures. Most important, the participants stressed the need for the integration of scientific and local indigenous knowledge to cope with climate variability and to move from study, assessment and discussion to the actual implementation of adaptation measures.

Consultations with scientists

Successful implementation of many adaptation options requires the support of and interaction with the scientific community to advance knowledge and understanding, develop new technologies and communicate information. Results from the local workshops were presented and discussed with leading scientists in the fields of animal husbandry and pasture management. The discussions focused on the following issues:

- What is the role of scientists in the implementation of adaptation measures?
- What should be done to introduce new varieties of drought-resistant pasture plants and to develop cultivated or irrigated pasture?
- What should be done to improve the productivity of animals?
- How should know-how on mechanical and automatic equipment or appliances to facilitate the manual labour of herders be transferred?
- What is the existing scientific, technical and financial capacity of the institutions to support implementation of adaptation measures?
- What should be done to improve the capacity of institutions and how should the barriers, if any, be overcome?

The discussions revealed serious concern among scientists about the problems of the livestock sector, awareness of the vulnerability of the sector to climatic stresses, and a willingness to cooperate in the development and implementation of adaptation measures. The scientists emphasized technologies for increasing livestock and pasture productivity, for example, with the development of new animal breeds and grass species through genetic engineering, and the use of experiments to test grassland adaptations for a warmer climate. Financial constraints and lack of collaboration among different institutions were identified as major obstacles for such work. The collaboration between local people and scientists established by our study team was acknowledged as a useful model and the creation of an institutional mechanism to continue this collaboration was suggested.

Consultations with policy and decision makers

The shortened list of adaptation options that passed the initial screening were evaluated by three different groups of stakeholders in a series of workshops and consultations. These included workshops with local community stakeholders and consultations with scientists, and with policy and decision makers from national ministries. The focus of the meetings with policy makers was to draw their attention to the impacts of climate change and urge their action in the implementation of adaptation measures and their integration in national planning.

Participants from the different ministries demonstrated good understanding of the current vulnerability of the livestock sector to climate extremes such as severe drought and *zud,* and also of the low technological advancement of the sector. Less well understood by the ministries were human caused climate change and the potential effects of climate change on the livestock sector. Knowledge of the design and implementation of adaptations to climate variability and climate change differs markedly across the ministries. Common to most is the perspective that adaptation measures can be synergistic with development policies in the livestock sector and that they should be implemented within that context. But the policy makers made no clear distinction between adaptation measures implemented at a project level and broader policy and institutional measures that could be enabling mechanisms for implementing adaptation across different sectors and different policy and decision-making contexts.

Priority Adaptation Measures

The adaptation measures that were discussed and evaluated in the stakeholder workshops can be classified into five main groups:

1 conserve natural resources;
2 strengthen animal biocapacity;
3 enhance capacities and livelihood opportunities of rural communities;
4 increase food security; and
5 improve understanding and forecasting of climate variations and extremes.

Mongolia's pastures are the key natural resource input to livestock production. Climate change threatens to reduce the production of forage grasses by this resource and may, in combination with heavy grazing pressures, degrade the land itself. Sustaining and improving Mongolia's pastures will be critical for future livestock production. A variety of options to conserve and improve pastures and pasture yields were identified.

Better management of grazing by limiting livestock numbers and returning to the traditional rotation of herds to different pasture each season would control grazing pressures, prevent degradation of pastures and help degraded pastures to recover. Degraded pastures can also be restored by reforestation and increasing vegetation cover. Drought-tolerant perennial species can be used to increase vegetation cover and increase pasture yields. Establishment of cultivated pasture would reduce dependency on nature and climate. A successful implementation of this measure would greatly reduce not only the expected impact of climate change but also the vulnerability to drought and harsh winters (*zud*). Expansion and rehabilitation of pasture water supply is also viewed as a promising measure for improved pasture utilization and stock survival, as well as ecosystem conservation and rural development.

Biocapacity is important for reducing the vulnerability of animals to drought and winter *zud*. Options identified for strengthening animal biocapacity include modifying the daily grazing schedule, increasing use of supplemental feeds, and increasing feed and pasture reserves. When temperatures get very high in the middle of the day in summer, livestock cease to graze and this can have a large impact on weight gain. Modification of the grazing schedule to put animals to pasture in early morning or late evening can partly compensate for the reduced grazing in the middle of the day due to high summer temperatures. But to fully compensate for the simulated effects of warming would require extending grazing at times other than the middle of the day by an impractical 6 hours per day by 2020 and even longer by 2050 and 2080. Thus modification of the grazing schedule needs to be considered in combination with other measures.

Supplemental feeds to increase daily feed intake could be given to livestock not only in winter but also in summer to fatten animals and increase their capacity to survive the winter. Simulation results show that 1.9–3.3kg/day of supplemental feed would be required per sheep in the summer of 2020 in order to compensate for the projected weight decrease. Feed reserves for emergency supply in case of drought or *zud* can be improved by increasing haymaking, sowing fodder crops, and increasing feed preparation and manufacture. Pasture can also be allocated as reserve not to be used except in the event of a harsh winter.

An important option for enhancing rural livelihoods is to promote traditional pastoral systems such as *khot ail* and *neg golynhon* that would revive traditional pasture management. The traditional way of life of Mongolian herders is in itself an effective indigenous measure to cope with climate extremes. Herders typically lived in groups of two or more households called *khot ail*, which served a variety of functions, including informal regulation of pasture use by member households and seasonal movement of herds. The *neg golynhon,* which means 'people from one river area,' is also a similar informal social network. Such community supervision was effective in protecting pasturelands and mitigating the impacts of droughts and *zud*. Demonstrations of community-based adaptation, in which the community decides on how best to share the limited common resource based on its indigenous knowledge about the livestock and the environment, hold promise as a potential adaptation measure for climate change.

A number of other options can complement collective organization of the pastoral system to enhance rural livelihoods. These include development and transfer of new technologies, education and training of herders, establishment of rural enterprises, access to credit for financing investments in land improvements, new technologies and enterprises, and establishment of insurance mechanisms for spreading risks. Extension services, education, training and other mechanisms are needed to enable herders to engage with experts and groups that can help find solutions and means to conserve natural pasture, improve livestock management, increase and improve the use of feed supplements, and increase the know-how of herders for managing and adapting to changing climate risks. Monitoring and research are needed to improve understanding of climate extremes and change and their consequences, and to develop forecasting and warning systems. Forecasts of drought and *zud* could be used to issue warnings and prompt actions by herders, local authorities and national ministries to prepare for and mitigate the impacts.

Barriers, Opportunities and Responsibilities

The implementation of adaptation measures to reduce climate change vulnerability in Mongolia faces many barriers. The legal framework for access to land and water treats these resources as common property and does not provide an effective means for regulating and managing their use. Lacking secure tenure to land and water, herders do not have appropriate incentives for conserving, improving and investing in pasturelands and water systems. Many of the identified adaptation options are therefore unlikely to be widely used by herders under the current framework. Privatizing ownership of pastureland and water would correct this failure, but would give rise to other problems and is widely and strongly opposed by herders, with the possible exception of small pastures for emergency use. Revival of a traditional collective system for regulation of access to pasture and water is attractive to many stakeholders but will require changes in legal and institutional frameworks to be successful.

The current institutional framework also poses barriers because responsibility for different sectors, resources, functions and programmes that are relevant to the livestock sector and the management of climate risks are divided among many different ministries and institutions. Successful implementation of a comprehensive adaptation strategy will require coordination and cooperation across these institutions. Such coordination does not occur at present. The participation of different ministries and institutions, as well as local stakeholders, in programmes such as AIACC and The Netherlands Climate Change Studies Assistance Program is helping to bring together the different stakeholders that are needed for successful adaptation. But more remains to be done.

Lack of modern technologies for animal husbandry, pasture improvement, cultivation, irrigation, processing, storing, packaging and transporting animal products results in low productivity of the sector, low incomes and inability to compete in the global market. The limitations include lack of infrastructure, machinery, improved seeds and animal breeds, and know-how. Rectifying these limitations is a priority for development of the livestock sector and would also enhance the capacity of the sector to adapt to climate change. Advances are being made but they are slow in coming and are sometimes reversed by events such as the droughts and *zud* of 1999–2002.

Many of the options have financial costs that are high relative to the financial means of most individual herders. Implementation at the level of the community, *sum, aimag* or national government allows for pooling of resources and may yield scale economies that reduce costs. But due to economic difficulties in Mongolia, financial resources are very limited at these levels, which would limit the implementation of adaptation measures.

A range of possible actors is likely to be involved in the implementation of the identified measures. Long-term concerns with respect to sustainable use of pasture resources are generally the responsibility of the national government, as the pasture is state owned. Hence the implementation of adaptation measures should begin at the level of national planning organizations. Planning for adaptation at the national or regional level is also justified when impacts and interventions cross local boundaries and/or economic and financial implications are beyond the capacity of local communities.

But most measures are to be implemented at the local level and by herders themselves. Thus success requires coordination between central and local levels of management to implement a combination of measures, allocate appropriate resources, provide support services, and create an institutional and legal framework that enables adaptation. Participation of national, provincial and local governments, scientists and herders is equally important in the implementation of any of the adaptation measures. Behavioural modifications with respect to pasture use and management, and livestock management among the herder communities could be a key factor in safeguarding the natural resource base.

Implementation of most of the identified adaptation measures requires substantial investment. At the moment Mongolia faces many other socioeconomic problems and financial constraints. Therefore, it is important that, at the national planning level, the available funds are more clearly prioritized and allocated based on the objectives of the selected adaptation strategies.

There are also many regional and global financing programmes related to the implementation of the United Nations Framework Convention on Climate Change (UNFCCC) goals. The Global Environment Facility (GEF) provides financial support to cover the incremental cost in developing countries and in countries with economies in transition of protecting and managing the global environment, including dealing with climate change impacts. Therefore, adaptation projects could be developed with GEF funding through its implementing agencies like the United Nations Development Programme, United Nations Environment Programme and the World Bank. Mongolia could also participate in regional, sub-regional and bilateral cooperation activities, and initiatives on climate change-related issues, so that it can gather more experience and knowledge on adaptation to its impacts.

Conclusions

The Mongolian livestock sector engages almost half the population, is a major contributor to national income and export earnings, and is highly vulnerable to climate change impacts. Observed increases in the incidence and impacts of severe weather events like drought and *zud*, and the potential for further increases as a result of climate change, are a particular concern for Mongolia.

Adaptation to reduce the impacts is vital for ensuring livelihood security and for promoting social and economic development.

Herders, technical and scientific experts, and representatives from local, provincial and national authorities give priority to adaptation strategies that would generate near-term benefits by improving capabilities for managing the extremes of drought and *zud*, and long-term benefits by improving and sustaining pasture yields. Specific measures identified as advancing these broad goals and warranting further consideration include:

1   improving pastures by reviving the traditional system of seasonal move-ment of herds, restoring degraded pasture, expanding and rehabilitating water supply, and developing cultivated pasture;
2   strengthening animal biocapacity by modifying grazing schedules, increas-ing use of supplemental feeds, and increasing feed and pasture reserves;
3   enhancing rural livelihoods by promoting traditional pastoral networks and herders' communities to regulate access and use of pasture and water, developing and transferring new technologies, educating and training herders, establishing rural enterprises, and providing access to credit and insurance;
4   improving food security by improving and diversifying food production and distribution systems; and
5   research and monitoring to develop and improve forecasting and warning systems.

Administrative decisions and actions are necessary to remove or ease barriers to the implementation of many of these adaptation measures. This includes revising the legal and institutional frameworks to provide incentives for conserving, improving and investing in pastureland and water systems, and promoting the organization of the herding population into local associations for managing resources and reviving practices of the traditional pastoral system. A sensible first step to initiate the process of adapting to longer-term climate change would be to facilitate existing adaptation strategies used by the herders to deal with climate variability and extreme events.

References

Batima, P. (2006) 'Climate change vulnerability and adaptation in the livestock sector of Mongolia', final report, Project AS06, Assessments of Impacts and Adaptations to Climate Change. International START Secretariat, Washington, DC, www.aiac-cproject.org

Batima, P., B. Bolortsetseg, R. Mijiddorj, D. Tumerbaatar and V. Ulziisaihan (2000) 'Impact on natural resources base', in *Climate Change and Its Impacts in Mongolia*, NAMHEM and JEMR publishing, Ulaanbaatar

Batima, P., L. Natsagdorj, P. Gombluudev and B. Erdenetsetseg (2005) 'Observed climate change in Mongolia', AIACC Working Paper No 12, International START Secretariat, Washington, DC, www.aiaccproject.org

Batima, P., B. Bat, S. Tserendash, D. Tserendorj, S. Shiirev-Adya, N. Togtokh, L.

Natsagdorj and T. Chuluun (2005) 'Adaptation to climate change', in P. Batima and D. Tserendorj (eds) *Climate Change Impacts on Extremes – Vulnerability and Adaptation Assessment for Grassland Ecosystem and Livestock Sector in Mongolia*, ADMON, Ulaanbaatar (in Mongolian)

Batima, P., L. Natsagdorj, N. Batnasan and M. Erdenetuya (2008) 'Vulnerability of Mongolia's pastoralists to climate extremes and changes', in N. Leary, C. Conde, J. Kulkarni, A. Nyong and J. Pulhin (eds) *Climate Change and Vulnerability*, Earthscan, London

Bayarbaatar, L., G. Tvuaansuren and D. Tserendorj (2005) 'Climate change and livestock', in P. Batima and B. Bayasgalan (eds) *Climate Change Impacts on Extremes – Vulnerability and Adaptation Assessment for Grassland Ecosystem and Livestock Sector in Mongolia: Impacts of Climate Change*, ADMON, Ulaanbaatar (in Mongolian)

Chuluun, T. and A. Enkh-Amgalan (2003) 'Tragedy of commons during transition to market economy and alternative future for the Mongolian rangelands', *African Journal of Range and Forage Science*, vol 20, no 2, p115

Mongolian Statistical Yearbook (2004) *Mongolian Statistical Yearbook 2003*, National Statistical Office, Ulaanbaatar, Mongolia

Natsagdorj, L. and G. Sanjid (2005) 'Climate change and drought', in P. Batima (ed) *Climate Change Impacts on Extremes – Vulnerability and Adaptation Assessment for Grassland Ecosystem and Livestock Sector in Mongolia: Livestock Sector Vulnerability to Climate Change*, ADMON, Ulaanbaatar (in Mongolian)

Natsagdorj, L., P. Gomboluudev and P. Batima (2005) 'Future climate change', in P. Batima and B. Myagmarjay (eds) *Climate Change Impacts on Extremes – Vulnerability and Adaptation Assessment for Grassland Ecosystem and Livestock Sector in Mongolia: Current and Future Climate Change*, ADMON, Ulaanbaatar (in Mongolian)

Tserendash, S., B. Bolortsetseg, P. Batima, G. Sanjid, N. Erdenetuya, T. Ganbaatar and N. Manibasar (2005) 'Climate change and pasture', in P. Batima and B. Bayasgalan (eds) *Climate Change Impacts on Extremes – Vulnerability and Adaptation Assessment for Grassland Ecosystem and Livestock Sector in Mongolia: Impacts of Climate Change*, ADMON, Ulaanbaatar (in Mongolian)

12

Evaluation of Adaptation Options for the Heihe River Basin of China

Yongyuan Yin, Zhongming Xu and Aihua Long

Introduction

The Heihe basin, located in northwestern China, is a region of predominantly arid and semiarid climate with extremely fragile ecological systems, few financial resources, poor infrastructure, low levels of education, and restricted access to technology and markets. The region suffers from climate variations and may experience more severe impacts of climate change on water resources, food production and ecosystem health in the future. Moreover, the region's adaptive capacity is much less than that in the coastal region of China. People in the Heihe region are facing substantial and multiple stresses, including rapidly growing demands for food and water, poverty, land degradation and other issues that may be amplified by climate change.

In the Heihe region, various water use policies and measures have been implemented or designed to limit or prohibit the utilization of water by different sectors or regions. Controversies have occurred, of course, as these policies are redistributive in nature, making some sectors or regions worse off and others better off. It is this redistributive nature of policies that often aggravates water use conflicts.

Measures for improving the adaptation of the water system to climate variability were introduced recently to the region, including water supply control, water permits, water right certificates, farmer water use associations, water pricing adjustment, and a better water allocation policy. What seems to be missing, however, is an overarching strategy that brings the climate change concern into water use decision-making process. For the most part, the impact of climate change on the water system has received scant attention from government agencies and others responsible for water resource management and planning. A partial explanation for the limited response to take consideration of climate change in water use management might be the lack of knowledge or awareness of the issue by policy makers and the general public.

Because of the limited knowledge and awareness, we undertook a study of the Heihe basin to improve understanding of the potential effects of climate change on water resources, the vulnerability of water users and options for adapting to the changes. Findings from the study about vulnerability to climate change are described in Yin et al (2008). Here we focus on our examination of adaptation, which applies multi-criteria evaluation methods. Our purpose is to provide decision makers with the information needed to improve the capacity of the water resource system for coping with and adapting to climate change in the Heihe region.

Current Status of Adaptation Science and Evaluation Tools

Research on developing well-designed adaptation strategies and options can provide the information and understanding necessary for establishing efficient adaptation options or policies to deal with climate vulnerability. In general, there are two broad approaches used for adaptation assessment. The first, developed by the IPCC (Carther et al, 1994) and applied within the context of climate change impact assessment, evaluates the effects of adaptation to lessen the impacts of climate change. The second is a policy analysis approach that seeks to understand the processes by which adaptation occurs, the multiple objectives of adaptation, the factors and conditions that enable or impede adaptation, the resources needed, and the consequences. Policy analysis of climate change adaptation aims to support decisions to improve the adaptability, resilience and sustainability of various systems with respect to climate change.

A wide range of tools has been used in the assessment of adaptation across and within different natural resource and socioeconomic sectors. The United Nations Framework Convention on Climate Change (UNFCCC) has collected and compiled information about methods and tools for evaluating climate change adaptation from member governments and other organizations (see Stratus Consulting, 1999). Their compendium provides basic information about adaptation evaluation tools, as well as related and complementary tools for constructing climate change scenarios and assessing climate change impacts and vulnerability. While the compendium is useful as a reference document, it is not a manual that describes how to implement each tool. It is rather a survey of possible tools that can be applied to a broad spectrum of situations.

We review below the main groups of tools included in the UNFCCC compendium that are relevant for policy analysis of adaptation. The compendium includes relatively few examples of such tools, indicating that there is a need for new research approaches and tools that can be used specifically in adaptation option evaluation, selection and decision making. Adaptation tools need to be able to evaluate alternative options or policies, a capacity, which many impact assessment methods lack. Appropriate decision tools exist, having been developed originally for policy evaluation in contexts other than climate change decision making. There are numerous applications of these tools in disciplines such as decision theory, management science, resource management and systems engineering. Introducing these tools into climate adaptation study can aid in developing more effective climate change adaptation strategies or policies.

Adaptation option evaluation tools in relation to impact assessment

As noted previously, the science of adaptation usually applies two approaches in evaluation of adaptations: impact assessment and policy analysis. Carter et al (1994) provided guidelines for climate change impact assessment in which impacts are estimated for cases in which agents are assumed to either adapt or not adapt. A comparison of cases with and without adaptation gives a measure of the potential effectiveness of the adaptation options considered, usually short-term or autonomous measures. A variety of assessment methods are presented by Carter et al (1994), but they are rather general and the guidelines do not offer specific recommendations or prescriptions.

Two concurrent research programmes, the US Country Studies Program and a parallel programme of the United Nations Environment Programme (UNEP), were developed to fill the gap. The former programme assisted more than 50 developing and transition economy countries to develop capacity for assessing vulnerability to climate change and evaluating adaptation options; it produced a guidebook of specific methods and approaches for impact assessment and adaptation evaluation (Benioff et al, 1996). The UNEP programme also sponsored the writing of a handbook on assessing climate change impacts and adaptation, to serve as a supporting resource for its own country studies programme. The handbook presents an overview of different methodologies and covers several sectors (Feenstra et al, 1998).

While these guidebooks and the approaches that they recommend have provided a useful framework for climate change research and assessment, they focus mainly on the impacts of climate change and give relatively little attention to the evaluation of adaptation options. It has been a common experience in applying these research approaches in climate change impact studies at country, regional and sectoral levels that the overwhelming part of the time, effort and resources were devoted to the selection and application of climate scenarios and impact assessments. It has been invariably noted that insufficient time and effort were left to develop the adaptation component.

There are some shortcomings associated with these scenario-driven approaches from the point of view of the need to improve our understanding and evaluation of adaptation in a policy analysis context. The problem is more fundamental than simply a matter of available project time and financial resources. There are many important reasons why a range of applications of the scenario-based approaches have not yielded useful results for the purposes of adaptation option evaluation and policy analysis (Lim et al, 2005).

The adaptation evaluation tools for policy analysis

The UNFCCC compendium (Stratus Consulting Inc, 1999) introduced a range of general decision tools that are applicable in evaluating adaptation policies in multiple sectors. The compendium groups these tools into three broad categories: initial survey, economic analysis and general modelling. The initial survey tools include expert judgement, screening of adaptation options and the adaptation decision matrix, which are useful for identifying potential adaptation strategies or narrowing down the list of appropriate options. They are relatively informal, inexpensive, and utilize qualitative judgement rather than quantitative data.

General tools of economic analysis that can be applied to policy evaluation of climate change adaptation include financial analysis, cost–benefit analysis, cost-effectiveness analysis and risk–benefit/uncertainty analysis. These are typically used to determine which options are most economically efficient and to assist the user in deciding which adaptation option is the most appropriate once a final list of adaptation options has been compiled. General modelling tools include TEAM and CC:TRAIN. These address different adaptation strategies across a number of sectors and are used to evaluate several sectors of concern in a particular region. More detailed information on these methods is available from Stratus Consulting Inc (1999).

To deal with the multi-criteria and multi-stakeholder nature of the adaptation evaluation process, multi-criteria evaluation technologies or tools can be adopted as effective evaluation instruments by which alternative adaptation policies can be compared and evaluated in an orderly and systematic manner. Given a set of adaptation policies available to deal with climate vulnerabilities or impacts on biophysical and socioeconomic aspects of our society, the evaluation tools can be used to help identify the policies that best satisfy selected decision criteria. A range of methods/tools developed in decision theory, multi-criteria evaluation and systems analysis can be adopted for adaptation option evaluation (Zadeh, 1965; Holling, 1978; Yin and Xu, 1991; Yin et al, 2000; Yin, 2001a).

An Integrated Assessment of Adaptation Evaluation in the Heihe Basin

The IPCC (2001) suggested a list of high priorities for narrowing gaps in vulnerability and adaptation research. Among these is integrating scientific information on impacts, vulnerability and adaptation in decision-making processes, risk management and sustainable development initiatives. In this section, an integrated approach is introduced as an example for discussion. The approach which was applied in the Heihe river basin case study, bringing together partners from the private sector, the public sector policy community and the academic research community, demonstrated how to meet the challenge of linking climate change adaptation and sustainable development. The analytic hierarchical process (AHP), a method of multi-criteria analysis, was applied in the case study to identify desirable adaptation options in dealing with climate change vulnerabilities.

Figure 12.1 illustrates the integrated assessment framework for adaptation option evaluation, linking impact assessment with local sustainability indicators, and using tools such as multi-criteria policy analysis and multi-stakeholder consultation in the Heihe region. In the following discussion, not all the components shown in Figure 12.1 are covered in the same detail. Rather, focus is on the two main concerns of this chapter: sustainability indicators/goals identification and adaptation policy evaluation. The case study presents the importance of indicator/goal setting in regional sustainability research and the approach to identify indicators/goals. The adaptation option evaluation system shown in Figure 12.1 represents a participatory approach to integrated assessment. Working in partnership with multiple stakeholders, alternative adaptation measures and water sustainability indicators are selected. Based on water vulnerability information provided by researchers of the project (Yin et al, 2008), the evaluation system identifies practical adaptation options to deal effectively with water vulnerabilities likely to become more severe in the Heihe river basin due to the impacts of climate change.

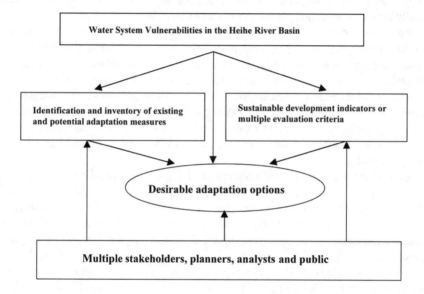

Figure 12.1 *Framework for multi-criteria evaluation of climate change adaptation options*

The first step in implementing the framework is to identify existing and possible adaptation options to deal with vulnerabilities of climate variation and change. This was done through multi-stakeholder consultations and using multi-criteria evaluation techniques. Numerous potential adaptation options are available for dealing with vulnerabilities to climate change. An initial screening process was conducted to reduce the number of options for further detailed evaluation. The multi-stakeholder consultation yielded a collective recommendation of adaptation options for further multi-criteria evaluation.

To link climate change impact analysis, adaptation option evaluation and sustainability evaluation, water system sustainability indicators must be set and the performance of adaptation options must be measured in a manner that integrates social, environmental and economic parameters that may be influenced by climate. In this study, indicators are evaluation criteria or standards by which the efficiency of alternative adaptation options can be measured.

There are several general frameworks that can be adopted for developing sustainability indicators. The first is a domain-based framework that groups indicators into three main dimensions of sustainability – economic, environmental and social. The three-dimensional nature of sustainability and the need to make trade-offs (for example, between economic growth and environmental quality) require maintaining these three components in a dynamic balance. Sustainability indicators thus should include economic, social and environmental information in an integrated manner.

Another important framework is a goal-based indicator system. Goals usually reflect the major development concerns of a nation or a region. For example, some concerns represent national or regional objectives of economic viability, maintenance of the resource base, and minimizing the impacts of climate change on natural ecosystems. Each goal is composed of a number of attributes or indicators which are measurable in most cases using existing sources of information. Other types of indicator frameworks include sector-based, issue-based, cause–effect and combination frameworks. No single indicator would be sufficient to determine sustainability or non-sustainability of a region or a system: a set of goals and/or indicators is required in sustainability evaluation.

Indicators are not a new concept and have been used to measure the performance of regional development policies or plans, to identify growth trends, to monitor the social and economic conditions of regions or nations, to inform the general public, to define planning goals or objectives, to guide strategic development options, and to compare different regions. For example, gross domestic product (GDP), housing price indices, unemployment rate and stock indices are commonly used to measure the social or economic performance of a society or economy. While these indicators strongly influence decision making by governments, other policy makers and the general public, they have shortcomings when used for measuring sustainable development. Recently, research has been initiated into developing indicators of societal sustainability.

Indicators may be conflicting in that the achievement of one target precludes the achievement of another. Possible trade-offs between indicators therefore need to be identified. Very often the trade-off relations are non-linear, creating situations of dramatic changes in the attainment of certain indicator levels once a threshold has been surpassed. Other indicators, however, are complementary – in other words by increasing the attainment of one indictor target it is possible to increase the attainment of other indicators. It has been suggested, for example, that development and environment are complementary up to some level of resource use. Indicators are also considered compatible when the attainment of one does not compromise the attainment of others.

Based on information gathered from stakeholders through householder surveys and consultation meetings, four indicators for evaluating adaptation options to reduce water system vulnerability were selected for our study of the Heihe basin (Table 12.1). The selected indicators corresponded to four broad water use sustainability goals. A multi-criteria options evaluation approach (MCOE) using these four indicators was implemented through stakeholder meetings and surveys to identify desirable adaptation measures by which decision makers can alleviate vulnerabilities and take advantage of positive impacts associated with climate change. The survey was mainly carried out in one-on-one interviews and in small group/workshop settings with a wide range of experts and stakeholders.

Table 12.1 *Indicators used to evaluate adaptation options in the Heihe river basin*

Sustainability goal	Indicator
Maximize water use efficiency	Reduce per unit production water use
Maximize economic return to society	Increase economic return per unit water
Minimize harm to natural environment	Ecological and environmental benefit
Minimize economic costs to society	Costs of adaptation options

Workshops were held to present climate impacts and vulnerability information to policy makers and stakeholder representatives for their review and comments. In the evaluation process, alternative options are evaluated by relating their various impacts to a number of relevant indicators. The results of impacts generated in various impact assessments are used as references for ranking the performance of each adaptation option against each sustainability indicator. Climate impact assessments of this project are discussed in detail in the project final report (Yin, 2006).

When multiple criteria are relevant to the evaluation of options, indicators associated with the different criteria must be combined if options are to be ranked. We use the analytical hierarchy process (AHP) method to combine our four indicators and rank adaptation options. The AHP method has been employed to evaluate alternative policies, allocate resources, conduct sensitivity analysis for resource-use planning, and select desirable project locations for both developed and developing countries (Saaty, 1980 and 1982). When applied to adaptation option evaluation, the AHP method requires decision makers to provide judgements on the relative importance of each option in relation to each criterion. In this exercise, a decision maker compares options two at a time (pairwise comparison). Then decision makers specify their judgements about the relative importance of each option in terms of its contribution to the achievement of the overall goal, in our case to alleviate the adverse consequence of climate change (Yin and Cohen, 1994; Yin, 2001b). The result of the AHP is a prioritized ranking indicating the overall preference for each of the adaptation options.

Applications and Results

The study area

The Heihe river basin covers an area of approximately 128,000 square kilometres and is located in a region with latitude of 35.4–43.5°N and longitude of 96.45–102.8°E. Figure 12.2 shows the location of the study region. The study area is the second largest inland river basin in the arid region of north-western China. The basin includes parts of two provinces (Qinghai and Gansu) and the Inner Mongolia Autonomous Region. The region is composed of diverse ecosystems including mountain, oasis, forest, grassland and desert. The River Heihe flows from a headwater in the Qilian Mountain area to an alluvial plain with oasis agriculture, and then enters deserts in Inner Mongolia, representing the upper, middle and lower reaches of the basin.

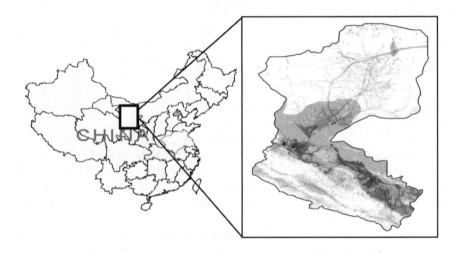

Figure 12.2 *Location of the Heihe river basin*

There is an increasing concern about water use conflicts in the Heihe region. The limited water supplies have to provide a number of economic sectors and communities in different jurisdictions with a range of different and often conflicting functions to meet the various demands. While the demands for resources increase as populations and economies grow, the availability and the inherent functions of water resources are being reduced by water pollution, environmental degradation and climate change. Competition over access to water resources in the Heihe region has led to disputes, confrontation and in many cases violent clashes. The growing water use conflicts have posed a big challenge for government agencies to implement effective water allocation policies.

Climate change trends in northwest China in the past 50 years have been investigated by analysing temperature and rainfall from 1951 to 2004 (Wang, 2005). The results of the analysis show a significant rise in daily mean temperature with linear trends in most areas ranging from +0.2°C to +0.4°C per decade. The Qilian Mountain glaciers are already undergoing rapid retreat, at a rate of about one metre per year. The region depends on the glaciers as important natural reservoirs for water supply. The water supply mainly comes from the spring melting of glaciers (Yin, 2006 and Yin et al, 2008).

One important water vulnerability indicator is the water withdrawal ratio, defined as the ratio of average annual water withdrawal to water availability. The WMO (1997) suggests that water withdrawal ratios that exceed 20 and 40 per cent be considered as indicators of medium and high water stress respectively. In northern China, however, where water withdrawal ratios typically exceed these thresholds, the government suggests that a 60 per cent threshold is a more practical indicator of high water stress given the severe water shortage situation in the area.

Table 12.2 lists water withdrawal ratios in the Heihe region under current climate conditions for the period 1990–2000. The current water withdrawal ratios in the region are extremely high (80–120 per cent), far exceeding the critical threshold levels set by both the WMO and the Chinese government. Conflicts over water use, including violent fighting for water, have been increasing in the basin over the past decade. The trend of this social indicator suggests that water shortage in the growing season is becoming more and more serious because of decreased water supply and increasing population and per capita water use (Yin, 2006).

Table 12.2 *Water availability, water withdrawals and water withdrawal ratio in the Heihe river region, 1991–2000*

Year	1991	1992	1993	1994	1995	1996	1997	1998	1999	2000
Water availability ($10^8 m^3$)	35.9	34.3	35.5	356	34.0	34.6	34.6	34.3	34.7	34.8
Total water withdrawal ($10^8 m^3$)	29.0	27.4	35.4	28.8	29.6	35.8	28.0	41.4	35.5	32.3
Water withdrawal ratio	81%	80%	100%	81%	87%	103%	81%	120%	102%	93%

Water stress in the region might intensify in the future because of growing water withdrawals related to population and economic growth, and decreasing water availability related to climate change. The National Climate Center of the Institute of Atmospheric Physics has developed climate change projections for western China for the 21st century using outputs from eight coupled global atmospheric and oceanic circulation models. These projections were calculated using the NCC/IAP T63 (National Climate Center/Institute of Atmospheric

Physics) model (Xu et al, 2003). Ding et al (2005) applied a regional climate model (Ncc/RegCM2) nested with a coupled GCM (NCCT63L16/T63L30) and Hadley Centre model (HadCM2) for climate change studies. Outputs of the Chinese regional-scale climate model were used to design scenarios of climate change for our study (Li and Ding, 2004). The climate change scenarios up to 2040 were combined with scenarios of socioeconomic change to calculate future water supply and demand for the study region. Results of projected water shortage under climate change and socioeconomic change show that water shortage in almost all the municipalities of the Heihe river basin will be worse than at present (Table 12.3).

Table 12.3 *Water shortage/surplus in Heihe river basin under climate change to 2040 ($10^8 m^3$)*

Municipality or county	Suzhou	Jinta	Jiayuguan	Shandan	Minle	Sunan	Ganzhou	Linze	Gaotai
Year 2000	−0.20	−0.07	−0.36	−0.12	−0.08	0.00	−0.80	−0.62	−0.77
Year 2010	−0.67	−0.41	−0.23	−0.27	−0.61	−0.01	−2.39	−1.99	−1.10
Year 2020	−1.26	−0.76	−0.18	−0.09	−0.24	0.04	−1.32	−1.33	−0.61
Year 2030	−1.01	−0.62	−0.05	0.09	0.15	0.09	−0.01	−0.52	−0.01
Year 2040	−0.19	−0.12	0.16	0.19	0.39	0.12	0.80	−0.26	0.29
Average	−0.67	−0.39	−0.13	−0.04	−0.08	0.05	−0.74	−0.94	−0.44

Note: Negative numbers indicate water demand exceeds water supply; positive numbers indicate water supply exceeds demand.

Most policy makers and communities across the Heihe river basin have very limited knowledge about the current adverse effects and impacts associated with climate change in water resource management and planning. It is also uncertain whether the region's water infrastructure and measures have the capacity to respond quickly and effectively to future climate change. Effects of climate change on water shortage may be so significant that a comprehensive adaptive action or strategy is required, involving the participation and coordination of national, provincial and local authorities, and other stakeholders engaged in water resource planning and management.

Water system adaptation options for evaluation

Numerous potential measures or options are available to alleviate negative consequences of extreme climate events or climate change for the water system. In general, the different options can be grouped into two categories: engineering and non-engineering measures. The former involve construction works that attempt to supply more water resources to various users. These structural measures include reservoirs, irrigation systems and wells. Options in the latter category do not involve construction and include demand management, water use policy, pricing, water trade and permits, and other institutional and governance measures.

In the Heihe region, both engineering and non-engineering measures have been implemented or designed to address the regions problems with water shortage and water stress. Controversies have occurred, of course, as a result of such policies. In 2001 the State Council of China enacted a compulsory water division regulation to limit water withdrawals by Zhangye City on the middle reach of the River Heihe, requiring the city to provide 0.95 billion m³ of water over a period of three years for the lower reach region in the Inner Mongolia Autonomous Region. The purpose of this government policy was to protect the extremely fragile ecological conditions in the lower reach, which had become a major source of sandstorm hazards affecting the Beijing region of China. While the policy helped improve the ecosystem condition in the lower reach, 40,000ha of cropland annually in Zhangye City could not obtain water for irrigation. As a result, farmers suffered considerable crop losses. Production from about 160,000ha of cropland was reduced due to water shortage, and some of the area produced no harvest at all. Farmers in the upper and middle reaches of the river argue that the policy's reduction of water for irrigation has led to detrimental consequences in the agricultural sector, while others have indicated that the new policy has enabled the revival of dried lakes located in the downstream region. Obviously, water policies or regulations may make some sectors or regions worse off and others better off because of their re-distributive nature. It is this re-distributive nature of policies that often aggravates water use conflicts.

There have also been a growing number of local and provincial initiatives and programmes to address various aspects of the water shortage problem. For example, farm water user associations were established to engage farmers in water use decision-making processes. Each association elects a chairperson based on a 'one family-one vote' basis. The association appoints an irrigation inspector to measure the volume of water inflow to the association, together with water managers from the local water resource bureau, and then to allocate water to each individual farm within the association. The municipal water resource bureau allocates water quotas to the farm water user associations. The chairperson organizes the 'collective activities' of constructing and maintaining on-farm irrigation systems and water tanks, protecting irrigation facilities, purchasing irrigation water and storing extra deliveries, catching return flow, and harvesting rainwater (Li, 2006).

Meanwhile, the government has established policies for water recycling, pollution control and water-efficient technology to improve water use efficiencies in industries and farming. For example, the Ministry of Water Resources introduced the first pilot project to incorporate practices of water conservation in the Heihe River region communities in 2001. Some adaptation measures were also introduced to the region, including overall water supply control, water permits, water right certificates, water pricing adjustment and better water allocation policy.

Effective water pricing mechanisms have been implemented to encourage water saving. Higher water prices reduce water demand and increase water supply. The Provincial Finance and Pricing Control Bureau determines the price per unit of water for each alternative water use, including irrigation, industry, domestic and hydropower. In Zhangye City, the water tariff per unit of water increases step-wise from one block to the next, depending on the amount of water used relative to a water quota. The municipal Water Resource Bureau sets a water quota to supply water users with their basic water needs at a price that is affordable. When water use exceeds the specified water use quota, water prices then shift to higher levels. The water price increases 50 per cent when water use exceeds 1–25 per cent of the quota and increases 100 per cent when water use exceeds 26–50 per cent of the quota. In addition, the water price is not constant in different seasons. The price increases during the dry season and decreases during peak rainy season. Preliminary effects of the pilot projects have shown some positive effects in dealing with the water shortage problem. The water price reform pilot projects in the region have provided water consumers with an incentive to reduce their water consumption (Liu et al, 2006).

Numerous potential measures or options are available to alleviate negative consequences associated with climate change. Based on government documents and existing literature on water resource management, the project researchers prepared a list of existing and potential options (Zhangye Government, 1998; Jia et al, 2004; Liu et al, 2006; Ministry of Water Resources of China, 2006; China Water Saving Irrigation Net, 2007). A primary screening process was conducted by the research team to select a limited number of adaptation options for further evaluation using the multi-criteria evaluation process. Eight options were selected for evaluation:

1   reform the economic structure to promote activities that are less water consuming (for example, promote drought-tolerant crops in place of high water consuming crops);
2   establish a water use permit and trading system;
3   construct water works (for example, irrigation systems);
4   establish water users' associations to engage farmers in water use decision making;
5   adopt advanced water use technologies (for example, water saving irrigation methods);
6   government pricing of water to control water demand;
7   involve multiple stakeholders in improving water allocation policies; and
8   increase awareness and education about water conservation.

The above list of adaptation options was evaluated using the AHP method to produce a ranking of options that reflects stakeholders' expectations of their effectiveness for reducing vulnerability as measured by four criteria: economic efficiency, environmental quality, equality and feasibility. The AHP application was facilitated by a series of workshops with participation of a broad range of stakeholders and policy makers from water sectors to identify sustainability indicator priorities, as well as a series of desirable adaptation policies. These policy workshops, which were carried out in Zhangye City, provided forums for the research team to present research findings to local officials, experts and policy makers, and for stakeholders to review the study results and provide comments on the findings and their relevance to government decision making in water resource management. During the policy workshops, participants completed evaluation forms for the set of eight adaptation options.

Survey questions on the evaluation forms were designed according to the principles of AHP so that the responses could be input into a software program, Expert Choice 2000, for compilation and analysis. The software computes an overall score for each alternative option by distributing the importance of the indicators among the adaptation options, thereby dividing each indicator's priority into proportions relative to the percentage rankings of alternative options. Using the four indicators listed in Table 12.1 and the above set of adaptation options, a decision hierarchy model was created. This decision hierarchy is quite simple because it includes a single overall goal, to reduce vulnerability, with two levels below it in the hierarchy: a set of criteria or indicators and a list of alternative adaptation options. Once the relative importance of individual criteria is determined, decision makers need only think about their preference for each alternative adaptation option in terms of achieving a single criterion.

The survey was designed as a series of tables. Respondents were given a pair of indicators or a pair of options, and asked to compare them using a numerical sliding scale. The comparison scale ranged from 1 to 5, with 1 representing options that are equally effective (or indicators that are equally important) and 5 representing options where one is very strongly more effective than another (see Table 12.4). The purpose of providing these comparison tables is to facilitate the AHP pairwise comparison process. Respondents select the relative effectiveness numeral based on their preferences and having been given certain impact information about the adaptation options. In this exercise, a stakeholder compares two options, given in the far left and far right columns of the table, against each criterion at a time.

In the example given in Table 12.4, the respondent considers the option 'reform economic structure' to be equally effective as 'adopt advanced water use technologies' (row 5: numeral 1). Comparing 'reform economic structure' with 'water use permit and water trade system' in row 3, the respondent's cross indicates a view that the second option is strongly more effective.

Results of the AHP analysis

Reform of the economic structure was ranked as the most desirable adaptation option for the Heihe region, with establishment of farm water users' associations ranked fairly high as well (see Table 12.5). The moderate performance levels for the options 'improve water allocation policies', 'establish water permits and trading' and 'increase awareness and education' are probably due to the fact that these are relatively new measures in water resource management in the study region. The scores for the options 'adopt advanced water use technologies' and 'implement water price system to control demand' are near the bottom of the list for most participants, especially from an economic

Table 12.4 *Example of AHP comparison table*

Adaptation option	Relative effectiveness scale									Adaptation option
	5	4	3	2	1	2	3	4	5	
Reform economic structure							X			Water use permit and water trade system
Reform economic structure		X								Construct water work
Reform economic structure						X				Establish farm water users' society or committee
Reform economic structure				X						Adopt advanced water use technologies
Reform economic structure			X							Government sets water price
Reform economic structure						X				Improve water allocation policies
Reform economic structure			X							Increase water saving awareness and education

Note: Relative effectiveness scale: 1 – equally effective; 2 – marginally more effective; 3 – moderately more effective; 4 – strongly more effective; 5 – very strongly more effective.

perspective, and are not considered to be desirable adaptation options. It appears that regional stakeholders consider the two options to be expensive alternatives for dealing with watershed management and farmers do not want to pay higher water prices. Constructing water works is judged to be the most inefficient option from an economic perspective and it is ranked at the bottom overall by regional respondents.

The above information is useful for government decision making in selecting efficient options, especially when considering the goal of Heihe river basin sustainability. The results suggest that institutional options (reform economic

Table 12.5 *Overall rank and score of adaptation options for the Heihe region*

Water resource adaptation option	AHP result	Rank order
Reform economic structure	0.26	1
Form farm water user society	0.18	2
Improve water allocation policies	0.14	3
Establish water permits and trade	0.13	4
Increase awareness and education	0.12	5
Apply water saving equipment and technology	0.08	6
Implement water price system	0.05	7
Construct water works	0.04	8

structure, water user associations and water permits) are considered by a wide range of stakeholders to be effective and desirable with respect to the four evaluation indicators. Implementation of these water-use adaptation options in the Heihe river basin might be able to reduce water vulnerabilities associated with climate change in the context of regional sustainability. These adaptation options can be incorporated into a comprehensive sustainable development plan for the basin.

Conclusions

Working in partnership with local, provincial and national governments, and other key stakeholders such as water use professionals, farmers and other organizations, the study has demonstrated the use of a multi-criteria evaluation approach to analyse adaptation options for the Heihe basin. The application has identified adaptation measures considered by stakeholders to be potentially effective for reducing vulnerability of water users by increasing economic efficiency of water use, improving environmental quality, promoting equality and reducing water costs. The highest ranked options, which emphasized institutional approaches, could become practical options to deal with water vulnerabilities likely to become more severe in the Heihe region due to the impacts of climate change as well as pressures from population and economic growth. Through public consultation activities, stakeholders' understanding of adaptation options and their possible effects was greatly enhanced. The increased awareness among local officials will certainly increase the effectiveness of implementing alternative adaptation policies.

A properly developed and implemented adaptation action plan consisting of various effective measures could have positive benefits to the well-being and productivity of all people living in the Heihe region. These effective adaptation measures could help reduce water system vulnerability and water use conflicts. Indirectly, a reduction in water system vulnerability will mitigate the impacts of climate change on agricultural systems and protect the livelihoods of farmers. Water system sustainability can also improve ecosystem health and reduce sandstorms, which have a global environmental impact. In addition, a successful adaptation action plan could become a useful model for communities across China and around the world.

References

Benioff, R., S. Guill and J. Lee (eds) (1996) *Vulnerability and Adaptation Assessments: An International Guidebook*, Kluwer Academic Publishers, Dordrecht, The Netherlands

Carter, T. R., M. L. Parry, H. Harasawa and S. Nishioka (eds) (1994) *IPCC Technical Guidelines for Assessing Climate Change Impacts and Adaptations*, Department of Geography, University College, London

China Water Saving Irrigation Net (2007) China Irrigation and Drainage Development Center, www.jsgg.com.cn/Index/Index.asp?ClientScreen=1024

Ding,Y. H., Q. P. Li and W. J. Dong (2005) 'A numerical simulation study of the impacts of vegetation changes on regional climate in China', *Acta Meteorological Sinica*, vol 63, no 5, pp604–621 (in Chinese)

Feenstra, J., I. Burton, J. Smith and R. Tol (eds) (1998) *Handbook on Methods for Climate Change Impact Assessment and Adaptation Strategies, Version 2.0*, United Nations Environment Programme, Nairobi, and Institute for Environmental Studies, Vrije Universiteit, Amsterdam

Holling, C. S. (ed) (1978) *Adaptive Environmental Assessment and Management*, John Wiley, Chichester, UK

IPCC (2001) 'Summary for policymakers', in J. McCarthy, O. Canziani, N. Leary, D. Dokken and K. White (eds) *Climate Change 2001: Impacts, Adaptation, and Vulnerability*, contribution of Working Group II to the Third Assessment Report of the Intergovernmental Panel on Climate Change, Cambridge University Press, Cambridge, UK, and New York

Jia, S. F., S. F. Zhang, J. Xia and H. Yang (2004) 'Effect of economic structure adjustment on water saving', *Journal of Hydraulic Engineering*, vol 3, pp111–118 (in Chinese)

Li, Q. P. and Y. H. Ding (2004) 'Multi-year simulation of the East Asian monsoon and precipitation in China using a regional climate model and evaluation', *Acta Meteorologica Sinica*, vol 62, no 2, pp140–153 (in Chinese)

Li, Y. H. (2006) 'Water saving irrigation in China', *Irrigation and Drainage*, vol 55, pp327–336

Lim, B., E. Spanger-Siegfried, I. Burton, E. Malone and S. Huq (eds) (2005) *Adaptation Policy Frameworks for Climate Change: Developing Strategies, Policies and Measures*, Cambridge University Press, Cambridge, UK

Liu, W., H. Huang, W. Zhang and M. Zhang (2006) *Evaluation Report of Water Saving Society Pilot Projects*, Research Center for Hydraulics Development, Nanjing, China (in Chinese)

Ministry of Water Resources of China (2006) 'Hydraulic works', available from water information website, www.cws.net.cn/

Saaty, T. L. (1980) *The Analytic Hierarchy Process*, McGraw-Hall, New York

Saaty, T. L. (1982) *Decision Making for Leaders: The Analytical Hierarchy Process for Decisions in A Complex World*, McGraw-Hall, New York

Stratus Consulting Inc (1999) *Compendium of Decision Tools to Evaluate Strategies for Adaptation to Climate Change Final Report*, FCCC/SBSTA/2000/MISC.5, UNFCCC Secretariat, Bonn, Germany

Wang, Z. Y. (2005) 'Climate change analysis for western China: 1951–2004', PhD thesis, China Meteorological Administration, Beijing

WMO (1997) *Comprehensive Assessment of the Freshwater Resources of the World*, World Meteorological Organization, Geneva, Switzerland

Xu, Y., Y. H. Ding, Z. C. Zhao and J. Zhang (2003) 'A scenario of seasonal climate change of the 21st century in northwest China', *Climatic and Environmental Research*, vol 8, no 1, pp19–25 (in Chinese)

Yin, Y. (2001a) 'Flood management and water resource sustainable development: The case of the Great Lakes Basin', *Water International*, vol 26, no 2, pp197–205

Yin, Y. (2001b) 'Designing an integrated approach for evaluating adaptation options to reduce climate change vulnerability in the Georgia basin', final report submitted to Adaptation Liaison Office, Climate Change Action Fund, Ottawa, Canada

Yin, Y. Y. (2006) 'Integrated assessments of vulnerabilities and adaptation to climate variability and change in the western region of China', final technical report of the AS25 Project, International START Secretariat, Washington, DC

Yin, Y. and S. Cohen (1994) 'Identifying regional policy concerns associated with global climate change', *Global Environmental Change*, vol 4, no 3, pp245-260

Yin, Y. and X. Xu (1991) 'Applying neural net technology for multi-objective land use planning', *Journal of Environmental Management*, vol 32, pp349-356

Yin, Y., S. Cohen and G. Huang (2000) 'Global climate change and regional sustainable development: The case of Mackenzie basin in Canada', *Integrated Assessment*, vol 1, pp21–36

Yin, Y. Y., N. Clinton, B. Luo and L. C. Song (2008) 'Resource system vulnerability to climate stresses in the Heihe river basin of western China', in N. Leary, C. Conde, J. Kulkarni, A. Nyong and J. Pulhin (eds) *Climate Change and Vulnerability*, Earthscan, London

Zadeh, L. A. (1965) 'Fuzzy sets', *Information and Control*, vol 8, pp338-353

Zhangye Government (1998) *Heihe Basin Ecological and Environmental Protection and Construction Plan: 1998–2010*, Zhangye, China

13

Strategies for Managing Climate Risks in the Lower Mekong River Basin: A Place-based Approach

Suppakorn Chinvanno, Soulideth Souvannalath, Boontium Lersupavithnapa, Vichien Kerdsuk and Nguyen Thuan

Introduction

Climate risks are not new to farmers of the lower basin of the Mekong river. For smallholder farmers of rain-fed rice, a dominant economic activity of the region, flood, drought and other climate hazards pose substantial threats to their livelihoods (Chinvanno et al, 2008). A variety of strategies and practices are employed to cope with and manage climate risks, which we document through field studies of farming villages in Lao PDR, Thailand and Vietnam. The strategies and specific measures for managing climate risks are broadly similar across the villages, but there are also important differences, despite the similar hazards being faced and the livelihood patterns held in common. In this chapter we examine these similarities and differences and their implications for promoting effective strategies for adapting to climate change.

Farmers' Concerns about Climate

Our study was conducted through household interviews and focus group meetings in farm communities of the Vientiane Plain and Savannakhet Province in Lao PDR, Kula Field and Ubonratchathani Province in Thailand, and the Mekong river delta of Vietnam. More than 1600 households plus local officials participated in the interviews and meetings, which were conducted in 2004 and 2005 and are detailed in Kerdsuk and Sukchan (2005) and Boulidam (2005). The locations of the study sites are shown in Figure 13.1.

The interviews and focus group discussions explored farmers' perceptions of climate hazards, the risks to their farming activities, observed changes in climate and the impacts, strategies and measures used to cope with climate risks, and options for improving the management of climate risks. The climate

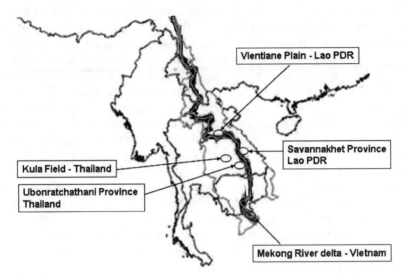

Figure 13.1 *Study sites in Lao PDR, Thailand and Vietnam*

risks found to be major concerns for farmers of the lower Mekong basin vary from location to location, depending on the geographical characteristics of the farmland, farming practices of the community and local features of the climate. However, two climate phenomena are identified by farmers at most of our study sites as significant threats to their livelihoods. These are prolonged midseason dry spells coming after sowing rice seeds or transplanting seedlings and flooding near the end of the crop cycle before harvest time.

With the limited extent of irrigated area in the region, most farmers rely mainly on natural rainfall for growing crops (Barker and Molle, 2004). In most parts of Thailand and Lao PDR, farmers of rain-fed rice practise single wet-season cropping, which normally starts in May and ends in October to November. These farmers start sowing rice at the beginning of the rainy season. Farmers who use a transplanting technique begin the process in mid-June to mid-July and harvest in October to November (Boulidam, 2005). The farmers of the Mekong river delta in Vietnam, where the rainy season is longer due to the influence of two monsoon systems, the southwest monsoon and northeast monsoon, are able to grow two rice crops per year (N. T. H. Thuan, personal communication, 2004).

A midseason dry spell typically occurs after seeding and/or transplanting. If prolonged, the midseason dry spell can seriously damage young rice plants. Such events increase the cost of production, as farmers may have to replant their rice. However, in some cases of delayed or prolonged dry spell, replanting may not be feasible because the rainy season would end before the replanted rice could reach maturity.

Floods that occur late in the rainy season, in October or November, pose serious risks for rice cultivation and farmers' livelihoods. The lower Mekong river basin experiences floods from the major tributary of the river, most commonly towards the end of the rainy season, when water flow is high and water from tributaries cannot flow into the main stem of the river. Sometimes the situation is made worse when water from the Mekong river is backed up into the tributaries (Mekong River Commission, 2005). This period of high flood frequency is close to harvesting time and for most farmers there would be no time to replant rice for that year if the crop were destroyed or damaged by a late-season flood. Only farmers cultivating areas close by the river or major tributaries and using short-cycle rice varieties have the possibility to replant after a late-season flood. In the discussions with farmer communities in Lao PDR and Thailand, the possibility of increasing flood risk in the future due to climate change raised high concerns among the farmers.

Direct and indirect impacts of floods and midseason dry spells reported to be major concerns by rice farmers in the lower Mekong are presented in Table 13.1. These have been categorized as first-order impacts (biophysical consequences of meteorological events), second-order impacts (crop production consequences of the biophysical impacts) and higher-order impacts that affect human well-being.

Table 13.1 *Multiple orders of climate impacts on rain-fed farms in the lower Mekong region*

Order of impact	Description	Impacts
First-order impacts	Biophysical consequences of meteorological events	Drying of soil due to midseason dry spell, particularly after seeding or transplanting Flooding due to heavy rain, particularly toward the end of the rainy season
Second-order impacts	Crop production consequences of the biophysical impacts	Damage to immature plants Reduced harvest Loss of harvest
Third-order impacts	Consequences of the second-order impacts	Increase in cost of production Food scarcity Decline in household income
Fourth-order impacts	Consequences of the third-order impacts	Degradation in household livelihood and socioeconomic condition (e.g. reduced financial and other wealth, reduced food reserves, malnutrition, increased debt) Migration of member(s) of the household (temporary or permanent) Migration of entire household and exit from farming Change in social status (e.g. change from independent farmer to contracted farmer or hired labour) Conflict among villages
Fifth order impacts	Consequences of the fourth-order impacts	Reduced labour force in farming communities Greater costs for hired labour, machinery to replace labour

Managing Climate Risks:
Current Practice and Potential Adaptation

Farmers surveyed in Lao PDR, Thailand and Vietnam identified numerous practices currently in use in their communities which they believe lessened their vulnerability to present-day climate variability and hazards. Some of the measures are motivated by climate risks. Others are primarily motivated by different concerns, yet nonetheless reduce climate risks by increasing the resilience of farmers' livelihoods to multiple sources of stress. They include on-farm and off-farm measures that are implemented at the household level (Tables 13.2 and 13.3), the community level (Table 13.4), and the national level (Table 13.5). Although none of the measures are motivated by perceived needs to adapt to human-induced climate change, many measures that are focused on near-term climate risks could be developed further for longer-term climate change adaptation (Kates, 2001). Implementation and the effectiveness of the measures in the different countries, some of the enabling and limiting factors that give rise to differences across the countries, and their potential as adaptations to climate change are examined below.

Vientiane Plain and Savannakhet, Lao PDR

Most farmers in Vientiane Plain and Savannakhet Province are subsistence farmers, producing rice mainly for their own consumption. They have farms of moderate but sufficient size for producing rice to support the annual consumption of the farm household. They produce a single rice crop each year, and their use of mechanized and advanced farm technology and formal institutional organizations (for example, cooperatives) is limited. The communities are still surrounded by intact natural ecosystems from which natural products can be harvested. This strengthens livelihoods by supplementing and diversifying the farm household's food and income sources (Boulidam, 2005).

Farmers of the Lao PDR study sites tend to rely mostly on farm-level measures for adapting to climate hazards and, to a lesser degree, on collective actions at the community level. Measures at the national level are very limited. Consequently, the capacity of the individual farm household to adapt is a key limiting factor at present for managing climate risks. The responses to climate hazards aim mainly at basic household needs, primarily food security of the household. Common measures implemented by rice farmers include seasonal changes in seed variety, cultivation methods, and timing of farm management tasks based on seasonal climate forecasts made with indigenous knowledge. Also common are raising livestock and harvesting natural products for additional food and income, which are considered major and primary adaptation measures in Lao PDR.

The use of indigenous knowledge to make seasonal climate predictions is still popular. Indigenous knowledge based on observations and interpretations of natural phenomena, for example, the height of ant nests in trees, the colour of frog's legs, the colour of lizard's tails and various indicators of the dry season weather pattern, is used to make forecasts of the

Table 13.2 *Household-level on-farm measures for managing climate risks*

Measure	Objective	Current Implementation	Effectiveness	Enabling and Limiting Factors
Change rice variety – seasonal	Avoid productivity loss from adverse climate conditions, improve food security	Lao PDR: Common practice. Rice grown for own and local consumption, market acceptance not a factor. Traditional knowledge used for seasonal forecasting. Thailand: Limited use. Local seed varieties not accepted by market. Vietnam: Moderate use. Short-cycle seed variety accepted by the market, but at a lower price	Moderate	Forecast accuracy; market acceptance of seed varieties; consumption preference
Change rice variety – permanent	Reduce variability of crop yield and income	Lao PDR: Limited use. Thailand: Common practice. Vietnam: Common practice – commercial farming	Moderate	Development of new seed varieties; market acceptance; consumption preference
Multiple, spatially separated farm plots	Diversify exposures to climate hazards	Lao PDR: Limited use. Thailand: Limited use. Vietnam: Limited use	High	Land availability and characteristics; population growth
Match method and timing of cultivation practices to seasonal climate	Avoid productivity loss from adverse climate conditions, improve food security	Lao PDR: Common practice. Use traditional knowledge, not constrained by market considerations. Thailand: Moderate use; change seedling technique. Crop calendar constrained by the market. Vietnam: Moderate use. Long rainy season allows more flexibility in crop calendar.	Low	Forecast accuracy; length of rainy season; market constraints on crop calendar
Manage water with small-scale irrigation, embankments	Water source during dry spells; control flooding	Lao PDR: Limited use. Thailand: Moderate use. Vietnam: Moderate use	Moderate to high if sufficient resources	Geographical features; financial resources for investment and operating costs

Table 13.2 (*continued*)

Measure	Objective	Current Implementation	Effectiveness	Enabling and Limiting Factors
Grow alternate crops between rice seasons	Increase and diversify food supply and income	Lao PDR: Limited to moderate use Thailand: Limited to moderate use Vietnam: Limited to moderate use; two crop seasons for rice is the normal practice	Moderate	Water availability in dry season; market for alternate crops; size and condition of farm land
Grow crops resilient to wider range of climate conditions than rice	Reduce variability of food supply and income	Lao PDR: Limited use Thailand: Limited to moderate use Vietnam: Limited use	High where feasible	Know-how; markets for other crops; financial reserves; farm size and soil condition; local culture
Livestock	Reduce variability of income, food security	Lao PDR: Common practice at a small scale Thailand: Common practice at a small scale Vietnam: Not available	High	Financial reserves; farm size and condition

onset and cessation of the rainy season, quantity of rain and other climate parameters (Boulidam, 2005). The forecasts are used for seasonally adjusting choices of seed varieties and time and methods for soil preparation, seeding, planting, fertilizing, weeding, harvesting and other tasks (Grenier, 1998). Because farmers in Vientiane Plain and Savannakhet Province grow rice mainly for their own consumption (and selling excess production to the local market for local consumption), they have flexibility to select the seed variety to match local climate conditions without regard for the requirements of the commercial markets of other regions.

Changing seed varieties in accordance with indigenous seasonal climate predictions is considered to be moderately effective by the surveyed farmers; adjusting the methods and timing of farming practices can be effective up to a point, but implementation has been patchy. Performance of these measures for adapting to climate change could potentially be enhanced by implementation of an early warning system based on modern inter-annual and seasonal climate forecasting, coupled with risk communication techniques to reach the populations at risk. Constraints on this measure include the precision of seasonal climate forecasts, ability and institutional network to communicate the forecasts in ways that are useful to farmers, acceptance of the forecasts by farmers, availability of suitable seed varieties, and flexibility for changing the crop calendar for their cultivation.

There is less flexibility for farmers in the Lao PDR sites to change the rice variety on a semi-permanent basis to one that is more climate-resilient or switching to an alternative crop. Constraints on these measures include lack of appropriate seed types, consumption preferences, national dependence on rice for food security, market conditions, lack of know-how and lack of required financial reserves. Consequently, these measures have limited current use. Where they have been used, these measures are considered by farmers to have moderate to high effectiveness for reducing vulnerability to climate and so are potential options for adapting to climate change. But the factors that constrain current use would need to be overcome. Growing a crop other than rice during the dry season is another moderately effective measure that is practised to a limited or moderate degree and can be an effective adaptation to climate change. But its use is restricted to areas where there is access to water and suitable markets.

The community still has an important role in the management of climate risks in the study areas of Lao PDR. For example, in the case of severe loss of rice production, the village leader would establish a cooperative network with other villages located near a river or stream or with irrigation systems, where supply of water is available for dry season crop. Shared farmland would be used for the cultivation of short-cycle rice varieties during the dry season to supplement the community's food supply. In addition, shared resources, such as a community rice reserve contributed to by households in the village or a community fish pond, also act as buffers to climate hazards that sustain the livelihoods and food security of the community. However, some of these collective actions are becoming obsolete, or will be in the near future, because of changes in socioeconomic conditions. Forces that have reduced the role of community-level

Table 13.3 *Household-level off-farm measures for managing climate risks*

Measure	Objective	Current Implementation	Effectiveness	Enabling and Limiting Factors
Harvest natural products	Increase and diversify food supply and income	Lao PDR: Common practice; Thailand: Limited use; Vietnam: Not available	High in Lao PDR; low to moderate in Thailand and Vietnam	Productivity, diversity and condition of natural ecosystems near villages
Produce and market non-farm products	Increase and diversify income	Lao PDR: Limited use; Thailand: Moderate use; Vietnam: Not available	Low to moderate in Lao PDR and Vietnam; moderate in Thailand	Know-how; access to market; market conditions
Seasonal migration for off-farm labour	Increase and diversify income	Lao PDR: Limited use; Thailand: Common practice; Vietnam: Not available	Low in Lao PDR and Vietnam; high in Thailand	Labour demand in urban areas; access to labour market; networks for job search
Permanent migration by family member	Increase and diversify income	Lao PDR: Limited use; Thailand: Common practice; Vietnam: Limited use	Low in Lao PDR and Vietnam; high in Thailand	Labour demand in urban areas; access to labour market; reduced farm labour for family

actions include population growth and expansion in the use of credit as an alternative to village rice reserves for coping with crop losses.

To date, national-level measures to manage climate risks are reported by surveyed farmers to be limited in scope and scale in Lao PDR. National action on climate risks has been constrained by local culture, lack of institutional arrangements to address climate risks, and limited know-how, resources and investment. Looking to the future, climate change is magnifying climate risks and increasing the amount of resources, technology and know-how that will be needed to manage the risks. Farmers have very limited capacity to adapt to the changes, and the diminishing role of communities is widening the gap between needs and capacities for managing risks. Consideration should be given to measures at the national level that would enhance capacity and enable actions for managing and adapting to climate risks at the farm level and at the community level.

Kula Field and Ubonratchathani Province, Thailand

Rice farmers in Thailand, particularly in the study areas in the northeast, are mostly commercial farmers who live in a monetary-oriented society and grow rice primarily for national and international markets. They have farms of moderate size on which they produce a single rice crop each year using mechanized and modern technologies, and formal organizations to support farm operations. The sale of rice is their main source of income, which is used primarily to purchase household basic needs, including rice for consumption, which could be cheaper in price and of different quality and texture than the rice the farm household grows. Only a small portion of farmers with larger farms are able to divide their farmland to grow both commercial rice variety for sale and a local rice variety for their own consumption or sale in the local market. The farming communities are closely linked to urban society. The surrounding land area is populated and used for settlements or is deteriorated natural forest that can provide only limited natural products as a supplement or alternative source of food and income (Kerdsuk and Sukchan, 2005).

According to the field assessment, farmers at the study sites in Thailand tend to rely on household and national-level measures for reducing climate risks, whereas the role of community-level measures has declined or been neglected. The household-level measures focus on income diversification, primarily from off-farm sources, which are not as sensitive to climate variations as income from rice (Kerdsuk and Sukchan, 2005). The main practice is seasonal migration to work in the cities, which can lead to the permanent migration of some members of the family in order to secure fixed income for the household. Wage income from city employment is less sensitive to climate and helps to insulate the farm household from climate-driven variations in farm income. Seasonal and permanent migration to diversify and supplement household incomes are more common in the Thai study sites than in Lao PDR and Vietnam and are made possible by close links between the rural villages and urban areas where there is demand for labour.

Table 13.4 *Community-level measures for managing climate risks*

Measure	Objective	Current Implementation	Effectiveness	Enabling and Limiting Factors
Shared resources – rice reserve/fish pond	Spread risks by creating food reserve; increase income for community	Lao PDR: Common practice Thailand: Limited use Vietnam: Not available	High in Lao PDR; low in Thailand and Vietnam	Cultural practices; strength of community institutions; guaranteed replenishment of rice reserve
Village fund	Finance investments to improve farms, livelihoods	Lao PDR: Limited use; use expanding under community management Thailand: Common practice; managed by government Vietnam: Not available	Moderate in Lao PDR and Thailand	Guaranteed repayment by borrower
Cooperative network among villages	Spread risks by sharing rice production, food supplies and labour with other villages	Lao PDR: Moderate use Thailand: Limited use Vietnam: Not available	Low to moderate in Lao PDR; low in Thailand and Vietnam	Relationship between village leaders; cultural practices
Cooperative processing and marketing of farm and natural products	Increase and diversify income	Lao PDR: Limited use Thailand: Limited use Vietnam: Not available	Moderate	Know-how; financial reserves; market access; market conditions

Unlike the studied communities in Lao PDR, where seasonal changes in rice variety and the crop calendar made in response to seasonal climate forecasts is common practice, these measures are little used by rice farmers in Kula Field and Ubonratchathani Province. Because they grow rice for national and international markets, they are limited in their ability to use local seed varieties, which fetch lower prices than commercial rice varieties, or to alter their crop calendar. In contrast, semi-permanent changes in seed variety to commercial varieties that are more resilient to climate stresses is common practice for farmers at the Thai study sites. This is made possible by the greater financial resources of commercial farming and by research and development programmes that provide new rice varieties that are both accepted in the market and more resistant to stress. This option could be moderately effective for adapting to climate change. Limitations on wider use are financial, technological and environmental.

Other on-farm measures for reducing climate risk practised by rice farmers in Thailand include changing seedling technique, using hired machinery, growing alternative crops between rice seasons and raising livestock. Some farmers make investments to increase and sustain the productivity of their farms in ways that make them more resilient with respect to climate variations and changes. For example, they construct small-scale irrigation systems to provide an alternative source of water for midseason dry spells or for growing a crop during the dry season. They may also build embankments to protect their fields from flood damage. Such measures are more common than in Lao PDR. But greater use is limited by financial requirements for investment and maintenance. A small number of farmers with large landholdings implement mixed-farming practices or switch part of their farmland from rice to a crop that is more resistant to climate stresses. Harvesting of natural products from forests, a common practice in Lao PDR, is limited at the study sites in Thailand because of high population densities and the degraded nature of forests that are adjacent to farm lands.

National-level policies and measures that serve to reduce vulnerability to climate hazards are more prevalent in Thailand than in Lao PDR and Vietnam. These policies and measures are not motivated by concerns about climate stress, especially climate change, but mainly by poverty reduction goals. Yet national measures in Thailand have supported financial needs, infrastructure development, transitions to more diversified farming systems, marketing of local farm products and farm planning, which have helped to improve the livelihoods of farmers and increase their resilience to climatic stresses. For example, an initiative of the Ministry of Agriculture and Cooperatives in 2004 (Department of Livestock Development, 2004) diversifies farming activity by promoting and providing support to farmers to raise livestock. Another initiative promotes transition from rice cultivation to other plantation crops that are more resistant to climate stresses, such as rubber trees. Research and development by government research facilities have provided new varieties of rice that are more resilient to climate variations, while maintaining the quality that is required by the market.

Table 13.5 *National-level measures for managing climate risks*

Measure	Objective	Current Implementation	Effectiveness	Enabling and Limiting Factors
Financial support to farmers	Assist investments to improve farms, livelihoods	Lao PDR: Limited use Thailand: Common practice Vietnam: Moderate use	Low in Lao PDR; moderate in Vietnam; high in Thailand	National financial condition; mechanism to allocate funds to the farmers; terms and conditions of loan
Support transition to other crops and more diversified farming systems	Diversify and improve farm livelihoods; increase resilience and sustainability of rural economy	Lao PDR: Limited use; rice farming central to food security Thailand: Limited to moderate use; market driven trend toward mono-cropping Vietnam: Limited use; rice farming central to food security	Low in Lao PDR and Vietnam; moderate in Thailand	National financial condition; know-how; mechanism to transfer know-how to farmers; markets; food security; soil properties
Support marketing of village products	Increase and diversify incomes	Lao PDR: Limited use Thailand: Moderate use Vietnam: Not available	Low in Lao PDR and Vietnam; moderate in Thailand	Markets; national financial conditions; know-how, mechanism to develop sustained market
Research and development of new seed varieties	Increase farm productivity and incomes; decrease variability of farm productivity and incomes; increase sustainability of farming	Lao PDR: Moderate use Thailand: Common practice Vietnam: Common practice	Low in Lao PDR; moderate in Thailand and Vietnam	National financial condition; time lag between research and availability of new seeds to farmers; technology; transfer knowledge to farmers
Develop rural infrastructure	Reliable water supply for irrigation; flood control	Lao PDR: Limited use Thailand: Moderate use Vietnam: Not available	Moderate	National financial condition; geographical conditions; technical feasibility
Provide information for farm management, including seasonal climate forecasts	Enable improved farm management	Lao PDR: Nonexistent Thailand: Limited use Vietnam: Limited use	Moderate	Communication channel; know-how to apply information; technology; forecast accuracy

Community-level measures are diminishing in Kula Field and Ubonratch-athani Province, with the exception of village funds for local investments to support farm livelihoods, which are managed by the government. But community or local administration units could play an expanded role to assist in planning as well as implementing future adaptation to climate change. The advantages of involving local institutions are that they are more aware of local risks, priorities and resources than national authorities and can be more flexible and timely in implementation.

The Mekong river delta, Vietnam

Rice farmers of the Mekong river delta in Vietnam are mainly commercial farmers. Unlike farmers at the study sites in Lao PDR and Thailand, they are able to grow two rice crops each year because of a longer rainy season. Farmers of the delta can grow sufficient rice to both supply the annual consumption of the household and sell rice to the market. They make moderate use of modern farm technology and formal institutional organizations in farming practice. The household relies heavily on income from rice production. The farm communities are surrounded by populated areas and are not tightly tied to the urban economic system (field interviews in Long An, Can Tho, Dong Thap and An Giang Provinces, Vietnam, 2004).

The farmer of rain-fed rice in Vietnam tends to rely on measures implemented at the household level and aimed mainly towards on-farm actions to protect against climate hazards. Community- and national-level measures play a very limited role in reducing their climate risks. The farm-level solutions include efforts and investments to increase and sustain the productivity of their farms, such as construction and maintenance of small-scale irrigation systems or embankments to protect their farmland from flooding. But investment costs and the limited financial capacity of farmers limit wider use of these measures. Using an alternative strategy, some farmers in the study sites have adapted to floods by accepting them as part of the ecosystem of their farmland, adjusting their crop calendar accordingly and allowing their lands to be flooded, thereby gaining advantages from nutrients being deposited that enhance soil fertility and pollutants being washed from their farmland. The use of alternative crops and seed varieties are also common adaptation measures used by farmers in the Mekong river delta.

Changing the variety of rice grown, both seasonally in response to climate forecasts and semi-permanently in response to markets and technological changes, is practised by Vietnamese farmers, even though they are commercial farmers and grow rice to match market demand. Because the rainy season in the delta region is usually seven months long, two crop cycles of rain-fed rice can be grown in one year. A two-crop cycle is also facilitated by the availability of short-cycle rice varieties that are suitable for growing in Vietnam and that are accepted by the market. This gives additional flexibility to farmers in Vietnam to select varieties of rice so as to balance the risk of losses from climate events against expected market returns according to their preferences regarding risk. Consequently, seasonal changes of rice variety are more commonly observed among rice farmers in Vietnam than in Thailand.

Community-level measures at the study sites in Vietnam are limited and have low effectiveness. Some measures that are implemented at a national level in Vietnam are considered by farmers to be moderately effective. National research and development programmes have facilitated changes in rice varieties by farmers that lessen vulnerability to climate extremes. Also being implemented, but on a limited scale, are national support for transition to alternative crops and provision of climate forecast information to farmers to assist with farm planning efforts.

Commonalities and Differences: A Matter of Context

Rice farmers are shown by the surveys to be experienced at managing climate risks, employing a variety of highly place- and time-specific measures to reduce their vulnerability. Many measures for managing climate risks are common to all of the study sites, at least in general characteristics. But there are also significant differences in the specific measures chosen and in the degree to which farmers rely on farm-level, community-level and national-level actions. These differences are apparent despite our focus on farmers who all make their livelihood primarily from growing rain-fed rice in a common river basin of Southeast Asia and who are exposed to similar climate hazards. The differences demonstrate the strong influence exerted by the local context on climate risk management. They arise from local differences in the specific climate hazards faced, physical and environmental constraints, available technologies, social and economic condition of the farm household and community, vitality of community institutions, degree of engagement in the market economy, market conditions, and the priorities and objectives of the farm households.

Even so, some commonalities do emerge from the experiences of farmers across the study sites. Some of the commonalities and differences are summarized in the following sections. In interpreting the findings, it should be borne in mind that the exploratory assessment surveyed farmers at only two sites each in Lao PDR and Thailand and only one site in Vietnam. While for convenience of exposition, we write about farmers in Lao PDR, Thailand or Vietnam, it would be misleading to extrapolate from farmers at the selected sites to characterize the condition and practices of farmers nationwide in any of the three countries. Differences in local context within a country can yield different risk management approaches and performance between communities of the country, just as they do in our comparisons of study sites from different countries.

At all of the study sites farmers rely primarily on their own capacity for implementing farm-level measures. But the context for farm-level action is shaped by what is done at community and national levels. Community-level measures are most prevalent in the farm communities of Vientiane Plain and Savannakhet Province in Lao PDR, where they play an important role in providing food security buffers and strengthening livelihoods. Farmers from the study sites in Thailand and Vietnam report that community-level measures are used only to a limited degree and are much diminished relative to the past. This too is the trend at the Lao PDR sites. The diminishing role of collective action at the community level may be an important deficit in the capacity of these communities to adapt to future climate change.

Our evaluation of national-level measures is based on the perspectives reported by farmers and community leaders at the study sites and does not reflect a comprehensive evaluation of national policies and programmes that are related to climate risks. But this is an important perspective, as it gives a sense of what is happening on the ground, at least in the communities surveyed. In none of the three countries can the national-level measures of which farmers are aware be described as constituting a national strategy for managing climate risks. The actions are not coordinated and typically are not designed specifically to combat climate risks.

National-level measures in Thailand, as perceived and reported by farmers in the Thai communities of Kula Field and Ubonratchathani Province, are greater than those reported by farmers surveyed in the other two countries and are an important complement to farm-level measures there. National-level actions in Thailand provide financial and other support for investments in farming infrastructure and expansion of farming technologies, including climate-resilient varieties of rice and other crops, sustainable farming practices, and diversified farm incomes. These efforts help to strengthen farm livelihoods and make them more resilient to climate and other shocks. In Vietnam, the national government supports research and development of seed varieties and provides financial support for investment in farm sector infrastructure, but other measures by the national government are reported by farmers to be limited. National-level measures are the least prevalent in Lao PDR and do not presently play a strong role in making farm households in the study areas climate resilient.

Farmers' objectives, priorities and capacities for using farm-level risk management measures vary between the study sites, and this influences their choice of measures. At the Lao PDR sites, most farmers practise subsistence agriculture and depend primarily on their own rice production for their food supply. Their choice of which rice variety to cultivate only needs to satisfy their own preferences and is not constrained by market requirements. They have access to healthy forests, from which they can harvest products to supplement their food supply. There are opportunities to earn monetary income, but these are little used. Consequently, their choices emphasize providing and protecting basic household needs, most particularly household food security, and employ strategies that have little financial cost and rely on household labour, indigenous knowledge and use of natural products.

Rice farmers in Kula Field and Ubonratchathani Province in Thailand are very much oriented to the market economy. They grow rice for cash income and have opportunities to participate in nearby urban labour markets. Their participation in commercial activities provides them with important financial resources and capacity, but their income can be volatile due to climate and market events, and market requirements for commercial rice can limit options for changes in rice cultivation. Consequently, their choices emphasize diversifying household income, particularly from off-farm labour, adoption of rice varieties that are more climate resilient and thus less variable in the income they provide, and investments such as small-scale irrigation and flood control that improve the productivity and resilience of their farmland.

In the Mekong River delta of Vietnam, farmers grow rice commercially but have little opportunity to participate in urban labour markets and so are highly dependent on the cash income from the sale of their rice. They have some financial resources and benefit from a longer rainy season than that at the Thai and Lao PDR sites, which allows them to grow two rice crops each year. The availability of short-cycle rice varieties that are suitable for growing on their farms and are accepted by the market also gives them greater flexibility to vary their rice cultivar and crop calendar if the season is expected to be unusually short or dry. Choices of the surveyed Vietnamese farmers emphasize varying cultivation practices to reduce the risk of damage or loss to the rice crop, and investments to improve the productivity and resilience of their farms.

Taken as a whole, the survey results suggest a pattern of climate risk management choices by farm households that is shaped by the socioeconomic condition of their surrounding communities. Farmers in communities with less developed socioeconomic conditions tend to pursue simple strategies targeted at increasing coping capacity and sustaining basic needs that can be implemented at the household or community level with limited financial and other resources. Farmers in communities with more developed socioeconomic conditions tend to pursue strategies targeted at reducing the variability of income and at improving the productivity and resilience of their farms. The measures that they adopt tend to depend more on market and other institutions, improved technologies and financial resources than is the case for farmers in less developed communities.

Climate Change in the Lower Mekong

Rain-fed agriculture is the dominant economic activity of countries in the lower Mekong basin and engages a high proportion of the population (Schiller et al, 2001; UN-ESCAP, 2006). Despite the efforts made to manage climate risks, farmers of rain-fed crops remain highly vulnerable to climate variations and extremes. Today, human-caused climate change threatens to magnify climate risks in the region and expose farmers to conditions that are outside of the range of current experience.

Many farmers over the age of 40 surveyed in Thailand and Lao PDR report noticeable changes in climate patterns over the past 25 to 30 years. These include increasing variability in the dates of onset and end of the rainy season, changes in wind direction, changes in the rainfall distribution pattern throughout the season, and an increase in thunderstorm activity. Thunderstorms, according to farmers at many of the study sites, have increased in frequency, and their occurrence has extended throughout the rainy season. In the past, they only occurred during the beginning and towards the end of the rainy season. This observed phenomenon may be an indicator of changes in the regional high–low pressure front during the rainy season, which no longer moves to a higher latitude after the beginning of the rainy season and southward again at the end of the rainy season. The front now seems to stay within the region throughout the rainy season. Some farmers also noticed a change in the wind direction pattern, which now varies throughout the season, unlike in the old days, when farmers observed that clouds and rain always came from a certain direction, which was thus more predictable.

The future climate of the lower Mekong, like much of the world, will be warmer. It is also likely to be wetter. Climate model simulations project trends toward greater precipitation and higher intensity precipitation in the region during the rainy season, which would increase the magnitude and possibly the frequency of floods in the Mekong basin. The greater precipitation during the rainy season suggests the potential for reduced frequency of prolonged dry spells in the middle of the growing season, but this will depend on how daily variability in rainfall changes.

Mathematical modelling simulations performed with the high resolution Conformal Cubic Atmospheric Model (CCAM) from McGregor and Dix (2001) are used to construct scenarios of future climate change for our assessment of impacts and vulnerability (Chinvanno et al, 2008). The scenarios correspond to increases in the atmospheric concentration of carbon dioxide from 360ppm to 540 and 720ppm, which are projected to be reached roughly by the 2040s and 2070s respectively in the IPCC's A1FI scenario (Nakicenovic and Swart, 2000). Projected changes in annual precipitation in subcatchments of the region range from no change to increases of more than 500mm per year (an increase of up to approximately 25 per cent), with the greatest increases projected for Lao PDR. Higher precipitation within a rainy season of approximately the same length as for the baseline scenario is projected by CCAM, implying potentially greater intensity of rainfall in the rainy season. Because of the potential for increased flood risk, as well as other changes in climate that would impact agriculture, there is a need to evaluate current practice for managing climate risks to the farm sector and strategies for adapting to future climate change.

Enabling Adaptation to Climate Change

The measures that are in use in the surveyed communities of Lao PDR, Thailand and Vietnam address current climate risks. They are not deliberate attempts to adapt to climate change. But they provide a basis of experience, knowledge and skills on which to build a climate change adaptation strategy. They also demonstrate a history of farmers in the region acting effectively, within their constraints, in their self-interest to reduce their vulnerability to climate hazards. Despite these efforts, however, farmers in the study area, particularly those who rely on rain-fed crops, are still strongly impacted by prolonged dry spells, floods and other climate events. They are highly vulnerable to climate hazards now and so can be expected to be highly vulnerable to climate change in the future.

Their vulnerability is partly due to lack of capacity of farm households, lack of capacity of rural communities, and lack of coordinated national strategies to support farmers and their communities in managing climate risks. An effective starting point for a national strategy of climate change adaptation would be to integrate policies to raise the capacities of farm households and rural communities for managing present climate risks into farm policy, rural development and poverty reduction efforts. Some national policies in the region already do this to a limited extent, though not explicitly.

Farm households need help with financial resources, opportunities for off-farm income, marketing of farm products, access to water and healthy ecosystems, information about current and changing climate hazards, know-how to diversify their farming practices and apply new farming methods and technologies, and access to improved varieties of rice and other crops. They also need buffers to protect their food security, health and livelihoods when they suffer severe crop or financial loss. Delivering this assistance to bolster the capacity of farm households requires community-level institutions with vitality and high capacity. Community institutions can also play a role in coordinating collective actions that require pooled resources to implement. Sadly, community-level institutions in the surveyed communities are in decline, and some community-level measures are becoming obsolete. A reversal of this trend will be important for maintaining existing capacity and raising capacity to the levels that will be needed to address the challenges of climate change.

An important concern for adaptation measures in the basin is that measures taken in one locality may have significant 'spillover' effects on neighbouring or downstream communities. A holistic approach to national policy and strategic planning for managing climate risks is needed in order to address concerns about potential spillovers. In addition, coordinated regional action by the countries of the lower Mekong river basin should also be considered as the countries share a common resource, the Mekong river, and some adaptation measures may only be feasible with regional collaboration.

Climate change will alter water availability, water quality, flood risks, and the performance and sustainability of river-dependent livelihood systems throughout the basin. The actions taken within any of the countries to adapt to these changes are also likely to have spillover effects that cross national borders. In this context, the countries of the lower Mekong river region should explore the potential for trans-boundary effects of their actions, options for reducing negative trans-boundary effects, and options for collective actions that may yield higher effectiveness of the adaptation measures and positive trans-boundary effects.

References

Barker, R. and F. Molle (2004) 'Evolution of irrigation in South and Southeast Asia', Comprehensive Assessment Research Report 5, Comprehensive Assessment Secretariat, Colombo

Boulidam, S. (2005) 'Vulnerability and adaptation of rain-fed rice farmers to impact of climate variability in Lahakhok, Sebangnuane Tai, Dong Khamphou and Khudhi villages of Songkhone District, Savannakhet, Lao PDR', Mahidol University, Salaya, Nakhon Pathom, Thailand

Chinvanno, S., S. Boulidam, T. Inthavong, S. Souvannalath, B. Lersupavithnapa, V.

Kerdsuk and N. Thuan (2008) 'Climate change risk and rice farming in the lower Mekong river basin', in N. Leary, C. Conde, J. Kulkarni, A. Nyong and J. Pulhin (eds) *Climate Change and Vulnerability*, Earthscan, London

Department of Livestock Development (2004) Website of Department of Livestock Development, Ministry of Agriculture and Cooperatives, Thailand, available at www.dld.go.th/transfer/Mcattle/index.php?option=com_content&task=view&id=15&Itemid=1, accessed 9 October 2006

Grenier, L. (1998) *Working with Indigenous Knowledge: A Guide for Researchers*, International Development Research Centre, Ottawa, Canada

Kates, R. W. (2001) 'Cautionary tales: Adaptation and the global poor', in S. Kane and G. Yohe (eds) *Societal Adaptation to Climate Variability and Change*, Kluwer Academic Publishers, Dordrecht, The Netherlands

Kerdsuk, V. and S. Sukchan (2005) 'Impact assessment and adaptation to climate change: The study of vulnerability and adaptation options of rain-fed farms in Tung Kula Ronghai', Khon Kaen University, Khon Kaen, Thailand

McGregor, J. L. and M. R. Dix (2001) 'The CSIRO conformal-cubic atmospheric GCM', in P. F. Hodnett (ed) *IUTAM Symposium on Advances in Mathematical Modelling of Atmosphere and Ocean Dynamics*, Kluwer Academic Publishers, Dordrecht, The Netherlands, pp197–202

Mekong River Commission (2005) *Overview of the Hydrology of the Mekong Basin*, Mekong River Commission, Vientiane, Lao PDR

Nakicenovic, N. and R. Swart (eds) (2000) *Emissions Scenarios*, Intergovernmental Panel on Climate Change special report on emissions scenarios, Cambridge University Press, Cambridge, UK

Rothman, D. S., D. Demeritt, Q. Chiotti and I. Burton (1998) 'Costing climate change: The economics of adaptations and residual impacts for Canada', in N. Mayer (ed) *The Canada Country Study: Climate Impacts and Adaptation*, vol VIII, Environment Canada, Montreal, Canada.

Schiller, J. M., S. A. Rao, Hatsadong and P. Inthapanya (2001) *Glutinous Rice Varieties of Laos: Their Improvement, Cultivation, Processing and Consumption*, Food and Agriculture Organization, Rome

United Nations Economic and Social Commission for Asia and the Pacific (UN-ESCAP) (2006) 'Annual indicators for millennium development goals', http://unescap.org/stat/data/goalIndicatorArea.aspx, accessed 9 October 2006

14

Spillovers and Trade-offs of Adaptation in the Pantabangan–Carranglan Watershed of the Philippines

Rodel D. Lasco, Rex Victor O. Cruz, Juan M. Pulhin and Florencia B. Pulhin

Introduction

Watersheds are a critical aspect of the economy and the environment in the Philippines. Approximately 18 to 20 million people inhabit the uplands of many watersheds and depend on their resources for survival. It is estimated that no less than 1.5 million hectares of agricultural lands presently derive irrigation water from these watersheds. However, despite their tremendous value, it has been observed that many watersheds are now in varying stages of deterioration (Cruz et al, 2000), largely due to population stresses. This could potentially serve to increase the vulnerability of their inhabitants to the impacts of future climate change and could have important implications for the viability of the economy and the environment that depend on them.

One of the most important watersheds in the Philippines is the Pantabangan–Carranglan watershed, which houses the Pantabangan Dam and supplies water for irrigation and power generation. It is also home to thousands of people largely dependent on agriculture as a source of livelihood. It has been projected that this watershed is likely to see important increases in temperature and precipitation by 2080 due to climate change, which could cause an increase in the frequency of extreme events such as floods. This could, in turn, significantly impact agricultural systems and endanger the survival of the local population dependent on farming. Additionally, such changes can also endanger the viability of the forest, grassland and brushland ecosystems and threaten the stability of this critical watershed.

With this in mind, in this chapter we evaluate potential adaptation options to address the vulnerability of natural and social systems in the Pantabangan–Carranglan watershed, namely forest and upland agriculture, water resources, and local institutions and communities. The shared water resource creates a high degree of interdependence among people, livelihoods and biophysical systems located within the watershed. Because of this interdependence, projects or policies implemented for the benefit of one sector often create spillover effects, both positive and negative, for other sectors. Yet cross-sector spillover effects or trade-offs are seldom considered. In our analysis of climate change adaptation strategies, we emphasize cross-sectoral effects and trade-offs to better understand how strategies implemented in one sector may conflict with or reinforce strategies of other sectors, and to identify potential win–win options. We anticipate that these findings will serve to better inform policymakers at the local and national levels in the Philippines and enable more effective decision making with respect to addressing the impacts of climate change on the Pantabangan–Carranglan watershed.

Characterization of the Study Site and Existing Vulnerability

The following description of the study site was partly based on the *Watershed Atlas of Philippine Watersheds* (Bantayan et al, 2000), as well as our primary data collection activities.

The Pantabangan–Carranglan watershed is located in central Luzon between the 15°44' and 16°88' north latitudes and the 120°36' and 122°00' east longitudes (Figure 14.1). It is bounded by the Caraballo Mountains on the north, northwest and northeast and by the Sierra Madre ranges on the south, southeast and southwest. The region has a complex topography varying from nearly level, undulating and sloping land to steep hilly and rugged mountain landscapes. It is largely characterized by the Philippine Climatic Type I, with two pronounced seasons – dry from December to April and wet the rest of the year (Bantayan et al, 2000). A small portion of the watershed falls under Climatic Type II, which has no dry season and very pronounced maximum rainfall from November to January. Average annual rainfall at stations in the watershed ranges from roughly 1780mm to 2270mm and the mean monthly temperature ranges from 25.7°C to 29.5°C. The watershed lies within the typhoon belt, with most typhoons occurring between September and October (Bantayan et al, 2000).

The major land cover types, shown in Figure 14.2, are natural forests (predominantly secondary forests), grasslands, reforestation areas, and alienable and disposable (A and D) lands, the latter including residential and *barangay* or community sites and cultivated areas. Grasslands occupy the largest portion, followed by secondary forests. Primary forests have largely disappeared since the logging boom of the 1960s (Saplaco et al, 2001). There has been a subsequent significant increase in reforested area, although it too is presently under increasing population pressure.

The Philippine government has classified the Pantabangan–Carranglan as a critical watershed since it houses the multipurpose Pantabangan Dam, built in 1974, which supports irrigation and hydroelectric generation and supplies domestic and industrial water needs. Construction of the dam and reservoir resulted in the submergence of the old Pantabangan town and seven adjacent

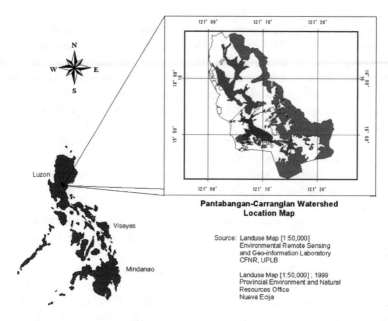

**Pantabangan-Carranglan Watershed
Location Map**

Source: Landuse Map [1:50,000]
Environmental Remote Sensing
and Geo-information Laboratory
CFNR, UPLB

Landuse Map [1:50,000] ; 1999
Provincial Environment and Natural
Resources Office
Nueva Ecija

Figure 14.1 *Location of the Pantabangan–Carranglan watershed*

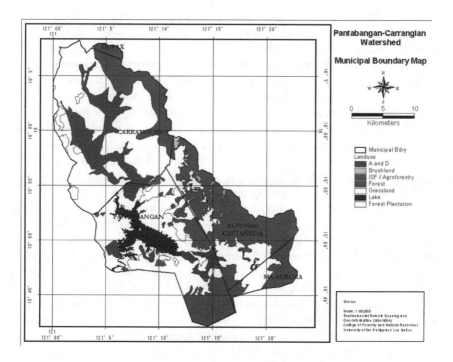

Figure 14.2 *Land use map of the Pantabangan–Carranglan watershed*

villages (Saplaco et al, 2001). All the displaced residents were resettled in the upper portion of Pantabangan and land grants were provided in place of submerged properties. The watershed is presently home to more than 60,000 people comprising of about 12,500 households (National Statistics Office, 2000a, b). Irrigation water from the reservoir is distributed to farmers outside the watershed in adjacent areas of central Luzon. Within the watershed, farmlands are largely unirrigated and dependent on rainfall. Rain-fed agriculture is the main occupation, with rice, vegetables, corn, cassava and onion as primary crops. Fishing is the second most important source of livelihood because of the reservoir and other occupations include grazing and reforestation activities, cottage industries and small businesses. Many residents are also employed in the labour force. Unemployment, however, remains a significant problem as a result of the limited opportunities, and many people resort to slash-and-burn farming and charcoal making at the expense of the local environment (Municipality of Pantabangan, undated; Municipality of Carranglan, undated).

The government has instituted several livelihood projects with support from various agencies and institutions to compensate for the residents' losses and to increase local productivity (Municipality of Pantabangan, undated; Toquero, 2003). These projects also include several reforestation programmes, integrated social forestry programmes and soil erosion control projects. One example is the RP-Japan reforestation project, with assistance from the Japanese government, which has reforested denuded portions of the watershed and created several local jobs. Another example is the World Bank-funded Watershed Management and Erosion Control project implemented by the National Irrigation Authority (NIA). However, despite these efforts, poverty remains a major issue and highlights the government's failures in providing viable resettlement alternatives. The government programmes have also had the negative effect of creating dependency such that once the programmes conclude, those relying on them for livelihoods resort to charcoal making for survival, thus destroying the very areas they reforested. It is estimated that more than 50 per cent of the residents in the watershed now follow such practices (F. D. Toquero, personal communication).

Added to these existing stresses, it is projected that climate change could cause important changes in temperature (about a 5 per cent increase as compared to 1960–1990) and precipitation (about a 13 per cent increase as compared to 1960–1990) in this area by 2080, which could increase the likelihood of floods in the wet season and the possibility of droughts in the dry season (Lasco and Boer, 2006). More than 25 per cent of the watershed is estimated to be highly vulnerable to the impacts of climate change, this including forests, grasslands and brushlands located on steep slopes, at high elevations and by roadsides. The moderately vulnerable areas constitute more than 65 per cent of the watershed and include grasslands, brushlands and forests in other locations. Of these ecosystem types, the dry forests are the most affected in simulation studies and face the possibility of elimination with a 50 per cent increase in rainfall. On the other hand, the wet forest and rainforest zones are expected to increase (Lasco and Boer, 2006).

From a socioeconomic perspective, the farming community is likely to be the most impacted by the changes in climate. Focus group discussions with vulnerable communities have led to the identification of areas most susceptible to the impacts of climate variability and change from the perspective of the local inhabitants. These include low-lying flood-prone and fire-prone settlements, agricultural areas prone to floods and droughts, dying streams and rivers, farmlands at the end of the irrigation canals, highly erosive areas along river banks, unstable areas with steep slopes that support infrastructure and grasslands, and forested areas and plantations near roadways (Lasco and Boer, 2006). Significant losses of life, property, infrastructure and livelihoods have been associated with past climate events in this area, with the more vulnerable population unable to ever fully recover from these setbacks. Incidences of illnesses such as diarrhoea, dysentery, dehydration, dengue, malaria and typhoid have also been blamed on climate variability and change.

Given this existing vulnerability of the population and ecosystems in the Pantabangan–Carranglan watershed, it therefore becomes extremely important to identify appropriate adaptive measures that can address the negative impacts of future climate change and also help sustain the local environment and economy. It is also critical that the selected sectoral adaptation measures are complementary and do not interact in a manner that reduces their effectiveness or causes undesirable outcomes in order to ensure their successful implementation.

Identification of Sectoral Adaptation Options

In attempting to determine and evaluate adaptation strategies for the Pantabangan–Carranglan watershed we focused on three sectors: the agro-forestry, water and institutional organizations of the region. Several different methods were used to identify climate change impacts, vulnerability and adaptation options, including GIS analysis, computer modelling, household surveys, focus group discussions, multi-stakeholder workshops and key informant interviews (for details see Pulhin et al, 2008 and Cruz et al, 2005).

Key informant interviews were conducted with government officials from different *barangays* or communities within the watershed and representatives from different local institutions, while focus group discussions targeted representatives from various institutions concerned with the Pantabangan–Carranglan watershed. A survey was conducted of 375 households from 25 *barangays* of the municipalities of Pantabangan and Carranglan in Nueva Ecija Province, Alfonso-Castañeda in Nueva Vizcaya, and Maria Aurora in Aurora. The overall objective of these participatory processes was to determine the services provided by the watershed to its different stakeholders; to obtain the perspective of participants on climate variability and extremes experienced in the area, their impacts and the coping strategies adopted; to assess the assistance provided by local institutions; and to solicit recommendations on improving responses to climate impacts.

Lastly, a major stakeholders' workshop was held in March 2004 to validate initial results of the study, conduct consultations on climate change impacts on forest, water and land use, and solicit feedback from various stakeholders. Thirty participants from different organizations within the Pantabangan–Carranglan watershed, particularly the National Power Corporation, the National Irrigation Authority, local government units, non-governmental organizations and people's organizations, were invited to attend this workshop. As anticipated, most of the adaptation options identified via the consultative processes described above were in response to the climatic variability observed in the study site. A brief description of these responses is provided below.

Tables 14.1, 14.2 and 14.3 show options for adapting to the impacts of climate variability and extremes that were identified during the multi-stakeholder workshop. The range of options identified by participants suggests that there is a high degree of awareness on adjusting to climate variability and extremes, which could serve to provide solid building blocks for adaptation to the impacts of future climate change. In general, the adaptation options identified are consistent with those recommended in the Philippines Initial National Communication (Government of Philippines, 1999).

Agro-forestry options for different land-use categories (Table 14.1) focus on the use of appropriate species and crops, scheduling of activities, technical innovations (for example, in water conservation), capacity building and law enforcement. Options for adapting water supply and use in response to different climate variations (Table 14.2) highlight water conservation practices, choice of crops/species, technical innovations and livelihood options. Options for institutional organizations (Table 14.3) include varied strategies used by the National Irrigation Administration, the Department of Environment and Natural Resources, the National Power Corporation and local government units.

Trade-off Analysis of Adaptation Options

The concept of trade-offs arises from the idea that resources are scarce. As a general principle, trade-off analysis shows that, for a given set of resources and technology, to obtain more of a desirable outcome for any given system, less of another desirable outcome is obtained (Stoorvogel et al, 2004a). Although there can be win–win outcomes in two dimensions, even such a win–win outcome must come at the expense of some other desired attribute.

Trade-off analysis has been used in exploring the effects of changes in land use, policies and scenarios in agricultural production systems in Ecuador and Peru (Antle et al 2003; Stoorvogel et al, 2004b). In these studies, the researchers developed a simulation model called TOA (trade-off analysis) for conducting an integrated analysis of trade-offs between economic and environmental indicators using biophysical as well as econometric process simulation models. Another form of trade-off analysis has been used in marine protected area management in the West Indies (Brown et al, 2000), which allowed decision makers to consider trade-offs between different criteria to evaluate alternative management options.

Table 14.1 *Adaptation options for agriculture and forestry by land use category*

Land Use	Adaptation Options
Lowland farms	Late rains: • Use of early maturing crop varieties • Shift to drought-resistant crops • Use of adaptable species • Supplemental watering Early rains: • Installation of small water impounding facilities
Upland farms	• Use of appropriate planting materials • Shift to more tolerant crops • Use of drought-resistant crops • Use of prescribed fungicides/pesticides • Installation of fire lines • Strict implementation of forest laws • Adoption of modern method of farming suited for uplands (e.g. sloping agricultural land technology (SALT)) • Greater visibility of enforcement agencies • Delay of planting
Tree plantation	• Adjust silvicultural treatment schedules and practices • Plant species that can adjust to variable climate situations • Proper timing of tree-planting projects or activities • Construction of fire lines • Control burning • Supplemental watering
Grasslands	• Supplemental feeding • Reforestation • Adaptation of SALT method of farming in combination with organic farming • Promote integrated social forestry and community-based forest management • Increase funds for forest protection and regeneration from national government • Increase linkages among local government, national government and non-governmental organizations • Introduction of drainage measures • Controlled burning • Introduction of drought-resistant species • Intensive information dissemination campaign among stakeholders
Natural forest	• Improve safety net measures for farmers • Coordination between local government units • Cancellation of the total ban on logging

In this chapter, we apply a less technically demanding approach for analysing the trade-offs between adaptation options for various sectors, which policymakers and stakeholders can use for first-order estimates. Specific adaptation options were initially analysed individually by sector (forest/agriculture, water resources and institutions and local communities), the details of which are reported in Cruz et al (2005) and Pulhin et al (2008). The trade-offs were

Table 14.2 *Options for adapting water resource supply and use in response to climate variations*

Climate Variation	Adaptation Options
General	• Adaptation of SALT method of upland farming • Implementation/intensification of reforestation programme • Strict implementation of forest laws • Programmes and research to increase groundwater utilization by households • More funds from national and local government • Construction of small water impounding facilities • Cloud seeding • Introduction of water conservation measures • Stabilization of watershed
Water shortage	• Use of shallow tube wells • Planting of new varieties of rice (e.g. Gloria rice) and other crops with less water requirements • Rotation method for scheduling irrigation • Planting early maturing varieties of crops and vegetables • Use the direct seeding method, which requires less water • Development and use of other water sources (e.g. the Atate and Penaranda rivers)
Floods	• None (wait for the next cropping season to respond) • Repair the damages • Close the main canal • Switch to other crops that can survive floods and heavy rainfall • Switch to early maturing varieties of crops (e.g. from palay to corn) • Explore other livelihoods through the Farmers' Business Resource Cooperative (e.g. pig rearing, squash and saluyot farming, canton (noodle) making and fruit juice making) • Construct fish ponds • Obtain training in farm management

then determined by means of matrix analysis to bring out the positive and negative interactions between sectoral adaptation options. The positive and negative ratings attributed were based on the expert judgement of the researchers. Next, mitigation measures were identified to minimize or eliminate adverse spillovers. Adaptation strategies common to all sectors were also identified.

Trade-off effects of adaptation strategies for forests and agriculture

Table 14.4 presents the analytical matrix describing the positive and negative impacts of the forest and agriculture sector on the other sectors. The effects of adaptation strategies for the forest and agriculture sector on water resources were generally found to be positive. This suggests a synergistic relationship

Table 14.3 *Adaptation strategies of different institutional organizations*

Institutions	Adaptation Strategies
National Irrigation Administration	• Reforestation and forest protection • Physical rehabilitation • Manage water releases from the reservoir to reduce flooding
Department of Environment and Natural Resources	• Reforestation and forest protection • Adjust programme priorities and schedules • Monitoring • Shading of seedlings in reforestation sites • Deploy forest guards to patrol the forest • Plant fire breaks • Information, education and communication • Hire additional manpower, especially casual labourers • Promote Integrated Social Forestry
National Power Corporation	• Reforestation • Information, education and communication • Proper choice of species • Adjust schedules and implementation
Local Government Unit	• Tree planting and reforestation • Information, education and communication, especially during typhoon season • Creation of El Niño/La Niña, task force formation of disaster brigade • Provision of relief goods • Provision of health services and medicines through extension workers • Bridge construction • Construction, maintenance and repair of roads • Construction, operation, maintenance and repair of water infrastructure (small impounds, wells, diversion canals, water tanks, pumps, etc) for irrigation and other uses • Provision of solar dryer facilities for drying palay • Obtain and distribute free seedlings from the Department of Agriculture • Visits by Barangay Tanod to the community to help with their problems • Small organizations that buy palay (rice) at higher prices and sell rice at lower prices • Training of People's Organizations

between adaptation for forest/agriculture and water. In contrast, adaptation strategies for forests/agriculture have a mixed effect on the various institutions in the Pantabangan–Carranglan watershed. Most of the recommended adaptation strategies require additional investments, which could pose a significant hurdle in their implementation given the tight budget constraints of many Philippine agencies. Likewise, the effects on local communities are mixed – in some cases there are positive effects but in others quite the opposite. For example, it is possible for farmers to obtain higher yields and incomes as a result of adaptation options such as the use of appropriate crop varieties; however, some

Table 14.4 *Cross-sectoral impacts of forest/agriculture sector adaptations on other sectors*

Adaptation Strategy for Forests and Agriculture	Effect on Water Resources	Effect on Institutions	Effect on Local Communities
Use of early maturing crops	+ Lower water demand	0	+ Higher income
Use of drought-resistant crops	+ Lower water demand	0	+ Higher income
Supplemental watering	− Higher demand for water	− Increase cost of developing alternative sources of water	− Greater labour demand + Higher income
Proper scheduling of planting	0	− Increase cost for training, technical assistance, research and development	0
Soil and water conservation measures	+ Conservation of water	− Increase cost for training, technical assistance, research and development	− Cash expenses
Establishment of fire lines	+ More vegetative cover promotes good hydrology	+ Less expense for fire fighting	− More labour demand + Less damage to crops from fire; more income
Construction of drainage structures	+ Better water quality (less sediment load)	− Increase cost of implementation	+ Less soil erosion in the farm; greater yield
Controlled burning	+ Less damage to watershed cover	0	0
Tree planting fuelwood	+ Better hydrology	− Increase cost of implementation	+ Steady supply of − Less area for farming
Enhance community-based organizations	+ Better conservation of water	+ Better participation in the political process	+ Better participation
Total logging ban	+ More forest cover	− Increase cost of enforcement and protection	− Less income from timber − Fewer sources of income
Use of appropriate silvicultural practices	+/− Could promote or impair hydrology depending on the practice	− Increase cost of implementation	− Increase cost of implementation
Better coordination between local government units	+ Promotes better watershed management	+ Greater collaboration among local government units	+ Better delivery of services to farmers
Information campaign	+ Better conservation of water	+ Increase awareness and competence	+ Increase awareness and competence
Better implementation of forest laws	+ Promotes better watershed management	− Increase cost of implementation	+/− Could adversely affect current livelihood of farmers that are deemed 'illegal'

adaptation activities such as supplemental watering could require more labour. Among other technical adaptation strategies, the use of early maturing crops and drought-resistant crops has the most positive effects on other sectors. The establishment of fire lines also has a generally positive effect but requires more labour time to establish. Similarly, social adaptation strategies (for example, community organizing) have positive effects on the other sectors. The negative effect in many cases is the additional cost required to implement adaptation strategies. This hurdle may prove daunting, considering the lack of resources of many Philippine government and non-governmental agencies.

In order to minimize the negative impacts, adaptation strategies for the forest/agriculture sector could be prioritized on the basis of their effects on other sectors (in addition to their effectiveness for forestry/agriculture). In general, those that have positive effects on other sectors should receive higher priority and attempts could be made to possibly alleviate the negative effects of the remaining options.

Trade-off effects of adaptation strategies for water resources

Adaptation strategies for water resources have overwhelmingly positive effects on forest and agricultural crop production as can be seen in Table 14.5. This is understandable considering that proper water management is essential to crop growth and development. Consequently, farmers can also earn a greater income if appropriate adaptation strategies for water are implemented. On the other hand, the effect on institutions is mainly negative as a result of the additional expenditures associated with the implementation of measures such as re-engineering, retooling, training programmes, technical assistance programmes and other related services. This implies that in the face of limited financial resources, adaptation strategies for water may not be fully implemented. One possible approach is to determine which of the recommended adaptation options are economically affordable given the current budget constraints of the implementing agencies.

Trade-off effects of adaptation strategies for institutions

Adaptation strategies identified by various institutions have generally positive effects on the other sectors (Table 14.6). This shows that the identified strategies are holistic in nature. (Of course, the underlying assumption is that financial resources are available to implement these strategies, which is the major constraint identified above.) There are a couple of exceptions to this. First, the strategy of stricter enforcement of forest protection rules could adversely affect farmers with no clear land tenure instruments. Many farmers in the watershed are informal settlers, and forest protection officers could compel them to leave their farms. Second, the strategy of releasing water from the Pantabangan Dam to prevent overflow could lead to flooding in downstream communities.

Overall, most of the sectoral adaptation strategies identified are found to have mixed interactions (both positive and negative), which suggests that generally, adaptation strategies are not neutral. Of the various adaptation

Table 14.5 *Cross-sectoral impacts of water sector adaptations on other sectors*

Adaptation Strategy for Water Resources	Effect on Forest Resources/ Agriculture	Effect on Institutions	Effect on Local Communities
Reforestation	+ Greater tree cover	– Higher investment cost	+ More income
Soil and water conservation measures	+ Increased yield	– Higher investment cost	+ More income
Water impoundment	+ Increased yield	– Greater expenses	+ More income – Greater expenses
Well construction	+ Increased yield	– Greater expenses	+ More income – Greater expenses
Cloud seeding	+ Increased yield	– Greater expenses	+ More income
Use of appropriate crops/varieties	+ Increased yield	– Greater expenses for research and development, technical assistance and information, education and communication	+ More income
Irrigation management	+ Increased yield	– Greater expenses for implementation	+ Increased income
Tap other water sources (e.g. rivers)	+ Increased yield	– Greater expenses	+ Increased income
Fishponds in flooded areas	+ Decreased pressure on forests and agricultural resources	– Additional expenses for technical assistance	+ Increased income – Greater expenses
Repair of damaged infrastructure	0	– Greater expenses	0
Shift in livelihood	+ Less use of land	– Additional expenses for technical assistance, training	+ Increased income
Strict implementation of forest laws	– Could affect crop production in areas deemed for forest	+ Strengthen role of regulatory agencies	+/– Promote peace but possibly lower income
Research on groundwater	0	– Greater expenses for research and development, technical assistance and information, education and communication	0
Capacity building activities	+ Build up of mass of competent players	– Greater expenses for research and development, technical assistance and information, education and communication	+ Build up of mass of competent players

Note: Key: + positive impact; – negative impact; 0 no effect.

Table 14.6 *Cross-sectoral impacts of adaptations by institutions on other sectors*

Adaptation Strategy for Water Resources	Effect on Forest Resources/ Agriculture	Effect on Institutions	Effect on Local Communities
Reforestation	+ Increased tree cover	+ Better watershed cover	+ Source of fuelwood/tree products
Forest protection	+ Reduce forest destruction	+ Better watershed cover	+/– Could affect source of forest products
Physical rehabilitation	0	0	+ Better facilities
Release of water from the dam	0	0	– Flooding in low-lying areas
Adjustment of schedule	0	0	0
Fire break establishment	+ Reduced fire loss	+ Better watershed cover	+ Reduced fire loss
Community-based management	+ Better forest land management	+ Better watershed cover	+ Empowerment of local people
Development of water sources	+ Increased crop yield	+ Stable water supply	+ Increased crop yield
Hiring additional personnel	+ Better forest protection	+ Improved water quality and regimen	+ Additional sources of income
Proper choice of species	+ Increased yield	0	+ Increased income
Provision of relief goods	+ Reduction of pressure on forest and agricultural resources	+ Reduction of pressure on water	+ Relief goods supplied
Creation of task forces	+ Better coordination	+ Better coordination	+ Better coordination
Infrastructure repair and construction	+ Increased farm yield	+ Stable and more efficient water supply	+ Increased income
Information, education and communication	+	+	+
Training of People's Organization	+ Better forest/farm management	+ Better water management	+ Skills developed

Note: Key: + positive impact; – negative impact; 0 no effect.

options recommended, four were common to all the sectors: tree planting/ reforestation, selection of appropriate species/crops and better implementation of laws and information campaigns (Table 14.7). This reflects the high degree of consciousness among stakeholders of the importance of forests in the watershed. Also, the use of soil and water conservation strategies was explicitly identified by the forest and water sectors and was also implied by the institutional sector. These results suggest that individual adaptation strategies could address more than one sector, allowing for greater synergy and cost-efficiency.

Table 14.7 *Adaptation strategies common to multiple sectors*

Adaptation Strategy	Forest/Agriculture	Water	Institutions
Tree planting/reforestation	X	X	X
Selection of appropriate crops/varieties	X	X	X
Better implementation of forest laws	X	X	X
Soil and water conservation measures	X	X	
Establishment of fire lines	X		X
Construction of drainage	X		
Controlled burning	X		
Enhance community-based organizations	X		X
Total logging ban	X		
Appropriate silvicultural practices	X	X	
Better coordination between local government units	X	X	
Information campaign	X	X	X
Water impoundment		X	
Well construction		X	
Irrigation management		X	
Cloud seeding		X	
Develop other water sources		X	X
Research		X	
Capacity building		X	X

While we have restricted ourselves to a qualitative approach to examine the trade-offs between various sectoral adaptation strategies, trade-offs can also be quantified. For example, the expense associated with reforestation, which has been identified as a desirable adaptation strategy by all sectors, is estimated by the Department of Environment and Natural Resources (DENR) to cost about US$900 per ha for three years (DENR, 1999). This level of investment may prove limiting for the organizations bearing the cost of reforestation such as the National Irrigation Administration, the National Power Corporation and the DENR. One possible way to reduce expenses is by encouraging more community participation to lower the labour cost, which constitutes the biggest fraction (about 70 per cent) of the total cost of reforestation. A somewhat less expensive adaptation strategy identified by stakeholders is fire line construction for forest/grassland fire prevention in the watershed, which costs about US$20/ha. Although this is typically a part of reforestation work, it can be undertaken as a separate project. Some of the other adaptation options are even less costly, for example, proper scheduling of planting to coincide with the late or early onset of the rainy season involves practically no financial cost to farmers and institutions and yet the positive effect on farm yields and income could be very high. While a detailed quantification of trade-offs in sectoral adaptation strategies is beyond the scope of this chapter, this is a potential area for further exploration in future studies.

Policy Implications

The findings discussed above have important implications for policy planning and management with respect to addressing the impacts of climate change in watersheds in the Philippines. By identifying the trade-offs between adaptation strategies for different sectors in the Pantabangan–Carranglan watershed, we have highlighted the significance of a cross sectoral analysis at the watershed scale, which can inform the design of effective policy and management responses to climate change. A cross-sectoral analysis of adaptation strategies will enable decision makers and managers to anticipate potential conflicts between sectors early on, thus providing greater opportunities for finding solutions.

Cost stands out as the most significant limiting factor in the implementation of adaptation options. The most common trade-off identified for all sectors is the additional cost of implementing adaptation strategies such as the construction of a water-impounding structure or tree planting. In developing countries like the Philippines, where priority for climate change adaptation is low, strategies that meet other, sometimes more important, goals may therefore stand better chances of implementation. For example, reforestation and tree planting programmes are already ongoing in the watershed, irrespective of climate change considerations, due to the other benefits they offer.

Besides highlighting the negative interactions, this type of trade-off analysis also helps to pinpoint those strategies that have synergistic effects across sectors and can be prioritized for implementation since they provide greater chances of stakeholder acceptance. A good example is tree planting/reforestation, which was identified as an adaptation strategy by all the sectors investigated.

It is also important that the positive and negative interactions between different adaptation strategies and their implications outlined above must be presented to policymakers in a manner that is easily comprehensible in order to enable effective decision making. One way of visually presenting the results of the study in a nutshell to policymakers is by means of the chart in Figure 14.3, which summarizes the impacts of adaptation strategies in one sector on other sectors and thus captures the potential synergies and conflicts in a simplified manner. For example, adaptation measures in water resources have mostly positive effects on forests and agriculture. These include tree planting and provision of irrigation water, which directly benefits forestry and agriculture. However, adaptation in water resources will have a largely negative effect on institutions. This can be attributed to the high cost of many of the adaptation measures identified, such as construction of shallow tube wells and impounding structures (Table 14.2). On the other hand, the effects of adaptation measures in forest resources and agriculture on local communities are mixed, in other words there are both positive and negative interactions. For example, use of more resistant varieties could lead to higher incomes but establishment of fires lines means higher labour costs.

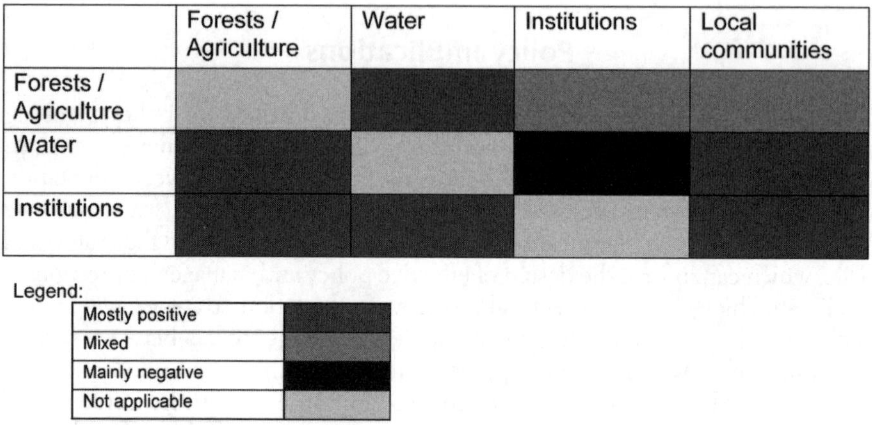

	Forests / Agriculture	Water	Institutions	Local communities
Forests / Agriculture				
Water				
Institutions				

Legend:

Mostly positive	
Mixed	
Mainly negative	
Not applicable	

Figure 14.3 *Summary of effects of adaptation strategies in one sector on other sectors*

This kind of visual presentation would thus enable policymakers and stakeholders to flag those adaptation measures that deserve attention and further study. Those with negative interactions could also be prioritized for quantitative analysis of trade-offs in order to determine the degree of impact and evaluate possibilities to minimize the impact.

References

Antle, J., J. Stoorvogel, W. Bowen, C. Crissman and D. Yanggene (2003) 'The trade-off analysis approach: Lessons from Ecuador and Peru', *Quarterly Journal of International Agriculture*, vol 42, pp189–206

Bantayan, N. et al (2000) *Philippine Watershed Atlas*, College of Forestry and Natural Resources, University of the Philippines, Laguna, Philippines

Brown, K., W. N. Adger, E. Tompkins, P. Bacon, D. Shim and K. Young (2000) 'Trade-off analysis for marine protected area management', CSERGE Working Paper GEC 2000-02, Centre for Social and Economic Research on the Global Environment, University of East Anglia, Norwich, UK

Cruz, R. V. O., R. D. Lasco, J. M. Pulhin, F. B. Pulhin and K. B. Garcia (2005) 'Assessment of climate change impacts, vulnerability and adaptation: Water resources of the Pantabangan–Carranglan watershed', Environmental Forestry Programme, College of Forestry and Natural Resources, University of the Philippines Los Baños College, Laguna, Philippines

Cruz, R. V. O., S. R. Saplaco, R. D. Lasco, M. M. B. Avanzado and F. B. Pulhin (2000) 'Development of water budget models for selected watersheds in the Philippines: Assessment of the impacts of ENSO on the water budget of selected watersheds', unpublished research report, Philippine Council for Agriculture, Forestry and Natural Resources Research and Development, Department of Science and Technology, Manila, Philippines

Government of Philippines (1999) *Philippines Initial National Communication to the UN Framework Convention on Climate Change*, Government of Philippines, Manila

Lasco, R. D. and R. Boer (2006) 'An integrated assessment of climate change impacts, adaptations and vulnerabilities in watershed areas and communities in Southeast Asia', final report, Assessment of Impacts and Adaptation to Climate Change Project No AS21, International START Secretariat, Washington, DC

Municipality of Carranglan (undated) *Development Master Plan of the Municipality of Carranglan, Nueva Ecija, 2003–2007*, Carranglan, Nueva Ecija, Philippines

Municipality of Pantabangan (undated) *Master Plan of the Municipality of Pantabangan, Nueva Ecija, 1998–2000*, Pantabangan, Nueva Ecija, Philippines

National Statistics Office (2000a) 'Census 2000: Philippines population by *Barangay*', CD-ROM, National Statistics Office, Makati, Philippines

National Statistics Office (2000b) *Philippine Statistical Yearbook*, National Statistics Office, Makati, Philippines

Pulhin, J. M., R. J. Peras, R. V. O. Cruz, R. D. Lasco, F. B. Pulhin and M. A. Tapia (2008) 'Climate variability and extremes in the Pantabangan–Carranglan watershed of the Philippines: An assessment of vulnerability', in N. Leary, C. Conde, J. Kulkarni, A. Nyong and J. Pulhin (eds) *Climate Change and Vulnerability*, Earthscan, London

Saplaco, S. R., N. C. Bantayan and R. V. O. Cruz (2001) *GIS-based Atlas of Selected Watersheds in the Philippines*, Department of Science and Technology, Philippine Council for Agriculture, Forestry and Natural Resources Research Development, and Environmental Remote Sensing and GIS Laboratory, University of the Philippines Los Baños College of Forestry and Natural Resources, Laguna

Stoorvogel, J. J., J. M. Antle, C. C. Crissman and W. Bowen (2004a). 'The trade-off analysis model: Integrated biophysical and economic modeling of agricultural production systems', *Agricultural Systems*, vol 80, pp43–66

Stoorvogel, J. J., J. M. Antleb and C. C. Crissman (2004b) 'Trade-off analysis in the Northern Andes to study the dynamics in agricultural land use', *Journal of Environmental Management*, vol 72, pp23–33

Toquero, F. D. (2003) 'Impact of involuntary resettlement: The case of Pantabangan resettlement in the province of Nueva Ecija', PhD thesis, Central Luzon State University, Science City of Muñoz, Nueva Ecija, Philippines

15

Top–Down, Bottom–Up: Mainstreaming Adaptation in Pacific Island Townships

Melchior Mataki, Kanayathu Koshy and Veena Nair

Introduction

Climate change, whether due to natural variability or human activity, is one of the most pressing issues for the Pacific island countries. The impacts of climate variability and extreme events such as cyclones, floods, droughts and sea level rise are rapidly pushing people beyond their coping range. The already strained economies are being drained trying to keep up with the impacts of these stresses on livelihoods. In the 1990s alone, the Pacific island region bore up to US$1 billion costs related to climate extremes (Campbell, 1999; Feresi et al, 2000), and the costs are expected to rise even further with a rise in the frequency and intensity of extreme events.

Climate projections for the South Pacific indicate warming of 0.8 to 1.8°C and precipitation changes that range from -8 to +7 per cent by mid-century (Ruosteenoja et al, 2003). By the end of the century, projected warming is 1.0 to 3.1°C and precipitation changes range from -14 to +14 per cent. Projections of globally averaged sea level rise range from 0.18m to 0.58m in 2090–2099 relative to 1980–1999, while tropical cyclones are likely to become more intense, have higher peak wind speeds and bring heavier rainfall (IPCC, 2007). Small islands share a number of characteristics that increase their vulnerability to climate variability and change, including small land area, proneness to natural disasters and climate extremes, limited water supplies, high concentrations of population and infrastructure close to coasts, open economies, low adaptive capacity, and adaptation costs that are high relative to national incomes (Mimura et al, 2007).

Analyses by Feresi et al (2000) predict up to 14 per cent loss of coastal lands in Fiji due to sea level rise and flooding by 2050, lands that are prime areas for economic activities and human settlements. In some areas, the demand for water resources is expected to outstrip supply by 5–8 per cent by 2050 (Feresi et al, 2000). Agriculture, human health and fisheries are also expected to be impacted negatively because of climate change, which, in turn, will have a negative impact on the economies of Pacific island countries. Following a 'do-nothing' option, a small island such as Viti Levu (Fiji) could incur a cost equivalent to 2–4 per cent of Fiji's gross domestic product, or US$23–52 million, by 2050 in damages associated with climate-related disasters (World Bank, 2000).

The capacity to mitigate the impacts of climate change and extreme events is beyond the island countries of the Pacific. Even if the developed nations reach the target of reducing emissions as proposed by the Kyoto Protocol, climate will continue to change, and the Pacific islands, among other poor countries of the world, will have to bear the consequences. The only logical option for Pacific island countries is to proactively learn to adapt to climate variability and extreme events.

Several adaptation options have been implemented in the Pacific islands through the actions of individuals, national governments and externally-funded climate change adaptation projects. The most common step being taken is the construction of seawalls to protect settlements against coastal erosion and storm surges.

However, some options have proven to be unsuccessful in solving the underlying problems. For example, in Qoma, Fiji, the community reported experiencing frequent inundation further downstream after the construction of a sea wall upstream (World Bank, 2000). Given the uncertainties regarding impacts and adaptation strategies, varying approaches have been experimented with in the islands. For example, the Secretariat of Pacific Regional Environment Program (SPREP) carried out the project Capacity Building for the Development of Adaptation Measures in Pacific Island Countries (CBDAMPIC), which focused on community and national capacity building, and identification and implementation of adaptation measures through community participation (Nakalevu et al, 2005). The Asian Development Bank- and Canadian Cooperation Fund for Climate Change-funded project Climate Change Adaptation in the Pacific (CLIMAP) has focused on developing case studies that demonstrate climate change adaptation through risk reduction. The case studies cover the spectrum from immediate project-level actions to longer-term national-level development planning.

This chapter presents lessons learned from a case study of vulnerability to river flooding and adaptation in Navua township of Viti Levu in Fiji, a typical community of the Pacific islands. The Navua case study is one component of a larger study implemented as part of the Assessments of Impacts and Adaptation to Climate Change (AIACC) programme. The project expanded an integrated framework for assessing climate change vulnerability and adaptation to incorporate both natural and human systems, and applied the framework to Viti Levu in Fiji and Aitutaki in the Cook Islands (Koshy, 2007).

Navua Township

Navua, characterized by rapid urbanization, meagre economic activities and low to middle incomes, is prone to recurrent flooding. The 1996 census recorded the residential population as 4220, 52 per cent higher than the population in 1986 (Sinclair Knight Merz, 2000) and the current population is estimated to be near

7000. Urbanization and resettlement of displaced sugar cane farmers following the expiry of their land leases has contributed to this rapid growth. Our surveys found that, on average, a Navua resident earns $US35–46 per week, which is comparable to the average weekly earnings recorded by a consulting firm in 2000 (Sinclair Knight Merz, 2000). This indicated that the socioeconomic status of average Navua residents has not improved over the past five years. Consequently, residents also rely on subsistence farming and fishing for sustenance and to supplement their incomes.

The land area of Navua is $16.7km^2$, with a maximum elevation of 31.4m and a minimum elevation of less than 0.6m above sea level. The Navua flood-plain is characteristically low lying, increasing the potential for flooding during intense and/or prolonged rainfall episodes. A section of the Navua river measuring about 163m wide and 5.8km in length runs along the town, with the central business district and some homes only a few metres from the river banks. The greater Navua area is crisscrossed by a network of irrigation channels and floodgates at the coast, previously used to distribute and control water needed for commercial rice farming.

Before 1990, commercial rice farming was an important economic activity in the area, but it was abandoned because of competition from cheaper rice imports from Asia, floods and pest infestation (Sinclair Knight Merz, 2000). Today, the main agricultural activities are small-scale commercial and subsistence farming of root crops such as cassava, dalo or taro, and vegetables, and raising livestock such as cattle and goats. Logging in the upper catchment of the Navua river and aggregate mining in the river are also significant activities.

In 2003, 113 properties comprising 45 business properties (private and government) and 68 residential properties in the project site were surveyed. All interviewees were adults present at the properties during the survey. Information and data were gathered concerning socioeconomic variables (population, economic activities and income level); building types, recollections of past floods, views about factors contributing to flooding, adaptive measures taken by residents to cope with flooding, the barriers to implementing adaptive measures, and perceptions of climate change in general. A second survey was carried out in 2004 following a flash flood that affected the study area in April of that year. Sixty-five per cent of properties initially surveyed in 2003 were surveyed again to record data relating to the recent flood. Interviews were also held with officials from the Navua Rural Local Authority and persons familiar with flooding in Navua. Results of these surveys are discussed below.

Flood Risks

Navua has experienced frequent flooding in recent decades (Fiji Meteorological Service, 2004), and flooding is the major threat to livelihoods of Navua residents. In the recent flooding episode of April 2004, the national government incurred costs of approximately US$65,000 for emergency food rations for a 30-day period for the greater Navua area. Damage to homes was estimated at over US$100,000 (SOPAC, 2003). More than 2700 people, representing about 40 per cent of Navua's population, were displaced from their homes because of the flooding and temporarily relocated to evacuation centres (Central Division Disaster Management Council Operation Centre, 2004).

Apart from businesses and the houses of a few middle-class residents, most of the properties in Navua are not insured because of financial constraints and inability to meet basic insurance requirements. And the full social and economic impacts of the most recent and previous floods are not known. However, the impacts are deemed substantial, taking into consideration destruction of crops, loss of income and properties, diseases and, in some cases, deaths. Many may not be able to improve their standard of living if they are to sustain significant damages to life and property on an annual basis.

In the surveys of residents, five significant local flood events were recalled since 1972, three of which flooded more than 80 per cent of the land area (Table 15.1). All five of the flood events came during the wet season of November to April, also the season of tropical cyclones, and four of the events were initiated by intense and prolonged rainfall associated with tropical cyclones. However, the most recent event, which caused the most extensive flooding of the five, was due to intensive rainfall associated with two consecutive tropical depressions. Because of strong variability in daily rainfall, there is also potential for flash floods in the dry season (May–October).

Table 15.1 *Flood extent, duration and rainfall in five recalled flooding episodes in Navua*

Climate Event	Dates and Duration of Event	Total Rainfall (mm)	Average Daily Rainfall (mm/day)	Area Flooded (% of study site)
Bebe	19 Oct–6 Nov 1972 19 days	652	34	86
Wally	1–6 Apr 1980 6 days	682	113	22
Oscar	28 Feb–2 Mar 1983 3 days	412	19	21
Kina	26 Dec 1992–5 Jan 1993 11 days	537	6	89
Two tropical depressions	6–15 Apr 2004 10 days	592	59	90

Flood frequency has been observed to have increased in the past decade compared to earlier periods. Survey participants were asked their opinion on the factors that contribute to increased potential for flooding in Navua. Contributing factors identified by the respondents include increased sediment input to the river, which raises the river bed; build up of sediments at the river mouth, impeding movement of water out of the basin; the presence of abandoned irrigation channels previously used in rice farming; non-functioning floodgates, especially those at the coast; lack of regular dredging of the Navua river; and changes in rainfall patterns.

While survey respondents report their perception that rainfall patterns have changed, this is not corroborated by our analysis of weather station data for Navua obtained from the Fiji Meteorological Service. Normal rainfall for Navua, based on the average for the 1961–1990 period, is 3500mm per year, but with large interannual variations of as much as +/-40 per cent (see Figure 15.1). Analysis of observed rainfall for the 43 year period 1960–2003 does not show any discernable increasing or decreasing trend. This result is consistent with similar analyses of rainfall patterns for Suva and Nadi carried out by Mataki et al (2006). The large interannual variations are driven mainly by movements of the South Pacific Convergence Zone, the main rain-producing system of the region, and the presence and absence of the El Niño Southern Oscillation and La Niña (Mataki et al, 2006). El Niño conditions are associated with droughts while La Niña episodes are associated with enhanced rainfall across the Western Equatorial Pacific, including Fiji.

We also analysed the average return periods for extreme rainfall events in the

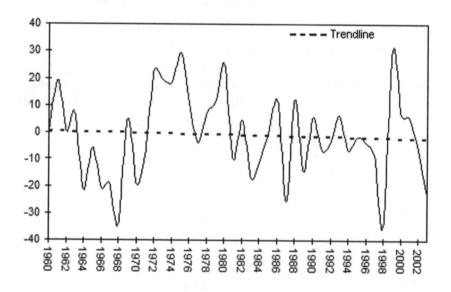

Figure 15.1 *Observed rainfall anomalies for Navua, 1960–2003 (percentage variation from normal rainfall); the trendline is not statistically significant*

two wettest months of the year, March and April, based on changes in the time period between peak rainfall events. The average return period of intense rainfall decreased from approximately 3 years in the period up to 1994 to 2 years since that time (see Figure 15.2). However, the reduction was found to be statistically insignificant. This further reinforces the conclusion that rainfall in the most recent decade has not diverged from the established norm of 1961–1990.

Nevertheless, the study site had been flooded in recent decades more

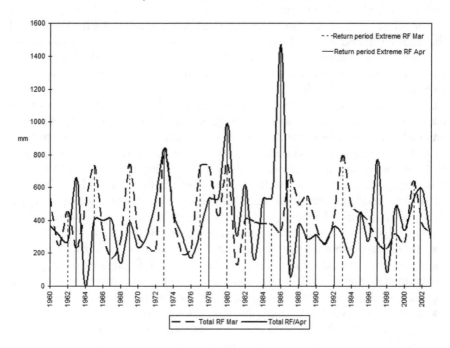

Figure 15.2 *Maximum daily rainfall during March and April, 1960–2003*

frequently than in the past. The explanation lies in the complex interplay between climatic factors and human activities in the Navua watershed. The Navua river is silted more intensely than before because of intensified human activities such as logging, aggregate mining and agricultural practices in the upper Navua river catchment (Sinclair Knight Merz, 2000; Ba, 1993; SOPAC, 2003). Siltation raises the riverbed and increases the river's potential to burst its banks during prolonged and/or intense rainfall episodes (Central Division Disaster Management Council Operation Centre, 2004; National Institute of Water and Atmospheric Research Ltd, 2004). The homeowners interviewed also recognized this as a major contributor to increasing flooding potential. Moreover, they stated the view that accumulated silt at the river mouth acts as a barrier to the free flow of water during floods.

Logging in the Navua catchment, mainly of mahogany, takes place in both natural and planted forests. Logging practices are regulated by the Fiji Forestry Decree of 1992 (Government of the Republic of Fiji, 1992) and the National Code of Logging Practice (NCLP), as well as by other environmental legislation and forestry regulations. These instruments have provisions to reduce the environmental impacts of logging, including soil erosion, blockage of waterways and sediment input to streams and rivers (Strehlke, 1996). Despite the regulations, however, logging operators often knowingly ignore requirements, engaging in excessive bulldozing and logging within waterway buffer zones and on steep slopes (Chand and Prasad, 2005). Such practices have been cited as contributors to recent floods in the Northern Division of Fiji that caused damages estimated at 10 million Fijian dollars (*Fiji Times*, 2007). Some of Fiji's largest mahogany forests are found in the upper catchment of the Navua river. Logging of these forests is contributing to problems of sedimentation of the river and is expected to increase in future years.

Aggravating the problem of sedimentation of the river is the failure of the national government to implement a programme of regular dredging of waterways. Dredging can facilitate the flow of rain-waters out to sea and dampen the severity and extent of flood events. However, the Navau river was last dredged in 1994 and funds that were earmarked for dredging the river in 2000 were diverted to other purposes (Auditor General of the Republic of Fiji, 2001).

Navua residents also attributed the extensive nature of flooding in April 2004 to dysfunctional irrigation channels and floodgates. Examination of the data in Table 15.1 indicates that the two most extensive floods occurred in 1993 and 2004, after commercial rice farming was abandoned. After the abandonment of rice farming, the irrigation channels and floodgates were poorly maintained, giving rise to blockages and uncontrolled movement of floodwater.

Elevation above the ground, location relative to the river and abandoned irrigation channels, other topographic features and structural strength all influence the vulnerability of buildings to flooding. Seventy-five per cent of surveyed homes were raised above ground level, while the remaining 25 per cent were built on ground level. Homes that were raised on pillars or had concrete porches were observed to be less affected by flooding, excluding factors such as the location of the home, the intensity of flooding and the strength of the building. The depth of water in homes that were flooded varied considerably. In most cases, both raised and unraised homes were flooded, but as expected, unraised homes experienced higher levels of water within them during floods, which also suggests that these homeowners generally sustain greater property losses. Nevertheless, even some homes raised nearly 2m above the ground level were also flooded during the five flooding episodes. These observations suggest that houses in Navua, especially those within a few metres of the river or confluence points of irrigation channels and those near previous flood water routes, may need to be raised by more than 2m to reduce the potential of being flooded in the future.

Adapting to Climate Change

Nearly 95 per cent of the interviewed residents and local authorities of Navua have heard about climate change, mainly through the media, though most are unaware of the human and natural factors responsible. They also associate the recent floods with climate change, which is an indication of their limited understanding of climate change science. Most are aware of measures that can be taken to reduce vulnerability to flooding and accept that as individuals, they have a role to play in reducing their vulnerability to flooding in collaboration with the government and its agencies. Nearly all survey respondents recommended dredging of the river as an important measure to reduce flood risks, in addition to building raised and sturdy homes. Other adaptive measures identified by respondents include taking insurance coverage, relocation, and maintaining irrigation channels and floodgates.

Adaptive capacity is defined by the IPCC as the ability of a system to adjust to climate change, including climate variability and extremes, to moderate potential damages, to take advantage of opportunities and to cope with the consequences (McCarthy et al, 2001). In the context of Pacific island countries, adaptive capacity is dependent on the net resources (financial, human and technological) available to national governments, communities and individuals to implement adaptation measures. Resource allocation and use are strongly influenced by factors such as governance, fiscal policies, tradition and culture, poverty, hardship, and prevailing socioeconomic and environmental conditions. On the basis of our studies, it appears that a majority of the residents in Navua lack sufficient net resources and the capacity to enable them to autonomously adapt to climate stresses and shocks to any significant extent without the government's intervention.

Adapting to present climate variability and extreme weather events is an early opportunity to enhance the resilience and the adaptive capacity of Navua residents to future climate change. Moreover, adaptation to climate change in a socioeconomically disadvantaged community such as the one in Navua is better approached from a broader development framework. Within such an approach, the government would oversee implementation of adaptation measures and incorporate adaptation measures in the development plans for the region. In addition, autonomous adaptation by individuals and communities should be encouraged with appropriate incentives and clear demarcation of responsibilities of government and communities in planning and implementing adaptation measures. Furthermore, government can also engage development partners (including funding agencies) through bilateral and multilateral arrangements to provide technical and financial support for the implementation of adaptation options. This is quite crucial for Pacific islands given the diminishing overseas development aid and the tight donor situation despite the continued requests of Pacific Islanders for donor contributions to satisfy various development objectives.

A lesson from our study is that a system embracing both top–down and bottom–up approaches to the adaptation process has the best chance of improving the adaptive capacity of towns in the Pacific with geographic features and socioeconomic backgrounds similar to those of Navua. This dual approach is also aligned with current regional efforts under the CBDAMPIC project to mainstream climate change adaptation into national development planning and concurrently engage and empower local communities and non-state actors to develop and implement effective and appropriate adaptation options. The two approaches are discussed separately in this chapter for clarity purposes only.

The top–down approach

The top–down approach recognizes the weak adaptive capacity of the Navua population and similar island communities, and places the onus on the national government and its agencies to mainstream adaptation by developing a framework for national adaptation policy within which to implement, promote and support adaptations that help to make society more climate-proof. National level actions should include developing and improving regulation of climate-sensitive sectors and geographic areas, investing in measures to reduce risks, and creating incentives for adaptation by local authorities, private entities and individuals. For example, incentives could be developed in consultation with local residents to enable them to afford flood-proof homes, take out insurance policies and relocate to sites that are less prone to flood risks.

Climate proofing of building codes, tourism and land-use plans can yield immediate positive results. As previously noted, Fiji has in place a variety of forestry regulations, but they are not rigorously enforced and compliance is imperfect. The result is that forestry activities have negative environmental impacts that could be avoided, including siltation of rivers that aggravate flooding. Improved enforcement of the regulations by the national government would help to climate-proof downstream communities.

Dredging of the river and proper maintenance of irrigation channels and floodgates would dampen the severity and extent of flooding in Navua. But these measures are beyond the financial capacity of the residents and require investment and action at the national level to bring the needed resources to bear. There is a national programme for dredging of rivers; however, dredging projects are implemented in an ad hoc manner and allocated funds can be diverted without proper consultation. The result is that some rivers are not dredged even several years after scheduled dates. The lack of adequate dredging equipment and faulty equipment are also partly responsible for delays, a situation that is prevalent in Pacific island countries for projects requiring specialized equipment. The government should revitalize the national dredging programme and prioritize and schedule dredging projects based on a comprehensive assessment of the vulnerability of communities and localities to flooding.

The bottom–up approach

A bottom–up approach to adaptation is a community-based approach that engages local stakeholders to identify and prioritize risks, select appropriate responses, and implement selected measures using local institutions, resources and knowledge. To be effective, individuals and local institutions need to be encouraged with proper support and incentives by the government. The approach should be underpinned by recognition and use of the positive aspects of cultural and communal-based traditions of Pacific island societies. The approach also recognizes the need to engage all stakeholders, including non-government organizations (NGOs) and intergovernmental regional organizations such as SPREP and the University of the South Pacific.

An important aspect of Pacific island communities, including those in Fiji, is the strong communal nature of living and working together, although at times this is also viewed by some as an obstacle to development (Duncun and Toatu, 2004). The communal nature of living in the Pacific, especially in the rural and urban hinterlands, brings forth an opportunity to pool and mobilize resources (for example, finance and local expertise) that are required for adaptation. Therefore, the adaptation process should be taken within the context of the community as a whole, although consideration should also be given to the needs of individuals.

NGOs and regional organizations with mandates in line with reducing climate-related risks have played significant roles in community-based development and advocacy for climate change in the Pacific. Their experience with community-level development, provision of technical advice and carrying out research on climate change issues will complement efforts by the national government and individuals to promote climate change adaptation. The engagement of regional organizations has been proven to be effective in the sharing of adaptation lessons learned elsewhere in the Pacific and to engender a consolidated stand on climate-related issues in international forums.

Autonomous adaptation has been observed within Navua town, especially in the construction of homes. Discussions held with officials of the Navua Rural Local Authority (RLA) indicated that the number of newly-approved buildings raised above the ground has been on the rise, especially since the recent floods of 2004. Residents were encouraged by Navua officials to build higher than the previous flood level. Such autonomous adaptation needs to be properly encouraged, as it will contribute to reducing the vulnerability of the Navua residents to flooding and reduce financial obligations of the national government during flooding disasters.

Our assessment of the situation in Navua, and the results from the CBDAMPIC project, indicate that local communities should be actively engaged in the full adaptation process, from planning to implementing and monitoring adaptation measures. Their involvement in this process is important, whether the technical advice on adaptation to climate change originates from local or international experts (Nakalevu, 2005). This approach will also contribute to heightening the community's responsibility to sustain adaptation to change and to proactively internalize the adaptation process. It is anticipated that by internalizing and sustaining the adaptation process, the communities' dependence on external assistance to implement adaptation options will progressively reduce over time.

Challenges to Implementing Adaptation in Pacific Island Countries

Four challenges to implementing adaptation to climate change in the Pacific are identified: (1) perceptions and competing government and individual priorities, (2) weak governance and institutional framework, (3) weak socioeconomic conditions, and (4) lack of technical capacity. We elaborate on each of these below.

Perceptions and competing government and individual priorities

Perceptions of the public and decision makers in the Pacific about climate change will influence the actions they take to deal with climate change risks. On the basis of our surveys, most Navau residents, local officials and national government officials have only a low level of awareness of climate change, in most cases influenced by media reports, which are seldom accurate. Only a few of them acknowledge the influence of human activities on the climate. This implies that many people are unable to perceive concrete links between climate change and the contribution humans make to aggravating climate change and variability.

Consequently, when the implementation of climate change adaptation is advocated, it is often perceived as an attempt to prepare for a future 'unlikely adversity', which is not as pressing as the need to meet basic daily needs such as food and shelter. Climate change is often viewed as a futuristic phenomenon and does not align well with the decision timeframes of individuals and governments, which are invariably short, at 1–5 years for governments, depending on the duration of the national parliament. Consequently, the notion of adapting to climate change is seldom regarded as a high priority and thus loses out in terms of funding and institutional support. In some cases, such perceptions are reinforced by the limited climate change awareness. A study in the Cayman Islands of the Caribbean also showed that policymakers seldom regard climate change as a priority environmental concern and therefore see little need to make policy responses to cater for it (Tomkins and Hurlston, 2003).

Perceptions that climate change is a distant and low-priority concern necessitate discussion of climate change adaptation in the context of climate variability and extreme weather events. People are better able to visualize the link between extreme weather events and climate variability and their livelihoods, and thus strategies for managing risks in this context. However, this approach to climate change discussion must be taken with care. Extreme weather events are frequent in Pacific islands and communities may perceive them as normal events.

For example, Fiji is affected by an average of two tropical cyclones per year and numerous tropical depressions. Consequently, if human-caused climate change is associated with climate extremes, climate change may come to be regarded as remaining within the norms with which islanders presently have to cope. This can lead people to downgrade the importance of adaptation to climate change.

The need for caution is also pertinent because national governments usually provide relief assistance during and after tropical cyclones and severe tropical depressions. Association of climate change with extremes could create expectations that individuals and communities should rely on government relief to cope with climate change. This could accentuate the local community's dependence on the national governments while also dissuading national governments from actively participating in the adaptation process. Proper public awareness about climate change adaptation should aim to unravel the above misconceptions.

Institutional framework and governance

Governments in most Pacific island countries are often challenged internally and externally to demonstrate good governance by establishing appropriate institutions with proper checks and balances to optimize the delivery of goods and services to the country as a whole. National governments can no longer afford to maintain rigid decision-making structures if they are to be effective and efficient in working towards the goal of enhancing the adaptive capacity of the population to climate change. The need to promote participatory approaches to planning and decision making in the context of climate change adaptation is pertinent to ensure the internalization and sustainability of the adaptation process. However, such changes by national governments towards participatory and decentralized decision making should be judiciously implemented with national interests at their core to avoid unnecessary delays and the continued dominance of decision making by a few stakeholders. Good governance is needed to enable climate change concerns to permeate all levels and sectors of the society, including the local communities.

The case of Navua demonstrates some of the problems with institutional frameworks in Fiji that hinder complementary decision making at local and national levels. The government activities within Fiji are undertaken through four distinct systems: the National Government Administration, the Fijian Administration (which exclusively looks after indigenous Fijian affairs), the Municipal Administrations (incorporated towns and cities) and Rural Local Authorities (RLA). Navua has not been incorporated as a town under the Local Government Act and is therefore governed as a rural local authority. The RLAs are essentially public health authorities responsible for public health, building construction and other matters coming under the Public Health Act. However, most functions and services are consolidated on a national basis for efficiency and economy of scale, and as a result RLAs have relatively limited powers.

Within this framework, as an RLA the residents of Navua do not have local-level political representatives, as is the case for incorporated town and city councils, although the Navua RLA officials work tirelessly to provide services and represent Navua residents with minimal financial and human resources. This ultimately means that local-level concerns about river flooding are often inadequately dealt with at the political and administrative levels. For example, the absence of a stronger political framework in Navua (Duncan and Toatu, 2004) made possible the easy diversion of funds earmarked for dredging in 2000 (Auditor General of the Republic of Fiji, 2001). Certain officials with the Navua RLA interviewed expressed their intention to legally incorporate Navua as a town as a means to have local-level political representation and improve the services and economic activities in Navua.

As mentioned earlier, the national government needs to establish a top–down or national framework of climate change adaptation policy and planning within which bottom–up strategies can be implemented in localities such as Navua. To drive this process, the institutional and governance structures need to be reinvigorated and strengthened. Awareness needs to be raised within these structures of the significance of adapting to climate variability and extreme weather events as preparation for changing climate patterns.

There is also a lack of communication and coordination between relevant government departments (Duncan and Toatu, 2004; Raj, 2004), an institutional setback apparent in many government departments in Pacific island countries. The fragmented jurisdictions over related areas reinforce the lack of communication in some circumstances. For example, in Fiji, the Land and Water Resources Management of the Ministry of Agriculture is responsible for river engineering, drainage and irrigation, while the Public Works Department is responsible for flood control, watershed management and flood forecasting. Although there is an amicable working relationship between the two government departments (Raj, 2004), regular communication on matters of mutual interest cannot be guaranteed as they go about their day-to-day operations. Furthermore, there is no central authority for flood management and the National Disaster Management Office only plays a coordinating role during disasters.

Weak socioeconomic conditions and lack of capacity

Large adaptation projects are often costly, especially for socioeconomically disadvantaged communities in the Pacific, such as that in Navua. An average income of $US35–46 per week barely meets basic needs let alone affords flood-proof homes, relocation or insurance. On the other hand, there are also certain individuals who are already implementing autonomous adaptation measures to the risks posed by flooding in Navua and coastal erosion and storm surges in Samoa (Nakalevu, 2005).

The lack of capacity for climate science to evaluate changing climate trends and risks and to predict future changes is pervasive throughout Pacific island countries. The lack of capacity is also evident at the systemic and institutional levels and therefore affects the ability to properly plan and implement climate change adaptation within a development framework. Also lacking is the technical capacity to formally evaluate the potential performance, costs and impacts of adaptation measures. For example, under the CBDAMPIC project, the cost–benefit analysis of adaptation options identified in the project sites had to be contracted out to a consultant because of the lack of expertise at the national level to carry out such analysis.

Conclusions

Climate-related disasters put a lot of strain on the sustainable livelihoods of communities in the Pacific islands. It is anticipated that with climate change, ongoing climate variability coupled with extreme weather events will increasingly threaten people's livelihoods. A significant and growing risk to livelihoods in the township of Navua is river flooding. The climate driver of river flooding, rainfall, has not shown any significant increase in its pattern or intensity to suggest that it is the dominant driver of vulnerability in the study region. Instead, non-climatic drivers such as increased sedimentation in the river from logging, mining and other activities, degraded infrastructure such as irrigation channels and flood gates, failure to regularly dredge the river, and locating housing and other structures in flood-prone locations have been responsible for increasing the vulnerability of Navua residents to flooding. In the future, however, climatic factors could amplify the risks as sea level rises and, possibly, rainfall from tropical cyclones intensifies.

Navua typifies local communities in the Pacific islands, which are locked in a vulnerable situation because of their poor socioeconomic conditions coupled with limited input to government decision-making processes and access to financial resources, expertise and technical knowledge. A way forward is to implement climate change adaptation embracing a connective top–down and bottom–up approach that provides a national framework for climate-proof development, and resources and incentives to enable local level action. Stakeholders from national to local levels, underpinned by lessons learned through experiences with climate variability and extreme weather events, should be involved in planning, implementing and monitoring adaptation measures.

References

Auditor General of the Republic of Fiji (2001) 'Report of the Auditor General of the Republic of Fiji Islands', Suva, Fiji

Ba, T. (1993) '*Wai-Magiti Na Kena Iyau*: Rivers and river basins in Fijian development', *Geography*, University of the South Pacific, Suva, Fiji

Campbell, J. (1999) 'Vulnerability and social impacts of extreme events', in *PACCLIM Workshop: Modeling Climate Change and Sea Level Change Effects in Pacific Island Countries*, International Global Change Institute, Waikato University, Hamilton, New Zealand, pp1–6

Central Division Disaster Management Council Operation Centre (2004) 'Report of the floods encountered in the Central Division on the 8th and 15th April 2004', Suva, Fiji

Duncun, R. and T. Toatu (2004) *Measuring Improvements in Governance in the Pacific Island Countries*, Pacific Institute of Advanced Studies in Development and Governance, University of the South Pacific, Suva, Fiji

Feresi, J., G. Kenny, N. Dewet, L. Limalevu, J. Bhusan and L. Ratukalou (eds) (2000) *Climate Change and Vulnerability and Adaptation Assessment for Fiji*, International Global Change Institute, Waikato University, Hamilton, New Zealand and Pacific Island Climate Change Assistance Programme, Government of Fiji, Suva, Fiji, p135

Fiji Meteorological Service (2004) *The Climate of Fiji*, Fiji Meteorological Service, Climate Division, Nadi, Fiji

Fiji Times (2007) 'Rehabilitation won't cost $10m', *Fiji Times Online*, www.fijitimes.com/story.aspx?id=57082, accessed 15 February 2007

IPCC (2007) 'Summary for policymakers', in S. Solomon, D. Qin, M. Manning, Z.

Chen, M. C. Marquis, K. Averyt, M. Tignor and H. L. Miller (eds) *Climate Change 2007: The Physical Science Basis*, contribution of Working Group I to the Fourth Assessment Report of the Intergovernmental Panel on Climate Change, Cambridge University Press, Cambridge, UK, and New York

Koshy, K. (2007) 'Modeling climate change impacts on Viti Levu (Fiji) and Aitutaki (Cook Islands)', final report of AIACC Project No SIS09, International START Secretariat, Washington, DC, www.aiaccproject.ort/FinalReports/final_reports.html

Mataki, M., M. Lal and K. Koshy (2006) 'Baseline climatology of Viti Levu (Fiji) and current climatic trends', *Pacific Science*, vol 60, pp49–68

McCarthy, J., O. Canziani, N. Leary, D. Dokken and K. White (2007) *Climate Change 2001: Impacts, Adaptation and Vulnerability*, contribution of Working Group II to the Fourth Assessment Report of the Intergovernmental Panel on Climate Change, Cambridge University Press, Cambridge, UK, and New York

Mimura, N., L. Nurse, R. McLean, J. Agard, L. Briguglio, P. Lefale, R. Payet and G. Sem (2007) 'Small Islands', in M. Parry, O. Canziani, J. Palutikof and P. J. van der Linden (eds) *Climate Change 2007: Impacts, Adaptation and Vulnerability*, contribution of Working Group II to the Fourth Assessment Report of the Intergovernmental Panel on Climate Change (forthcoming)

Nakalevu, T. (2005) 'CIDA/SPREP CBDAMPIC project in the Pacific', paper presented at the Samoa Training Institute on Climate and Extreme Events, Apia, Samoa, Secretariat of the Pacific Regional Environment Programme, Apia, Samoa

National Institute of Water and Atmospheric Research Ltd (2004) 'The island climate update no 44', www.niwa.co.nz/ncc/icu/2004-05/icu-2004-05.pdf/view_pdf, accessed 20 May 2005

Raj, R. (2004) 'Integrated flood management (Case study: Fiji islands flood management – Rewa river basin)', www.apfm.info/pdf/case_studies/cs_fiji.pdf, accessed 28 June 2006

Ruosteenoja, K., T. R. Carter, K. Jylha and H. Tuomenvirta (2003) *Future Climate in World Regions: An Intercomparison of Model-based Projections for the new IPCC Emissions Scenarios*, The Finnish Environment 644, Finnish Environment Institute, Helsinki, Finland

Sinclair Knight Merz Pty Ltd (2000) *Environmental Impact Assessment for the Navua River Mouth Dredging Project*, Ministry of Agriculture, Fisheries and Forests, Suva, Fiji

SOPAC (South Pacific Applied Geoscience Commission) (2003) *Proceedings of the 4th National Fiji Consultations: Reducing Vulnerability of Pacific ACP States*, South Pacific Applied Geoscience Commission, Suva, Fiji

Tompkins, E. L. and L. A. Hurlston (2003) 'Report to the Cayman Islands' Government. Adaptation lessons learned from responding to tropical cyclones by the Cayman Islands' Government, 1988–2002', Tyndall Centre Working Paper No 35, Tyndall Centre for Climate Change Research, www.tyndall.ac.uk/publications/working_papers/wp35.pdf, accessed 24 April 2005

World Bank (2000) *Cities, Seas and Storms: Managing Pacific Island Economies: Vol IV, Adapting to Climate Change*, East Asia and Pacific Region, Papua New Guinea, and Pacific Island Countries Management Unit, World Bank, Washington, DC, p48

16

Adapting to Dengue Risk
in the Caribbean

*Michael A. Taylor, Anthony Chen, Samuel Rawlins, Charmaine
Heslop-Thomas, Dharmaratne Amarakoon,
Wilma Bailey, Dave D. Chadee, Sherine Huntley,
Cassandra Rhoden and Roxanne Stennett*

Dengue – A Caribbean Health Problem

Dengue fever is a potentially serious vector borne viral disease transmitted by the *Aedes aegypti* mosquito.[1] It is generally associated with common viral symptoms but may progress to the more severe dengue haemorrhagic fever, which can prove fatal, particularly for children, young adults and the elderly. Dengue is endemic in several tropical and subtropical countries, particularly affecting Southeast Asia and the Western Pacific (WHO, 2002). In the Caribbean the existence of dengue has been reported for well over 200 years (Ehrekranz et al, 1971) with sharply decreasing intervals between subsequent epidemics since 1970.

Four closely related viral strains (dengue serotypes 1–4) are responsible for the dengue epidemics in the Caribbean region either due to their introduction or reintroduction. For example, a devastating pandemic reported in 1977 and lasting until 1980 (PAHO, 1997) was caused by the reintroduction of the dengue-1 serotype. Similarly the dengue-4 strain emerged in this region in 1981, and currently all 4 serotypes coexist here (Rawlins, 1999)[2], with the number of recorded cases continuously increasing since 1981 (Figure 16.1), drastically so since 1991. Between 1981 and 1996, 42,246 cases of dengue hemorrhagic fever and 582 deaths were reported by 25 countries in the Caribbean and wider Americas (CAREC, 1997).

The various factors that influence dengue outbreaks in the region include the presence of the disease in a territory, the immunity of the population, and socioeconomic conditions that impact (among other things) on vulnerability and the ability of the disease to spread and do so quickly. The free movement of people between and beyond Caribbean borders also aids the spread of the disease. For example, the reintroduction of dengue-1 in the region in 1977 was

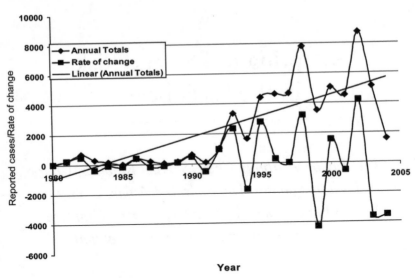

Figure 16.1 *Annual variability of the reported cases of dengue and the rate of change (increase or decrease from previous year) for the Caribbean*

initially detected in Jamaica, from where it spread to virtually all other islands of the Caribbean.

Presently, however, the factors that have emerged as the most significant are vector abundance and the frequency of vector–human contact. There is reason to believe that climate change is creating favourable conditions that are contributing to the increasing frequency of dengue epidemics in the region. Warmer temperatures result in shorter viral cycles within the vector, thus causing faster transmission to humans. This is aided by the increased vector abundance brought on either by excess rainfall (breeding sites resulting from pools of stagnant water) or by the lack of it (breeding sites resulting from stored water). Variability in these two climate stimuli, temperature and rainfall, probably accounts for some of the outbreak patterns seen since 1981 (Figure 16.1), and further changes in both variables, for example, a warmer wetter scenario, would similarly influence future outbreaks. Added to these risk factors is the fact that there is currently a deficit in the capacity of regional public health programmes to deal with the existing level of threat. A future increase in the frequency of dengue outbreaks due to climate change would only further overwhelm these already inadequate and resource-constrained programmes.

Keeping these issues in mind, in this chapter we attempt to examine the potential for adaptation to the increased future risk of dengue fever due to climate change in the Caribbean. The dengue–climate link is first explored to determine whether experience from past outbreaks can establish the increased future threat of dengue epidemics. Thereafter, the effectiveness of current adaptive strategies to contain dengue at present risk levels and their ability to cope with the possible increased future risk is also evaluated. Finally, options for expanding, strengthening or combining existing programmes or implementing new ones to better respond to the challenge of increased risk, given the constraints of the Caribbean reality, are suggested.

Data and Methodology

Dengue records were obtained from the Caribbean Epidemiological Center (CAREC). Climate data were obtained from the Workshop Database of the Climate Studies Group, Mona (CSGM), and consisted of daily and monthly station data (maximum and minimum temperatures and rainfall) for 27 Caribbean territories. Based on the available data the study period was selected as 1980–2001; climate and dengue incidence data from this period were then statistically analysed to indicate associations between the variables and to identify target countries for detailed analysis.

Geographic and socioeconomic data and survey instruments were used to identify vulnerable subgroups within target countries and to isolate common characteristics among the subgroups. The identification and classification of potential adaptive strategies was done in collaboration with CAREC and the health ministries of the target countries; these were then assessed using matrix analysis. Next, a limited set of best practices was isolated with the help of expert judgement in order to recommend a course of action that can inform appropriate policy formulation to address the increased risk of dengue fever within the region.

Dengue and Climate – A Cause for Concern

Our analysis indicates that there is a clear seasonality to dengue outbreaks within the Caribbean, with a tendency to peak towards the end of the year (see Figure 16.2; the pattern shown here, for Trinidad and Tobago, can be generalized for the entire region[3]). On average, outbreaks were observed to lag the second and principal maximum in rainfall by 3–4 weeks and the maximum in temperatures by 6–7 weeks. Similar observations have been made for Trinidad and Tobago (Campione-Piccardo et al, 2003; Wegbreit, 1997) and for Barbados (Depradine and Lovell, 2004). The climatological sequence is therefore as follows: warm temperatures (with the climatological maximum being attained) → abundant rainfall (in most cases the late rainfall season has begun to end) → dengue outbreaks.

The association between dengue outbreaks and climate was confirmed by statistically significant lag correlations between weekly dengue cases and temperature and precipitation in given years, with the relationship with temperature being stronger than that with precipitation. Strong associations with temperature are not unexpected, as studies by Koopman et al (1991), Focks et al (1995) and Hales et al (1996) have all shown that there is a shortening of the extrinsic incubation period of the dengue virus within the mosquito with warmer temperatures (in other words a shortening of the interval between the acquisition of the virus by the mosquito and its transmission

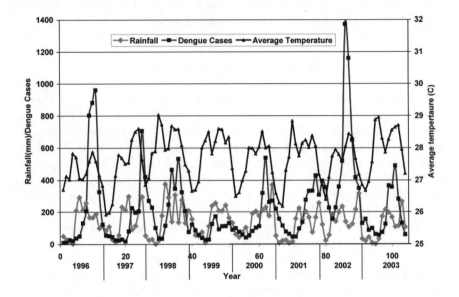

Figure 16.2 *Monthly variability of the reported dengue cases, rainfall and temperature from 1996 to 2003 in Trinidad and Tobago*

Source: Chen et al (2007).

to other susceptible vertebrate hosts). An increase in the intake of blood meals by the vector has also been associated with warmer temperatures (McDonald, 1957), which means an increased vector–human contact. The linear associations with rainfall (though weaker) are also not unexpected since both excess or lack of rainfall can contribute to increased vector abundance due to either rainwater accumulation or water storage. Note, however, that despite being weaker, the link between rainfall availability and vector abundance should not be dismissed and rainfall variability must be monitored and/or accounted for in adaptation strategies.

Besides seasonality, there are also changes on longer timescales, as well as an interannual variability in the number of dengue cases reported in the Caribbean. Reported cases increased dramatically in the 1990s, and there is marked variability from year to year superimposed on the increase. Stratifying dengue epidemics according to the El Niño Southern Oscillation (ENSO) phases reveals a bias towards El Niño or El Niño+1 years (Amarakoon et al, 2003); this is shown in Table 16.1. Similar links between upsurges of dengue fever and ENSO events have been noted for the South Pacific (Hales et al, 1996) and Colombia (including its Caribbean coast) (Poveda et al, 2000). The correlation between reported four-weekly cases and monthly temperatures was also found to be significant over the period of analysis, with a stronger correlation for El Niño and El Niño+1 years (correlations are found to be weaker and insignificant with rainfall).

Table 16.1 *Distribution of epidemic peaks among ENSO phases, 1980–2001*

Region	Total	El Niño and El Niño+1	La Niña	Neutra
Caribbean	8	7	–	1
Trinidad and Tobago	8	6	–	2
Barbados	6	5	–	1
Jamaica	5	4	–	1

Source: Chen et al (2007).

We suggest that the link with ENSO events results from their impact on both the temperature and rainfall regimes of the Caribbean. The rainy season is drier and hotter during El Niño occurrences, with warmer temperatures continuing into the El Niño+1 year (Chen and Taylor, 2002; Taylor et al, 2002). The early part of the rainy season also tends to have wetter than normal conditions during El Niño+1 years (Taylor et al, 2002). These climatic factors create conditions conducive to the propagation of the dengue virus.

Past weather records for the region show that average temperatures (maximum and minimum) have increased over the past 40 years (Peterson et al, 2002; Taylor et al, 2002), and this trend is projected to continue in the near future. Projections for 2080 suggest a Caribbean that is warmer by 2 or 3°C (Santer, 2001), with statistical downscaling confirming increases of this magnitude for Trinidad and Tobago, Jamaica and Barbados (Rhoden et al, 2005). Additionally, the El Niño phenomenon is also expected to respond to climate change (the exact manner of the response is still unclear), and some models indicate an increase in the frequency of ENSO events in a climate forced by future greenhouse warming (Timmermann et al, 1999). Both the projected temperature increases and the possible increase in El Niño frequency would impose additional stresses on an already fragile and vulnerable health system by favouring increased vector abundance and vector–human contact. This poses a serious cause for concern in the region and underscores the urgent need to develop appropriate policies and programmes to address this significant health threat.

Retirement, Pitfour and John's Hall – Profiles of Vulnerable Communities

In order to determine appropriate adaptation strategies to the increased threat of dengue fever, it is first important to identify the communities that are the most vulnerable. A local level study was conducted for a section of the parish of St James in western Jamaica, based on the pattern of dengue occurrences in 1998, to isolate the common characteristics of those most vulnerable to dengue fever (Heslop-Thomas et al, 2006). In that year, there was a concentration of cases within the parish capital city of Montego Bay and sporadic cases along a permanent stream with associated seasonal streams and gully banks. Three communities, Retirement, Pitfour and John's Hall (Table 16.2) were selected along this hydrological feature, and a questionnaire was administered to a 10 per cent sample comprising heads of households. The questionnaire solicited information on socioeconomic conditions, support systems, knowledge of the disease and cultural practices that might have implications for the spread of dengue fever.

Table 16.2 *Socioeconomic characteristics of three communities in Western Jamaica and survey sample size*

	Granville/Pitfour	Retirement	John's Hall
No of households	1507	485	572
Infrastructure	Mixture of formal and informal structures (ratio 50:50)	Few informal dwellings	Rural squatters
Economic bracket	Low income Heads of families self-employed or in the service sector in Montego Bay	Lower middle income Heads of families in the service or public sector in Montego Bay	Poor Rural squatters; primarily female heads of household engaged in domestic service or petty trading
Sample/households	151	49	57

Survey responses were then used to construct a vulnerability index based on a number of indicators identified in the literature, which include group immunity; knowledge of symptoms and vectors of the disease; the use of protective measures; source of water and water storage devices; distance from the nearest health facility; chronic illnesses; and measures of resilience and stress – education, employment, income, female household headship, room densities, coping strategies and integration into the community. The index was determined for the community in general and for the surveyed households and subgroups consisting of the most and least vulnerable within each community.

Some of the important findings from the assessment of index values and survey responses are as follows:

• Seventy-eight per cent of the respondents felt that dengue control should be the responsibility of the government, in sharp contrast to the Jamaican Ministry of Health's view that dengue control be a community responsibility. More than half the people interviewed in the communities were unaware of the causes of dengue fever, and the overwhelming majority had no knowledge of its symptoms.

• The most vulnerable subgroups, comprising 14 per cent of the entire sample, had the largest representation in John's Hall and Granville/Pitfour – the two communities with squatter settlements. Prominent characteristics of the most vulnerable group were that 95 per cent of the household heads had no skills, 87 per cent earned the minimum wage or less and 84 per cent had female heads. The least vulnerable were the employed urban dwellers (97 per cent), those who possessed marketable skills (78 per cent) and those who earned wages that were above the national minimum (81 per cent).

- In the surveyed communities, 23 per cent had no access to piped water and water supply was irregular for all respondents. Fifty-four per cent of the respondents stored water in drums, which in most cases were uncovered to facilitate rainwater collection and/or easy access to the stored water. Such outdoor drums have been found to be the most productive *Aedes* mosquito breeding containers (Focks and Chadee, 1997). Significantly, 97 per cent of the least vulnerable subgroup had water piped into their dwellings.

In summary, the study suggested that the most vulnerable to dengue fever were the poor, who lived in informal or squatter settlements and who, as a result, lacked basic community infrastructure, including access to piped water and adequate garbage disposal. The common practice of storing water in open drums creates breeding grounds for the mosquito vector. These settlements were also found to lack any community structure to facilitate collective action on any issue, including the threat of a dengue fever outbreak. The most vulnerable were also, to a large extent, unaware of the cause of dengue fever and how their actions might contribute to the disease and felt that it was the responsibility of the government to contain the disease or any threat of its outbreak.

Current Adaptation Strategies Assessed

A marked similarity can be observed in the strategies adopted for handling dengue fever in the islands of the English-speaking Caribbean. This is probably due in part to a common heritage that has led to common social structures, the active involvement of the PAHO/WHO in supporting anti-dengue programmes in member countries and the work of CAREC in defining regional strategies. Because of the similarities, it is possible to generalize about the strategies used, highlighting, wherever necessary, the exceptions. In this study we have assessed the measures of the health ministries of Trinidad and Tobago and Jamaica since both countries had a high percentage of dengue cases in our data set and both have expended resources in the recent past on the management of the disease. The strategies generally fall into one of three categories: health promotion and education, surveillance, and vector adult control.

Health promotion and education: This strategy generally targets the entire population but only after the risk is detected. Media resources, posters, printed flyers, booklets and, in limited instances, health education teams in outbreak communities are used for this purpose. Various messages regarding the identification and elimination of breeding sites, the description of symptoms, and disease treatment options are conveyed. However, despite the existence of such education and promotion units within both the Trinidad and Tobago and Jamaican health ministries, neither has staff specifically dedicated to dengue fever and nor are media resources employed before the occurrence of an outbreak. Effectiveness of promotion strategies is measured by the reduction/increase in reported cases rather than an assessment of knowledge gained or changed behavioural practices. The assumption is that the former implies the latter.

Surveillance: As with education and promotion, the health ministries of both Jamaica and Trinidad and Tobago possess surveillance units staffed by government-employed public health inspectors. In Jamaica, staffing is deemed inadequate and there is no routine surveillance of communities and dwellings except in response to outbreaks or to directives from senior health officials. Trinidad and Tobago, on the other hand, employs in excess of 600 inspectors who conduct surveillance on a quarterly cycle, but these cycles are not necessarily synchronized with dengue outbreaks, nor do they help anticipate or reduce them. Surveillance programmes in both countries emphasize the identification and treatment of breeding sites, as well as recording epidemiologically important indicators such as vector abundance – adult, larval or pupal – particularly in regions with recorded or suspected cases.

Control of adult mosquitoes: This is carried out on request or directive and involves the use of ultra low-volume (ULV) or thermal fog sprays of an appropriate insecticide. Although the consensus of the health ministries is that its effectiveness is short term (a few days), it is an often demanded solution to intolerable adult mosquito levels. Fogging is an expensive exercise and is limited to areas where dengue outbreaks have occurred or where risk is high because of the abundance of adult mosquitoes. Other vector adult control practices are usually initiated by individuals, largely to alleviate the nuisance of mosquito bites. These include the use of repellents to reduce vector–human contact or the use of physical methods such as the installation of mesh screens on buildings. In Jamaica, there is no sustained programme aimed at the reduction of larval or pupal abundance, although on request there is limited chemical control with Abate granules. In Trinidad and Tobago, inspectors from the surveillance unit carry a supply of temephos (Abate) for the active treatment of breeding places on discovery.

As is obvious, a primary weakness of current adaptation strategies is that they are reactive rather than anticipatory. In general, reactive strategies fail to engender long-term behavioural change and instead help to institutionalize the idea of the government being solely responsible as opposed to fostering community or personal responsibility. Even the best education programmes therefore fail to translate into sustained dengue preventive actions at the community level since their suggested strategies for vector control are merely viewed as necessary only when mosquito levels become intolerable. The general lack of anticipatory action suggests that the Caribbean population has developed a high tolerance for mosquitoes and mosquito bites, and any existing risk-reducing activities therefore lose some of their inherent effectiveness because they are often deployed late in the vector and/or viral development cycle.

Cuts in budgetary allocations to the health ministries and an overall resource problem are the primary reasons for the reactive approach to dengue control and prevention. In Jamaica, dengue fever is classified as a Class 2 disease and is given significantly less priority than Class 1 diseases, especially

HIV/AIDS (Class 1 diseases are those which must be reported immediately to the Ministry of Health and include HIV/AIDS, malaria and diseases preventable by immunization; Class 2 diseases are reportable weekly in line listing). There is correspondingly a lack of adequate manpower and facilities to tackle the problem on an ongoing basis; for example, there is only one understaffed virology lab in Jamaica and excess samples must therefore be sent to Trinidad and Tobago, resulting in delays in outbreak identification.

Despite the financial constraints, the Jamaican health system does possess a strong network, which can be mobilized in emergency situations. Primary health care is well organized and based on a nested system of health centres offering different levels of care. There is a health centre within five miles of every community, and the decentralization of health services has resulted in the division of the island into four health regions. This enhances the delivery of primary care by allowing autonomy in meeting the identified health needs of a region. There is, therefore, greater sensitivity to local needs and the potential for greater responsiveness in the event of epidemic outbreaks.

In summary, the current adaptive strategies are clearly ineffective in addressing the needs of the most vulnerable. Limited financial resources constrain the ability of these strategies to address the prevention of dengue as opposed to providing a reactive response after an outbreak has occurred. They are also unable to rally or facilitate collective action, nor can they assist in transferring responsibility from governments to communities. The relatively well-organized health system is the only plus, and though it does not serve as an adaptive strategy, it can allow for targeted resource allocation to the affected community's health centre in the event of an epidemic and ensure some level of primary care to all affected.

Assessment of Potential Future Adaptation Options

Table 16.3 presents a matrix of possible adaptation options available for coping with an increased threat of dengue fever. The methods listed include those currently employed in the Caribbean region (as discussed in the previous section), other options practised elsewhere in the world or on a very limited scale within the region, and options that present themselves as future (though not too distant) possibilities, as a result of ongoing research in the region.

These options are assessed by six characteristics rated high, medium and low. For example, cost is a serious adaptive constraint and so each proposed adaptation option is rated on the likely cost of implementation within the context of the Caribbean region. The assessments are a best guess (expert opinion) and are guided by the views and knowledge of the region's environmental health officers. The six assessment characteristics are:

1   cost of implementation;
2   effectiveness (as measured by its long-term ability to reduce risk or address vulnerability);
3   social acceptability;

Table 16.3 *Adaptation strategies matrix*

Measures	Cost	Effectiveness	Social Acceptability	Friendly for Environment	Neighbour Effects	Technical Challenges and Socioeconomic Change	Score
Short term							
1. Adulticide (ULV or thermal fog sprays) in truck or air	H	L	L	L	L	H	6
2. Education (disease symptoms, sanitizing the environment)	M	M	H	H	H	M	24
3. Surveillance for vector or larval/ pupal control	H	M	M	M	M	L	18
Long term							
1. Surveillance for vector or larval/ pupal control and environmental sanitation	H	M	M	M	L	L	16
2. Community education and involvement	M	H	H	H	H	M	26
3. Chemical control	H	M	M	L	M	L	16
4. Biological control	H	H	M	H	M	M	20
5. Adult control							
– Mesh windows	M	H	H	H	H	H	24
– Personal protection	M	M	M	M	M	H	16
6. Use of physical control; low-cost secure drums	H	H	M	H	H	H	20
7. Granting security of tenure to squatters	H	H	H	M	H	H	20
8. Early warning system	M	H	H	H	H	H	24

Note: Columns 2 through 7 indicate assessment criteria. Column 8 gives a composite score based on the ranking in columns 2–7.

4 environmental friendliness;
5 promotion of neighbourliness; and
6 technical and/or socioeconomic challenges to implementation.

A simple composite score is offered in the final column for comparison purposes. In compiling the score, high is given a score of 5, medium a score of 3 and low a score of 1, except for categories 1 and 6, in which the scoring allocation is reversed. The maximum possible score is 30. The adaptation strategies are again categorized under the headings of health education and promotion, surveillance and adult vector control. They are also divided into short-term and long-term practices, in other words, by whether their intent is to immediately alleviate the threat associated with dengue fever or to do so gradually.

The short-term strategies are those currently adopted in the region (see previous section), namely public education aimed at encouraging individuals to identify and eliminate current breeding sites and to identify dengue symptoms; surveillance in outbreak communities for the purpose of environmental sanitization; and adult mosquito control with an appropriate insecticide (fogging). Of the three, public education received the highest composite score, whereas adulticidal fogging received the lowest score. Education benefits from the fact that in the present framework it is generally medium to high ranked in each category. Its effectiveness is medium ranked because of the seasonal nature of the campaign, while the presence of established units to handle education accounts for the medium (as opposed to high) ranking with respect to cost and technical challenges. Insecticidal fogging, though often demanded and practised, received a low score because of limited long-term effectiveness, inability to promote neighbourliness (people shut their windows during fogging) and limited social acceptability (the commonly used insecticide, malathion, has an unpleasant odour and needs specialized equipment for distribution).

Of the long-term strategies assessed, the education strategies again achieved the highest composite ranking (though only marginally), with the focus on sustained campaigns aimed at community education (as opposed to targeting individual behavioural practices) and community involvement. Chemical control, surveillance practices and strategies relying on individuals to personally protect themselves received the lowest scores. Surveillance as a long-term approach does not engender neighbourliness (general suspicion) while the best personal protective measures come at a cost, thereby limiting their possible use by the poorest, who are the most vulnerable.

Generally, however, most strategies fall in the mid range of scores (16–24), suggesting that relative advantages in one area are offset by disadvantages in another. Physical control of mosquitoes via the use of low-cost covered drums would address vulnerability issues associated with water storage, but such drums or drum covers are yet to be designed and would have to be subsidized or made available free to the most vulnerable. Even then, much would depend on householders being vigilant in covering containers. Granting security of tenure to squatters would promote community structure and increase the possibility of the eventual implementation of appropriate infrastructure for regular water supply. Such a move, however, is costly and fraught with social tensions. Biological control, for example, using fish to control mosquito larvae, is an environmentally friendly option, but is not suited for community practice unless the community could be persuaded of the benefits of proper implementation. Finally, an early warning system for action would require the coordination of a number of agencies (for example, climate research and monitoring agencies and health ministries) and necessitate the development of appropriate thresholds for action and coordinated action plans.

Best Practice Recommendations

At present there is no single 'best' adaptation measure to counteract the threat of increasing dengue fever within the Caribbean. As suggested by Table 16.3, the various strategies have their relative merits and demerits. In light of this, we offer three potential options for tackling the dengue adaptation problem. Each option represents a combination of selected strategies outlined in Table 16.3, with due consideration given to their relative strengths and weaknesses. The options also give primacy to the need to address vulnerability issues, namely the lack of knowledge about dengue fever, the lack of community structure to facilitate collective action and the issues of water storage. The options are in increasing order of human and economic investment required and all assume that the currently practised strategies (outlined above) are at least maintained.

Option 1 – Refocusing current strategies

Option 1 advocates that currently used strategies at least be maintained at their present level of activity and funding, but that approaches to them be refocused and relatively minor modifications be made. The key emphasis would be on education with a leaning towards personal and community education derived from the environmental sanitation and vector control strategies proposed in the campaign. This is as opposed to merely providing information about the disease and the steps to be taken to reduce mosquito abundance. A proposed modification would also be to engage communities before the rainy season through organized activities in local churches, schools, and youth and service clubs, and using competitions to test knowledge and community cleanliness. Involvement before dengue onset would promote long-term behavioural change (not restricted to the dengue season) and community responsibility. Behavioural change strategies have been advocated by the WHO as an important mechanism for prevention and control of dengue fever (WHO, 2000) and community mobilization has shown recent success in reducing cases of dengue fever during outbreaks (WHO, 2005). Vector surveillance in its current form would provide added support for the educational activities, particularly approaching the dengue season.

Option 1 calls for the least additional investment, though a capacity upgrade of the education and promotion units of the health ministries would be required to initiate and sustain activities outside the dengue season. The possibility of cost sharing with the engaged community groups should also be explored.

Option 2 – Plus proper water storage

Option 1 does not address the vulnerability issues surrounding proper water storage. The proposed adaptation strategies for this purpose given in Table 16.3 (design of drums and covers and security of tenure) are, however, costly; consequently, option 2 would require an even greater investment by the ministries of health.

In option 2, the refocusing actions of option 1 are still undertaken, as they address education deficiencies and community involvement and responsibility, which are the two identified characteristics of the most vulnerable. In addition, however, the design of a suitable low-cost water storage drum or drum cover would be actively pursued. Currently, water is stored in discarded oil drums that are left open to catch rainwater runoff from roofs, which encourages mosquito abundance by creating breeding places. A covered low-cost unit which allows water in and whose cover is easily removable but secure, or from which water can easily be otherwise removed, would be ideal. There is also the option to design only a drum cover that meets the above requirements since the commonly used storage drums are fairly standard in size. Such storage units/covers do not currently exist and might be costly to design and manufacture with little guarantee of their eventual use by the community. To ensure the latter, incentives would have to be offered (for example, in the form of subsidies) and an intensive public education campaign emphasizing the value of covered drums/drum covers would have to be undertaken. Incentives may also have to be offered to ensure the appropriate use of drum covers and efforts would have to be made to ensure that other habitats are made vector-free.

Option 3 – Plus an early warning system

Like option 1, an early warning system has the advantage of anticipatory action. However, whereas option 1 promotes education simply based on the knowledge that there is a dengue season, an early warning system attempts to gauge the severity of any possible outbreak. Consequently, responses can be appropriately tailored based on the anticipated level of threat.

Option 3 therefore proposes the actions of option 1 but coupled with an early warning system. An example of the structure of a simple early warning system is given in Figure 16.3 and would involve multi-sectoral cooperation. Monitoring of climatic indices would be undertaken by the meteorological services, the regional universities and/or the regional climate research institutes. Monitoring would involve the continuous close tracking of dengue-related temperature indices (for example the MAT index discussed in Chapter 2 of Chen et al, 2007), especially immediately following the onset of El Niño, when the risk of a dengue epidemic is greatest. If climate (tempera-

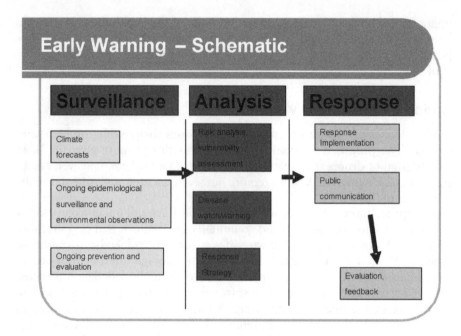

Figure 16.3 *Schematic of a possible early warning system*

ture) thresholds are being quickly approached, then an alert or *watch* would be issued. On this basis, epidemiological surveillance by the ministries of health would be initiated or its frequency altered, and the education campaign tailored to meet the perceived level of threat. If the surveillance data confirm the presence of the pathogen or an increase in its abundance and if climate thresholds are exceeded then subsequent *warnings* would be issued, as needed. One benefit of this multi-staged early warning approach is its cost effectiveness since continuous monitoring of climate threshold is less costly than continuous epidemiology surveillance. With this approach response plans can be gradually ramped up (for example, by the inclusion of other strategies such as chemical or biological control) as forecast certainty increases, allowing public health officials several opportunities to weigh the costs of response actions against the risk posed to the public.

The implementation of option 3, however, requires a memorandum of understanding between the cooperating institutions, a definition of roles, a focal point, some investment in research and the possibility of staging a pilot project. A similar framework for a health early warning system has been provided in US National Research Council (2001).

Conclusions

There has been a significant increase in dengue cases in the Caribbean since 1991, and presently there is concurrent circulation of all four dengue serotypes in the region. The increase can be linked to climate since both the abundance of vectors and the transmission rate are modulated by temperature and rainfall. A marked seasonality in dengue outbreaks has been observed and extreme changes in climate stimuli (for example, as occur during an El Niño or El Niño+1 year) also appear to increase the risk of severe outbreaks. Current adaptation strategies within the region are limited as they are reactive rather than anticipatory, primarily due to cost constraints, and give priority to reducing vector abundance over reducing vector–human contact. Though both strategy foci are important, the current strategies also fail to address the needs of the poor, who are the most vulnerable to the disease. Water storage is a critical issue for this vulnerable population, which lacks both an adequate knowledge of the disease and a sense of neighbourliness that can promote community action. Adaptive strategies that target these issues, particularly appropriate water storage that discourages vector breeding, would be ideal, though not necessarily easy or practical to implement within the Caribbean context.

Admittedly, however, even if proper water storage facilities were provided, there would still remain the uphill task of convincing people to remove the additional breeding sites in the vicinity of residences, including pools of water collected in old tyres, in plant pot bases and in other domestic receptacles lying around. Consequently, of the three adaptation options offered in the previous section, the implementation of an early warning system (option 3) might hold significant potential. As its basis, option 3 suggests that it might be easier to achieve behavioural change during an emergency than its induction as an everyday practice. Because it is possible to predict the likelihood of outbreaks on the basis of climatic changes, for example, El Niño events, public health authorities could try issuing emergency warnings and urging prompt action during such critical periods that are favourable to vectors. If this temporary behaviour change is successful, it may eventually become widespread and permanent. Over time, option 3 could then facilitate the transfer of information, the transformation of behavioural practices, (hopefully) the engendering of community spirit and action, and the gradual shift of responsibility for alleviating the dengue threat to government–community partnerships. Of course, the meteorological services and public health authorities must first be persuaded to cooperate in operationalizing this option.

We finally note again that even at current levels of threat, there is an inability to cope, and reported cases are continuing to increase. Inaction is therefore not an option for the Caribbean, especially in the light of increased future threat due to anticipated climate change.

Notes

1   The *Aedes egypti* mosquito acquires the dengue virus when it bites a sick person. The virus then undergoes a 8–10 day period of incubation within the mosquito after which it is can be transmitted to any other human the mosquito bites for the purpose of obtaining a blood meal. A mosquito carrying the dengue virus is capable of transmitting it for life (WHO, 2002).
2   Dengue 2 and dengue 3 were previously introduced during the early 1950s (Downs, 1964) and 1960s (Ehrekranz et al, 1971) respectively.
3   The Trinidad data also show a bimodality to the dengue case pattern, which, however, is not consistent in other territories.

References

Amarakoon, D., A. A. Chen and M. A. Taylor (2003) 'Climate variability and patterns of dengue in the Caribbean', *AIACC Notes*, vol 2, no 2, p8, Assessment of Impacts and Adaptation to Climate Change, International START Secretariat, Washington DC

Campione-Piccardo, J., M. Ruben, H. Vaughan and V. Morris-Glasgow (2003) 'Dengue viruses in the Caribbean. Twenty years of dengue virus isolates from the Caribbean Epidemiology Centre', *West Indian Medical Journal*, vol 52, pp191–198

CAREC (1997) 'Epinote: An update of dengue fever in the Caribbean', Caribbean Epidemiological Centre, Trinidad, West Indies

Chen, A. A. and M. A. Taylor (2002) 'Investigating the link between early season Caribbean rainfall and the El Niño +1 year', *International Journal of Climatology*, vol 22, pp87–106

Chen A. A., D. D. Chadee and S. C. Rawlins (eds) (2007) 'Climate change impact on dengue: The Caribbean experience', Climate Studies Group, Mona, Kingston, Jamaica

Depradine, C. A. and E. H. Lovell (2004) 'Climatological variables and the incidence of dengue disease in Barbados', *International Journal of Environmental Health Research*, vol 14, pp429–441

Downs, W. (1964) 'Immunity patterns produced by arthropod borne viruses in the Caribbean area', *Annals of the Institute of Tropical Medicine*, (Lisbon), vol 16, no 9, pp88–100

Ehrenkranz, N. J. et al (1971) 'Pandemic dengue in Caribbean countries and the southern United States – Past, present, and potential problems', *New England Journal of Medicine*, vol 285, pp1460–1469

Focks, D. A., E. Daniels, D. G. Haile and L. E. Deesling (1995) 'A simulation model of the epidemiology of urban dengue fever: Literature analysis, model development, preliminary validation, and samples of simulation results', *American Journal of Tropical Medicine and Hygiene*, vol 53, pp489–506

Focks, D. A. and D. D. Chadee (1997) 'Pupal survey: An epidemiological significant surveillance method of *Aedes aegypti*: An example using data from Trinidad', *American Journal of Tropical Medicine and Hygiene*, vol 56, pp159–167

Hales, S., P. Weinstein and A. Woodward (1996) 'Dengue fever in the South Pacific: Driven by El Niño Southern Oscillation?' *Lancet*, vol 348, pp1664–1665

Heslop-Thomas, C., W. Bailey, D. Amarakoon. A. Chen, S. Rawlins, D. Chadee, R. Crosbourne, A. Owino, K. Polson, C. Rhoden, R. Stennett and M. A. Taylor (2006) 'Vulnerability to dengue fever in Jamaica', AIACC Working Paper No 27, International START Secretariat, Washington, DC, www.aiaccproject.org

Koopman, J. S., D. R. Prevots, M. A. V. Marin, H. G. Dantes, M. L. Z. Aquino, I. M.

Longini Jr and J. S. Amor (1991) 'Determinants and predictors of dengue infection in Mexico', *American Journal of Epidemiology*, vol 133, pp1168–1178

McDonald, G. (1957) *The Epidemiology and Control of Malaria*, Oxford University Press, London

National Research Council (2001) *Under the Weather: Climate, Ecosystems and Infectious Disease*, Commission on Geosciences, Environment and Resources, Board on Atmospheric Sciences and Climate, National Research Council, National Academy Press, Washington, DC, US

PAHO (Pan American Health Organization) (1997) 'Re-emergence of dengue and dengue hemorrhagic fever in the Americas', *Dengue Bulletin*, vol 21, pp4–10.

Peterson, T. C., M. A. Taylor, R. Demeritte, D. Duncombe, S. Burton, F. Thompson, A. Porter, M. Mercedes, E. Villegas, R. Semexant Fils, A. Klein Tank, R. Warner, A. Joyette, W. Mills, L. Alexander, B. Gleason (2002) 'Recent changes in climate extremes in the Caribbean region', *Journal of Geophysical Research*, vol 107(D21):4601, doi:10.1029/2002JD002251

Poveda, G., N. E. Graham, P. R. Epstein, W. Rojas, M. L.Quiñones, I. D. Valez and W. J. M. Martens (2000) 'Climate and ENSO variability associated with vector-borne diseases in Colombia', in H. F. Diaz and V. Markgrtaf (eds) *El Niño and the Southern Oscillation*, Cambridge University Press, New York

Rawlins, S. (1999) 'Emerging and re-emerging vector-borne diseases in the Caribbean region', *West Indian Medical Journal*, vol 48, pp252–253

Rhoden, C. L., A. A. Chen and M. A. Taylor (2005) 'Scenario generation of precipitation and temperature for the Caribbean', in *Proceedings of the 7th Conference of the Faculty of Pure and Applied Sciences*, Kingston, Jamaica

Santer, B. D. (2001) 'Projections of climate change in the Caribbean basin from global circulation models', in *Proceedings of the Workshop on Enhancing Caribbean Climate Data Collection and Processing Capability*, Kingston, Jamaica

Taylor, M. A., D. B. Enfield and A. A. Chen (2002) 'The influence of the tropical Atlantic vs the tropical Pacific on Caribbean rainfall', *Journal of Geophysical Research*, 107(C9):3127, doi:10.1029/2001JC001097

Timmermann, A., J. Oberhuber, A. Bacher, M. Esch, M. Latif and E. Roeckner (1999) 'Increased El Niño frequency in a climate model forced by future greenhouse warming, *Nature*, vol 398, pp694–697

Wegbreit, J. (1997) 'The possible effects of temperature and precipitation on dengue morbidity in Trinidad and Tobago: A retrospective longitudinal study', PhD thesis, University of Michigan, Ann Arbor, MI

WHO (2000) 'Strengthening implementation of the global strategy for dengue fever/dengue haemorrhagic fever prevention and control', report of informal consultation, 18–20 October 1999, World Health Organization, Geneva, Switzerland

WHO (2002) 'Dengue and dengue haemorrhagic fever', World Health Organization Fact Sheet No 117, revised April 2002, available at www.who.int/mediacentre/factsheets/fs117/en/

WHO (2005) 'Dengue outbreak in Timor-Leste', *Communicable Disease Newsletter*, vol 2, no 1, p12

17

Adaptation to Climate Trends: Lessons from the Argentine Experience

Vicente Barros

Introduction

Adaptation to climate change is not usually a priority issue in most countries; this has largely been attributed to a lack of adequate and unambiguous information on the specific future impacts of climate change, especially at regional and local scales. In developing countries, adaptation is further hindered by the relative absence of social and institutional practices of long-range planning. The only exceptions are cases where there is some certainty about the direction of future trends; for example, sea level rise. In contrast, the issue of climate variability is better understood and so far much scientific research has been directed towards understanding adaptation to climate variability in the hopes of drawing important lessons that can inform adaptation to climate change in the future.

Another aspect of the global climate system that is even less well understood is that of long-term climate trends (of 20 to 50 years) that have in the past substantially altered the climate in some regions of the world. These trends could be the result of either global warming or interdecadal natural variability or both. Attribution of these trends to a specific cause is a complex issue because they may also be influenced by many natural or local human-driven processes. However, irrespective of their relation to global climate change, these trends do produce important social and economic impacts, adaptation strategies for which could also provide important insights into adapting to future climate change. Since such trends result in entirely new climatic conditions, adapting to them is much more complex than adapting to interannual climate variability. Sometimes it may be decades before there is public or even technical awareness of such trends and their associated costs or opportunities, while in other cases responses may be relatively fast.

Over most of southern South America, there were important long-term trends in precipitation during the second half of the last century (Barros et al, 2000), which were also reflected in the trends in mean river streamflows (Camilloni and Barros, 2003). These trends were positive on the eastern side of the continent and negative on the Andes Cordillera and west of it (Giorgi, 2003). As a result, there are a number of areas and socioeconomic systems in southern South America that are now under entirely new climatic conditions. This region therefore offers very recent experiences in adaptation to such long-term climate trends and could potentially provide valuable lessons for adapting to the impacts of climate change in the future.

In this context, this chapter analyses five such experiences: the rapid adaptation of the agriculture system; the increasing frequency of floods along the flood valleys of the great rivers of eastern Argentina (Camilloni. and Barros, 2003; Barros et al, 2004); the increasing frequency of extreme precipitation events in the central and eastern part of that country (Re et al, 2006); the decreasing trend in the discharges of rivers fed by the melting of snow and ice from the Andes and rainfall in central Chile; and the recurrent floods on the coastal zones of the Río de la Plata caused by wind storms (Barros et al, 2003). These five case studies vary in terms of climate aspects, socioeconomic systems, and responses to the new climate conditions or climate threats. However, all of them indicate that awareness about the climatic changes plays a central role in the adaptation process. They also highlight the role of scientific knowledge in enabling timely and effective adaptation by building awareness and guiding appropriate responses. An in-depth discussion on each of these case studies is offered in the following sections, categorized on the basis of the nature of responses to climate trends; the nature of the trends themselves; and changing attitudes in adaptive responses.

Autonomous and Planned Adaptation Responses to Current Long-Term Trends

Autonomous adaptation is usually triggered by climate-driven changes in natural or human systems (McCarthy et al, 2001). It is typically the result of reactive responses to current climate impacts (rather than preventive measures) by individuals or groups acting independently that could nonetheless result in large cumulative effects.

On the other hand, planned or anticipatory adaptation takes place before impacts of climate change are observed and is also referred to as proactive adaptation (McCarthy et al, 2001). This adaptation is necessarily based on scientific knowledge that anticipates how changes in climate will evolve and is usually undertaken under collective and organized social policies or measures, more likely directed by a government agency or by a large public or private organization.

An example of each of these adaptation types is offered in the two case studies that follow.

A case of autonomous adaptation: The expansion of the agriculture frontier

Southern South America, east of the Andes Cordillera, is one of the regions of the world that has seen the largest positive trend in mean annual precipitation during the 20th century (Giorgi, 2003), especially after 1960 (Figure 17.1).

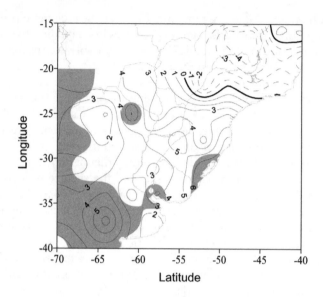

Figure 17.1 *Linear trends of annual precipitation (mm/year), 1959–2003*

Until recently, technical knowledge about such long-term climate trends was quite modest (Castañeda and Barrros, 1994; Barros et al, 2000), with almost no public awareness. As a result, adaptation responses to the changing precipitation trend in southern South America have so far primarily been reactive in nature and limited to the agricultural sector, where the effects of these changes could be more easily perceived and managed. The increased precipitation therefore prompted a rapid expansion of agriculture into the former semiarid areas of this region, towards the west of what was known as the humid plains, aided by new technologies in crop production. Figure 17.2 shows this increased precipitation zone as a 100-km westward shift of the isohyets that are considered the boundary of extensive agriculture, namely those of 600mm annual rainfall in the south and 800mm in the north.

This positive trend in precipitation was first observed in the provinces of Buenos Aires, La Pampa and Córdoba, south of subtropical Argentina, in the 1960s (Barros et al, 2000). In response, there was significant agricultural extension in the provinces of La Pampa and Córdoba during the 1970s and in Buenos Aires province there was an increase in cultivated area from 8.3 to 9.6 million hectares between 1971/1972 and 1982/1983. Between 1992 and 2004, the total cultivated area in Argentina expanded from about 20 million hectares to about 29 million hectares, 4.6 million hectares of the newly added 8.7 million hectares being located in the provinces of La Pampa, Córdoba, Santiago del Estero and Chaco, in other words in the western and northwestern border of the traditional humid plains. Much of the change in the La Pampa region happened prior to 1982/1983, but in the Córdoba, Santiago del Estero and Chaco provinces a steady increase in cultivated area continued,

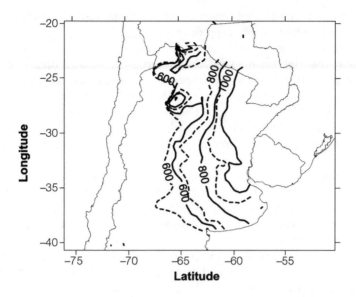

Figure 17.2 *Isohyets in mm: 1950–1969 (solid line) and 1980–1999 (dashed line)*

with soybean being the primary crop of choice. These newly extended agricultural areas also saw an increase in crop yields, which were partly due to the positive precipitation trends and partly due to the geographic extension of the growing area (Magrin et al, 2005). Table 17.1 shows the cultivated areas for these four provinces.

Table 17.1 *Cultivated areas*

	1971/1972	1982/1983	1992/1993	2003/2004
La Pampa	1.6	2.2	2.1	1.9
Córdoba	3.5	4.6	3.9	6.6
S. del Estero	0.3	0.3	0.3	1.2
Chaco	0.7	0.7	0.7	1.5
Total four provinces	6.6	7.8	7.0	11.2
Country	20.1	23.9	20.2	28.8
Share of the four provinces in %	32.8	32.6	34.7	38.9

Note: Values are in millions of hectares.

Initially, crops other than soybean were cultivated, but from the 1970s onwards much of these new agricultural areas were taken up by soybean farming. Besides the positive trend in precipitation between the second half of the 1970s and the 1990s that made soybean cultivation possible in the former semiarid regions, other factors such as favourable international prices and new technological packages that included minimum or no tilling practices contributed to the selection of soybean as the crop of choice.

This adaptation of agriculture to changing climatic trends was entirely autonomous and not planned by the government or any organization. It resulted from a large number of individual decisions that were taken even before technical specialists became aware that new climate conditions allowed successful crop production in lands that previously did not support farming. The lag time prior to this adaptation or adjustment was about one decade, this being the period before the farmers realized that the new climate conditions were persistent.

The economic returns from the agricultural expansion were, however, not always positive. In northern Argentina, some farmers suffered severe losses after switching from livestock to crop farming because there has not only been an increase in the mean annual precipitation in this region over the past 30–40 years, but also an increase in its interannual variability. Figure 17.3 shows this increase in interannual variability as the intensification of the rate between the interannual standard deviation and the mean of annual precipitation. This climatic trend has thus simultaneously increased the vulnerability of farming, especially in the case of modern agriculture, where the costs of inputs are considerable. While the rise in the mean annual precipitation was rapidly noticed by farmers, the increasing interannual variability was not always perceived, leading to huge losses in some cases.

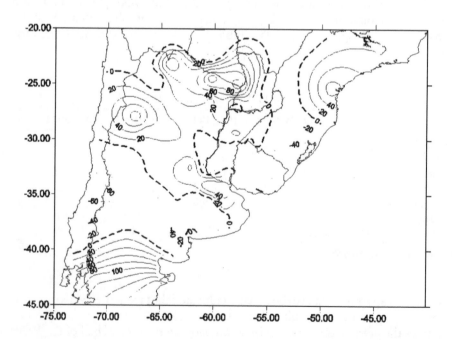

Figure 17.3 *Percentage change in the rate between standard deviation and mean value in the 1980–1999 period with respect to 1950–1969*

The financial benefits to the farmers were additionally constrained by the lack of an adequate network of rural roads in the region, which prevented them from fully exploiting the newly available cultivable area under the new climatic trends. Table 17.2 shows the low density of rural roads in the meridional axis of the expansion of the agriculture frontier that runs from La Pampa in the south, through Córdoba, to Santiago del Estero and Chaco in the north. The lack of roads has deterred the further expansion of agriculture in these provinces despite favourable climatic conditions.

Table 17.2 *Density of rural roads in six provinces*

Province	Density of Rural Roads (km/km²)
Santa Fe	0.90
Buenos Aires	0.48
Córdoba	0.34
Chaco	0.22
La Pampa	0.16
Santiago del Estero	0.11

Source: Adapted from Escofet and Menendez (2006).

Besides the economic issues, the rapid expansion of agriculture has also negatively impacted the natural environment in certain regions. Deforestation to convert land to crop farming use is causing ecosystem losses and affecting biodiversity in the northern region of the country. Moreover, if this long-term precipitation trend is associated with global warming impacts, it is possible that it could reverse in the future, since its relation with temperature may not necessarily be linear. In such a situation, the loss of the natural vegetation cover due to deforestation would favour a desertification process and further add to the economic losses of farmers. These considerations have now led to a government-issued moratorium forbidding further deforestation in the province of Santiago del Estero.

Yet another drawback of this agricultural expansion is associated with the very nature of the Argentine Pampas, which are characterized by large plains with very small slopes, hindering water runoff. As a result, the positive trend in precipitation here has been accompanied by a greater frequency and extension of floods. However, public initiatives to address this flooding were absent for decades and only recently have some projects been undertaken to facilitate and manage the runoff. In the meantime, a chaotic network of unauthorized private channels for water drainage has developed, leading to numerous conflicts as water drained away from one field often floods the neighbouring one. In addition, existing roads also lack adequate drainage provisions and further add to the problem of waterlogging.

These undesirable consequences of the rapid adaptation in the Argentine agricultural sector indicate that autonomous adaptation to climate change is a process that requires more attention and research. The lesson learnt here is that adaptation in the form of anticipatory and inexpensive regulation, and its effective enforcement is not only convenient but highly recommended as it is difficult to impose restrictions after the impacts of climate change have occurred. This must be supported by technical advice that can help moderate the negative impacts and guide the adoption of better choices, both from the environmental and the economic point of view.

A case of planned adaptation: The great river floods

Consistent with the precipitation trends described above, streamflows and flood frequency of the great rivers of the Plata basin, namely the Paraná, Paraguay and Uruguay rivers, have also considerably increased since the mid 1970s (García and Vargas, 1998; Genta et al, 1998; Barros et al, 2004). The percentage amplification of their streamflows was two to three times greater than that of precipitation over their respective basins. Berbery and Barros (2002) have shown that this amplification of streamflows was primarily due to the features of the basin and that streamflow trends were largely a result of the precipitation trend. This has also been shown, with the help of various examples of sub-basins, by Tucci (2003), who estimated that at least two-thirds of the increasing trends in the Plata basin streamflows were caused by the precipitation trends. This large amplification of the streamflow response to changes in precipitation implies that activities based on or affected by these streamflows would likely be highly vulnerable to climate variability and to climate change.

While the positive consequence of this increase in streamflows has been the increase in hydroelectric power generation, which was greatly favoured by this regional climate trend, the negative consequence was the increased frequency of major floods, which caused huge social and economic damages. Though the large streamflows originate in the Brazilian and Paraguayan territories, the greatest floods tend to occur in Argentina along the banks of the Paraná. In 1998, the flooded area reached 45,000km², and similar areas were affected by floods of comparable magnitude in 1992 and 1983.[1]

There are also clear indications of a recent change in the frequency of occurrence of these floods: four out of the five greatest discharges of the 20th century of the major river in the Plata basin, the Paraná, occurred in the past 20 years (Table 17.3). A similar situation occurred with the Uruguay river (Table 17.4), although the floods from this river extended over less territory and had smaller socioeconomic impacts than in the case of the Paraná. It can be seen that during the second half of the last century, there was a consistent increase in the frequency of large discharges: one in the 1950s, two in the 1960s and 1970s, five in the 1980s, and six in the 1990s.

The magnitude of the area flooded by the great rivers and the large number of people affected resulted in the generation of rapid awareness about the change and the need to cope, which in turn created the conditions for government action. Moreover, the worst flood events were found to be related to the El Niño phenomena (Table 17.3), which helped to identify a cause and facilitate the availability of international credit (Table 17.5). As a result, a public

Table 17.3 *Major monthly streamflow anomalies (m³/s) at Corrientes*

Date and ENSO Phase	Streamflow Anomaly (m³/s)
June 1983 El Niño	38,335
June 1992 El Niño	26,787
June 1995 El Niño	24,153
May 1998 El Niño	22,999
September 1989 Neutral	16,698

Note: Mean flow is 18,000m³/s.

adaptation policy was implemented after the great flood of the great tributaries of the Plata river in 1983 and its recurrence in 1992. After the 1983 flood, a hydrologic alert system was also implemented with a focus on the floods of the great rivers of the Plata basin. This system was improved after the 1992 flood, and several programmes related to reconstruction and the building of structural measures (defences) were executed with credit from international banks (Table 17.5).

Table 17.4 *Largest daily discharge anomalies (larger than 3 standard deviations) of the Uruguay river at the Salto gauging station, 1951–2000*

Date	Discharge Anomaly (m³/s)
9 June 1992	31,784
17 April 1959	30,575
21 July 1983	27,831
7 January 1998	27,677
16 April 1986	26,779
5 May 1983	25,678
8 March 1998	25,302
15 June 1990	24,355
24 October 1997	23,967
20 June 1972	20,660
24 April 198 7	20,187
9 September 1972	18,664
1 May 1998	18,089
16 November 1982	17,317
19 November 1963	16,867
20 September 1965	15,913

Note: Mean flow is 4500m³/s.

Table 17.5 *Programmes funded by international banks to ameliorate and prevent damages from floods in Argentina*

Programme	Funds in millions of US$	Purpose
Programme of reconstruction for the emergency caused by floods (PREI)	293.4	434 works of infrastructure and housing
Programme of defence of floods (PPI)	420	155 works of infrastructure and improvement of the hydrologic alert system
Programme El Niño Argentina (defences)	60 (25 for the great rivers area)	Works of defence
PREI (second phase)	17.3	Works of defence, housing and infrastructure studies

The benefits of this adaptation process are obvious from a comparison of the flood events of 1983 and 1998. Although streamflow and flood duration were different in the two cases, the areas affected were about the same. However, because of the defences built, the number of people evacuated was considerably lower in 1998, about 100,000 versus 234,000 in 1983. Unfortunately, this is the only overall objective measure of the successful, although yet incomplete, adaptation process, besides the qualitative assessment made after the 1998 flood. A national assessment of economic losses was conducted for the 1983 flood but a similar assessment for the 1998 flood is still lacking.

What this case shares in common with the preceding example was awareness. This was favoured by the fact that in both cases the changes were perceived by the key sectors. In the first example, the farmers initially became aware of the changes and their reactions came before the government's response, which is still lagging behind in many cases. On the other hand, in the example of the great river floods, the change was easily perceived by the entire society due to the nature of its impacts, and this led to prompt public action.

Masked Climate Trends

It is evident that the first requirement for adaptation to climate change or a changing climatic trend is awareness and perception of the associated advantages or threats. However, certain climate trends may be masked due to various natural and socioeconomic factors and their impacts may not be easily perceptible to the public. In this section two such examples of masked climate trends are explored: slow but very long trends and changes in extreme but infrequent events.

Slow trends

Climate trends can be slow enough to pass unperceived or get masked by interannual or interdecadal climate variability. An example is the situation with water resources in the regions of Cuyo and Comahue in western Argentina and in central Chile.

Western Argentina is mostly arid and the region of Cuyo (between 27° and 38°S), near the Andes Mountains, receives an annual precipitation of about 100mm, with population and economic life largely concentrated along the river oasis. The rivers are primarily fed by snow melting in the mountains and therefore have a pronounced annual cycle with a maximum in summer. In the Comahue region (between 36° and 42°S), annual precipitation is also quite small, less than 200mm in the plains, although considerably greater over the Andes. Population and economic activity in the plains, as in Cuyo, are dependent on the rivers, which in this case are fed by both snow melt and rainfall in the Andes.

A long-term decreasing trend in river flows exists in these two regions, but for various reasons the potential dangers of this trend have so far not been noticed by the public or by the key sectors involved in water administration. One problem is the lack of adequate monitoring of precipitation over the Andes and the absence of long series of snowfall data. Therefore precipitation trends over the Andes can only be indirectly assessed using data from central Chile (the same synoptic systems are responsible for producing precipitation in Chile and snowfall over the Andes). Figure 17.4 shows annual precipitation for two stations that are at the extremes of the latitudes between which Cuyo and Comahue are located. Both stations show definite downward trends that are also present in all the other stations (not shown) at intermediate latitudes. In addition, most climate models project a continuation of this downward trend during the present century, with some variation in intensity. This agreement in the direction of climate trends between different socioeconomic scenarios and models suggests that this signal is robust and very likely to occur.

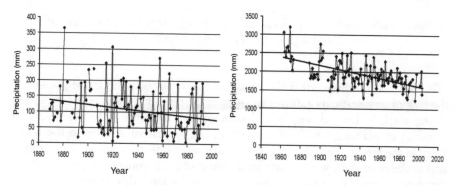

Figure 17.4 *Annual precipitation in Chile: La Serena (29.9°S, 71.2°W) (left); Puerto Montt (41.4°S, 73.1°W) (right)*

Consistent with the precipitation trends in Chile, river flows in Cuyo and Comahue have a general downward trend that is illustrated by two examples in Figures 17.5 and 17.6. In the case of Cuyo, there were two downward trends during 1920–1938 and 1945–1970, followed by periods of recovery (Figure 17.5). During the second negative trend, there was widespread public concern about the future viability of the oasis economy that led to a better administration of water. However, given the recovery after each negative trend, particularly after 1970, the key sectors have become sceptical about the dangerous implications of this long-term trend. This is significant in view of the fact that a new downward trend began in the mid-1980s (Figure 17.5). The interannual and interdecadal variability in this case tends to mask the significance of the long-term climate trend in public memory and contributes to the lack of risk perception.

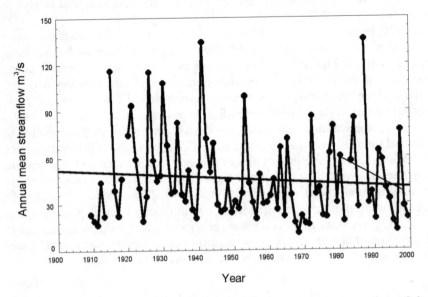

Figure 17.5 *Mean annual streamflow (m³/s) of a representative river of the Cuyo region, Los Patos river, 1900–2000*

This low level of concern is the cause (and also partially the consequence) of the lack of systematic monitoring of snow in the mountains and of the evolution of the glaciers. Indeed, the little documentation available indicates a general receding of glaciers, which, if correct, would indicate a decline in the water stock in the mountains. This, together with the downward trend in precipitation and in the river flow, would indicate a matter of great concern, especially given that future trends are projected to continue in the same direction.

In the case of Comahue, the downward trend in river flows was ignored until recently, when technical specialists began to analyse climate change data. The reason behind this attention on Comahue is that it possesses 25 per cent of the installed hydroelectric power projects in the country, providing 15 per cent of the country's generated power. A technical analysis of the generation capacity found that if the existing dams were to be operated with 1940 river flows, they would produce about 30 per cent more energy than they do now,

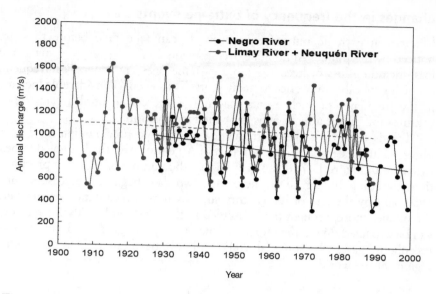

Figure 17.6 *Mean annual streamflow (m³/s) of rivers of the Comahue region, 1900–2000; note that the Negro river starts at the junction of the Limay and Neuquén rivers*

given the same level of investment. This clearly pointed to a marked decrease in river flows since 1940 (Devoto, 2005). The other major user of water in this region is the farming sector, but since it uses only a small fraction of the available water resource, no shortage has yet been experienced.

What sets Cuyo and Comahue apart is the fact that Cuyo has a long-established population, whereas Comahue has been occupied relatively recently and most of the inhabitants are immigrants from other provinces or countries. There is thus a weak collective memory about natural conditions and hazards in this area, which could account for the lack of awareness among the general public about long-term trends in the river flows.

In comparison to Cuyo and Comahue, central Chile is more developed and populated. The lack of rainfall during summer makes irrigation necessary for most crops. In addition, power generation is also largely based on hydropower. However, despite Chile's heavy reliance on its water resources, little public and scientific attention has been accorded to the slow but persistent downward trend in precipitation during the past century (Figure 17.4) and its projected continuation over the present century, as predicted by almost all climate scenarios. This is possibly because, once again, this long-term declining trend in precipitation has been masked by interannual and interdecadal variability.

It can be concluded that slow but steady trends, which could be related to climate change and are likely to continue in the future based on the projections of climate change scenarios, can be masked by the interannual and interdecadal variability that mislead the population and the key sectors, thus preventing awareness about the impacts of such trends.

Changes in the frequency of extreme events

Changes in extreme but infrequent events can take time before they are noticed by the public. Although extreme events can cause severe damage and loss, including loss of lives, the more extreme ones are generally infrequent, occurring possibly only once in many years. Thus even a considerable increase in the frequency of such events may not be perceived until a catastrophe captures the public attention, as happened in Argentina.

Trends towards a greater frequency of extreme precipitation have been observed in the central and eastern part of Argentina over the last few decades, as shown in Figure 17.7. It can be seen that the number of events of precipitation exceeding 100mm in no more than two days began to increase around 1980, and by the end of the century such extreme precipitation events were three times more frequent than observed in the 1960s and 1970s. Such trends can be expected due to the impacts of increased atmospheric concentration of greenhouse gases (Watson et al, 2003) and have been observed in many other regions of the world.

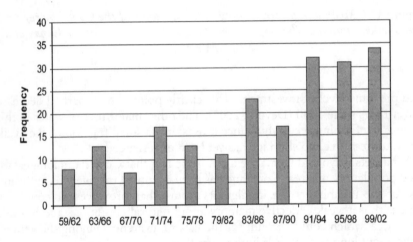

Figure 17.7 *Number of events with precipitation greater than 100mm in no more than two days in periods of four years (16 stations in central and eastern Argentina)*

In Argentina, these trends emit a robust signal, and one that does not depend on the threshold. For instance, for a 150mm threshold, Figure 17.8 shows the annual frequency of events in the 1983–2002 period and the ratio of the annual frequencies between the 1983–2002 and the 1959–1978 periods. In the eastern part of the country, this ratio is greater than 1:1 almost everywhere, and in some areas in the northeast the ratio is as high 4:1 or even 7:1. A ratio of 7:1 means that where extreme precipitation events could be expected once every 7 years in the past, they now occur every year.

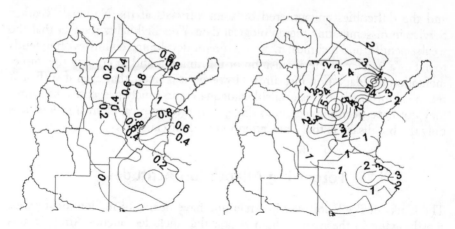

Figure 17.8 *Annual frequency of cases with precipitation over 150mm in less than two days (left); for the same threshold (150mm), the ratio of the annual frequencies between the 1983–2002 and 1959–1978 periods (right)*

Such extreme precipitation events tend to produce devastating local floods in Argentina that affect both rural and urban environments. Cultural habits and infrastructure developed during a period with a different climate are now proving to be a burden under these new climate conditions. The inadequate infrastructure (drainage, bridges, roads and so on), which was not designed to withstand such circumstances, tends to further enhance the damages caused by these flood events. In spite of this, new infrastructure continues to be developed without accounting for the changed situation.

The population in the area where the 150mm precipitation events increased twofold is 2.5 million, while in the area where these events are now four times more frequent it is 1 million. This population has only a diffuse and unclear knowledge of this change. The poor population, which bears much of the negative impacts, was probably aware of the changing precipitation trends but assumed the burden fatalistically without demanding adaptive measures. This attitude has, however, begun to rapidly change of late, especially after the event of April 2003 that flooded half the city of Santa Fe and took many lives. The people have now begun to press for solutions and no longer accept the standard excuses from public officials about the event being extraordinary or unexpected. Thus national and provincial governments have, in response, begun to institute some adaptive measures, including implementing new early-warning systems at the provincial level (in Santa Fe) or enforcing land zoning (in Chaco). In addition, the Institute for Water Research has also started a programme to develop new standards for the design of water management.

The delay in the initiation of adaptive activity in this case has been attributed to several factors, including the lack of technical knowledge, the lack of an appropriately dense network of pluviometers (to measure precipitation), and the difficulties encountered by some officials at the National Weather Service in disseminating meteorological data. This highlights the fact that the weak scientific and technical infrastructure of developing countries often tends to act as a barrier to the acquisition of information and to guiding the implementation of adaptation measures. However, it is likely that even if sufficient technical information was available, adequate and timely preventive measures may not have been implemented since, as is often typical of such events, it is only the big disasters that draw attention to their causes.

Forgetting Adaptation Attitudes

The banks of the Plata river in Argentina have nearly 14 million inhabitants, mostly living in the metropolitan region that includes Buenos Aires. In this area, high tides associated with the inward drag of strong winds from the southeast are common, especially if they occur simultaneously with high astronomic tides. These events are locally known as sudestadas and cause floods along the low coasts of the Argentine margin. Because of the shape and other characteristic features of the Plata (Figure 17.9), the sea level rise resulting from climate change is also expected to propagate inwards into the river. The results of modelling studies[2] indicate that higher sea levels are likely to cause an increase in the frequency of storm surge floods along the coast of metropolitan Buenos Aires.

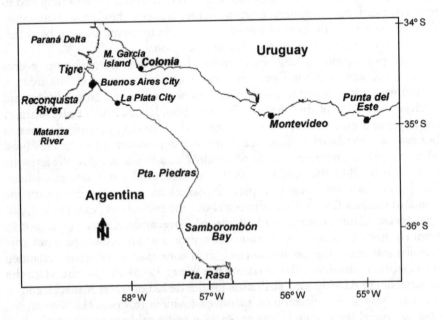

Figure 17.9 *The Plata river estuary*

In coastal neighbourhoods such as La Boca and Avellaneda, there exists a long tradition of coexistence with floods. Here, informal networks among neighbours support local practices and strategies that aid in anticipating the arrival of floods (including an early warning system and self-help and evacuation strategies), and tend to diminish local vulnerability. However, it has lately been observed that, in both these areas, the increasing influx of newcomers is gradually eroding this collective cultural adaptation to floods.

Another factor that tends to compromise the shared knowledge of cultural adaptation to floods, despite its importance, is the construction of structural defences against flooding. The coastal defence structure in the city of Buenos Aires, built in 1998, has successfully mitigated recent floods and cultural adaptation practices no longer assume as much importance. This defence was, however, designed without taking into consideration future river level rise, which may reduce its effectiveness in the coming years. By this time the knowledge of practical flood prevention and coping strategies will probably have been diminished or forgotten, and even institutional mechanisms of response to floods dismissed.

One traditional adaptation strategy employed until recently was the avoidance of living in areas exposed to frequent floods. These areas are typically located away from the city and such an adaptation strategy would also have ensured minimal social impacts from the greater frequency of future floods in this region. However, since the 1980s people have increasingly begun to favour living in private gated communities over living in the metropolitan area. The rising demand for private gated towns is making the natural areas near the water attractive for the upper middle class (Ríos, 2002). It has thus become very common for gated communities to be sited on lands that are frequently flooded, though often at an artificially elevated level of 4.4m above sea level, which is the assumed secure height against flood risks. Besides the fact that this massive modification of the environment affects ecosystems and creates drainage problems, this height of 4.4m above sea level may not be adequate protection against future more intense flood threats (Barros et al, 2005).

This trend favouring an increased number of gated communities near the coast, in the margins of the Paraná or even in the front of the Paraná delta, is expected to increase in the coming years (Ríos, 2002). At the same time, a growing occupation of the low lands in the valleys of some tributaries of the Plata river by low income people is also observed. This current tendency of occupying lands at risk of flood, by both very poor settlements and upper middle class gated communities, thus negates customary strategies for adaptation to present and future scenarios of recurrent floods.

The changing relationship of society with the Plata river and its hazards in this case study indicates that new habits, greater resources or a large percentage of newcomers may lead a social group to forget its collective adaptive attitudes towards climate hazards. In this context, the dissemination of the findings of the AIACC project (Barros et al, 2005) in collaboration with a local non-governmental organization, Fundacion Ciudad, proved to be opportune, as these results can serve to guide governments in more effective decision making regarding potential adaptation strategies for climate change impacts in this region. They can also assist individuals and private developers in making appropriate decisions regarding future property development. This case study thus also helps to highlight the importance of scientific research in helping develop and maintain the collective awareness of both present and future climate hazards.

Discussion and Conclusions

The case studies examined in this chapter show that adaptive strategies to current regional climatic trends can offer important lessons for designing socioeconomic responses to future climate change. The important trends in climate that were observed over the past 40 years in Argentina provide a wealth of valuable experiences that could help direct future adaptation planning.

It is noted that in some sectors like agriculture, autonomous adaptation can be relatively fast, as has happened in Argentina over the past few decades. This adaptation was facilitated by the relatively short production cycle in agriculture and the independent process of decision making of farmers, which produced quick, results-oriented experiences and choices. However, autonomous adaptation may also have undesirable implications for the environment, other sectors, society as a whole and sometimes even the very people who take the adaptive actions. Therefore, autonomous adaptation needs to be guided by the public and research sectors to improve its benefits and reduce its negative impacts.

When decisions cannot be individually taken, but depend on large entities like governments or big organizations, public awareness of the changing climatic conditions becomes a key issue in initiating the process of adaptation. This can be observed in the case of adaptation to the increasing flood frequency of the great rivers of the Plata basin, where public awareness of the impacts led to the implementation of a public adaptation policy by the government. Often this kind of awareness initially begins in the technical spheres, but its dissemination among the wider public is important to ensure the feasibility of political and economic decisions, even where the required funds come from international agencies, as is often the case in developing countries.

Sometimes, climate trends may have features that make them difficult to be noticed by society. This lack of awareness is more likely to occur when the local climate observation system lacks the capacity to provide the necessary information. Two cases in this chapter show that slow but steady trends, which could be related to climate change, can often be masked by interannual and especially by interdecadal variability. This variability can confuse the population and key sectors, thus preventing their perception of the risks associated with the long-term trend. The first case showed that a downward trend in precipitation and river streamflows has been taking place for about a century in two regions of Argentina and in central Chile without much public awareness, largely because it is masked by the effects of interannual and interdecadal variability. In the second case, there was a trend towards an increasing frequency of extreme, but occasional, precipitation events in eastern Argentina about which the local population was largely unaware. It was only after a catastrophic flood event in the city of Santa Fe in April 2003 that the attention of the public, authorities and specialists was drawn to the issue.

Various factors, such as changes in social attitudes and habits, new resources like structural defences or highways, a large percentage of immigrants, and even new technologies, may also reduce local adaptive capacity by diminishing communal memory of collective adaptation attitudes to climate hazards as can be observed in the final example of the coastal area of the Plata river. This process could have significant social and economic repercussions, especially if these hazards are further enhanced by climate change.

All the above experiences on adaptive responses to long-term climate trends in Argentina offer important lessons that can serve to guide the process of adaptation to climate change, especially in developing countries. What is common to these cases is that they all point to the significance of awareness about the changing climatic trends in order for adaptive strategies to be effectively implemented. However, various factors, such as the lack of technical knowledge, lack of an appropriate monitoring system, and difficulties in the dissemination of data and information, may act as barriers in the development of public awareness and delay adaptation in developing countries. This could have important implications in terms of future ability to adapt to the impacts of climate change. Scientific and technical organizations therefore have a key role to play in providing the necessary information for designing and implementing effective adaptive responses when necessary. Nonetheless, the weak scientific and technical capacity of developing countries could prove to be a big drawback for such countries and increase their vulnerability to the impacts of climate change.

Notes

1   A study made for the World Bank indicates that Argentina ranks 14th among the countries affected by floods, with economic losses that sometimes reach more than 1 per cent of annual GNP (World Bank, 2001).
2   These were developed as part of the Assessments of Impacts and Adaptations to Climate Change (AIACC) project 'Global Climate Change and the Coastal Areas of the Río de La Plata'.

References

Barros, V. (2005) 'Global Climate Change and the Coastal Areas of the Río de la Plata, final report', Assessment of Impacts and Adaptation to Climate Change Project No LA 26, International START Secretariat, Washington, DC

Barros, V., E. Castañeda and M. Doyle (2000) 'Recent precipitation trends in South America East of the Andes: An indication of climatic variability', in P. Smolka and W. Volheimer (eds) *Southern Hemisphere Paleo and Neoclimates: Key Sites, Methods, Data and Models*, Springer-Verlag, Berlin and Heidelberg, Germany, pp187–206

Barros, V., L. Chamorro, G. Coronel and J. Baez (2004) 'The major discharge events in the Paraguay river: Magnitudes, source regions and climate forcings', *Journal of Hydrometeorology*, vol 5, pp1061–1070

Barros, V., I. Camilloni and A. Menéndez (2003) 'Impact of global change on the coastal areas of the Río de la Plata', *AIACC Notes*, vol 2, pp9–12

Barros, V., A. Menéndez and G. Nagy (eds) (2005) *El Cambio Climático en el Río de la Plata [Climate Change in the Plata River]*, CIMA, Buenos Aires

Berbery, E. and V. Barros (2002) 'The hydrologic cycle of the La Plata basin in South America', *Journal of Hydrometeorology*, vol 3, pp630–645

Camilloni, I. and V. Barros (2003) 'Extreme discharge events in the Paraná river and their climate forcing', *Journal of Hydrology*, vol 278, pp94–106

Castañeda, E. and V. Barros (1994) 'Las tendencias de la precipitación en el Cono Sur de América al este de los Andes', *Meteorológica*, vol 19, pp23–32

Devoto, J. (2005) personal communication

Escofet; H. and A. Menéndez (2006) 'Vulnerabilidad de campos productivos a mayor intensidad y frecuencia de grandes precipitaciones' ['Vulnerability of farming to the greater intensity and frequency of extreme precipitations'], in *Vulnerability to Climate Change in Argentina*, Di Tella Foundation, Buenos Aires, pp411–465

García, N. and W. Vargas (1998) 'The temporal climatic variability in the Río de la Plata basin displayed by the river discharges', *Climate Change*, vol 38, pp359–379

Genta, J., G. Perez Iribarne and C. Mechoso (1998) 'A recent increasing trend in the streamflow of rivers in southeastern South America', *Journal of Climate*, vol 11, pp2858–2862

Giorgi, F. (2003) 'Variability and trends of subcontinental scale surface climate in the twentieth century. Part I: Observations', Climate Dynamics, vol 18, pp675–691

Magrin, G. O., M. I. Travasso and G. R. Rodríguez (2005) 'Changes in climate and crop production during the 20th century in Argentina', *Climatic Change*, vol 72, pp229–249

McCarthy, J., O. Canziani, N. Leary, D. Dokken and K. White (eds) (2001) *Climate Change 2001: Impacts, Adaptation and Vulnerability*, Contribution of Working Group II to the Third Assessment Report of the Intergovernmental Panel on Climate Change, Cambridge University Press, Cambridge, UK, and New York

Re, M., R. Saurral and V. Barros (2006) 'Extreme precipitations in Argentina', in *Proceedings of the 8th International Conference on Southern Hemisphere Meteorology and Oceanography, Foz de Iguazú, Brazil*, American Meteorological Society, Boston, MA, pp1575–1583

Ríos, D. (2002) 'Vulnerabilidad, urbanizaciones cerradas e inundaciones en el partido de Tigre durante el período 1990–2001' ['Vulnerability, gated communities and floods in the Tigre District during the period 1990–2001'], Tesis de Licenciatura en Geografía [Thesis of Degree in Geography], Facultad de Filosofía y Letras, University of Buenos Aires

Tucci, C. E. (2003) 'Variabilidade climática e o uso do solo na bacia brasileira do Prata' ['Climate variability and land use in the Brazilian Plata basin'], in C. Tucci and B. Braga (eds) *Clima and Recursos Hidricos no Brazil [Climate and Water Resources in Brazil]*, ABRHA, Porto Alegre, Brazil

Watson, R. T. and the Core Writing Team (eds) (2001) *Climate Change 2001: Synthesis Report*, Contribution of Working Groups I, II and III to the Third Assessment Report of the Intergovernmental Panel on Climate Change, Cambridge University Press, Cambridge, UK

World Bank (2001) 'Gestión de los recursos hídricos: Elementos de política para su desarrollo sustentable en el siglo XXI' ['Water resource management: Elements for a sustainable development policy in the 21st century'], Report no 20729-AR, World Bank, Washington, DC, US

18

Local Perspectives on Adaptation to Climate Change: Lessons from Mexico and Argentina

Mónica Wehbe, Hallie Eakin, Roberto Seiler, Marta Vinocur,
Cristian Ávila, Cecilia Maurutto and Gerardo Sánchez Torres

Introduction

The *municipio* of González, Tamaulipas, in northern Mexico and the South of Córdoba Province in the Argentinean *Pampas* are regions strongly dependent on agriculture and therefore highly vulnerable to climate variability and change. Adverse climatic events such as floods, droughts and frosts can negatively impact the economy of these regions and also affect social composition and stability. In the absence of conscious efforts to adapt, potential increases in the frequency or magnitude of adverse climate events or changes in climate averages (Houghton et al, 2001) may make it more difficult for some producers to participate in the agricultural market. This may be particularly true for small-scale commercial farmers with limited capital, who are not always able to recover from recurrent crop failures.

The centralized nature of sector policy in Mexico and Argentina also implies that adaptation strategies may be assessed and planned without taking into account specific local needs. On the other hand, the availability of technology, information and other resources at the local level are what determines the socioeconomic characteristics of production and the performance of farmers and communities, and their capacity to cope with adverse climate impacts. Understanding the existing coping strategies of farmers in specific geographic contexts is thus the first step towards the identification and prioritization of appropriate options to increase the adaptive capacity of particular farmer groups to future climate change and facilitate the creation of a more sustainable and equitable production environment (IUCN/IISD, 2004; Wehbe et al, 2005a and b).

With this objective of focusing on the local context, we therefore present two case studies of grain and cattle producers in the above mentioned regions of Mexico and Argentina. The cases are distinguished by important differences in their respective socio-productive structures, which point to the significance of the geographic context and socioeconomic circumstance in understanding the challenge of adaptation to climate change. Yet, in presenting the cases together, we illustrate that farmers in both locations have experienced similar processes of institutional and policy reforms that have had important implications for adaptive capacity. We therefore suggest that any interventions intended to enhance adaptation to climate risk need to be considered in the context of the opportunities and constraints posed by the broader institutional environment and, conversely, that there is also a need to examine how institutions and policy explain differential adaptive capacities at the farm level.

Geographic Background

González, Tamaulipas (Mexico)

The *municipio* of González lies in the watershed of the River Panuco (or Guyalejo) in southern Tamaulipas and covers an area of 3491km². It receives an annual rainfall of about 850mm, largely concentrated in the months of June to September, with a midsummer period of diminished rainfall (*la canícula*) in July and August. Drought is the most common natural hazard, although there has been occasional flooding due to cyclones and even hurricanes.

Crops and pasture together cover 50 per cent of the *municipio*'s land, and nearly 60 per cent of the economically active population is involved in these primary sector activities (INEGI, 2000). In the southwest of the *municipio*, irrigated production is supported by the Las Animas Dam, which allows for the cultivation of vegetables in addition to grain crops. In the rain-fed area, sorghum is the principal crop, followed by safflower, maize and soybeans (according to the Secretary for Agriculture, Livestock, Rural Development, Fisheries and Food, SAGARPA, these crops represented 47 per cent, 17 per cent, 13 per cent and 12 per cent respectively of the planted area in 2002). Many farmers manage two annual harvests: sorghum during the summer rainy season and safflower or additional sorghum in winter with the residual soil moisture.

The majority of the farmers in the *municipio* are *ejidatarios*, or farmers who received land as part of the land distribution programme after the 1910 Agrarian Revolution. Approximately 20–30 per cent of the farmers have a form of private tenure that generally permits larger landholdings (*pequeños propietarios*). Although sorghum is one of Tamaulipas's most important commercial crops, the local economy is not particularly prosperous. Approximately 71 per cent of the economically active population reported receiving less than two minimum salaries in 2000, 10 per cent less than in 1990 (one minimum salary in 2000 in González was approximately US$100/month) (INEGI, 2000). Over half of the adult population in the latest population census reported having had no or incomplete primary school education, and 13 per cent were illiterate (INEGI, 2000).

South of Córdoba Province (Argentina)

The South of Córdoba Province in central Argentina occupies approximately 9 million hectares or (90,000km²), with 75 per cent dedicated to agricultural activities. It lies in the western portion of the Argentinean Pampas and is a transitional area between the humid and arid regions. The average annual precipitation ranges from 700 to 900mm (SMN, 1992) and is mostly concentrated in the months of September to March (Ravelo and Seiler, 1979).

This region is home to more than 800,000 inhabitants but there exists an established trend of rural–urban migration. In the period between the last two National Agricultural Censuses (1988 and 2002), the number of farm units here declined from 21,645 to 14,299. An accelerated process of land concentration, particularly during the 1990s, has left 50 per cent of the agricultural land in the province in the hands of 9 per cent of landowners (each with more than 1100 hectares). The other 50 per cent of the land is distributed among the remaining 90 per cent of farmers; this is a highly heterogeneous group with a wide range of landholding sizes (INDEC, 2002) and varying levels of prosperity. The majority are family farmers that primarily depend on agricultural activities for their livelihoods.

Much of the commercial grain and cattle production in the area is rain-fed, although a few farmers have incorporated groundwater irrigation systems. Because the soils rarely freeze, most farmers manage two annual harvests: wheat and other fodder crops in winter, and soybean, maize, peanuts, sorghum and, to a lesser extent, sunflower (among other less important cash crops) in summer. The area is historically characterized by mixed cash crop and livestock production; however, declining relative prices of beef have resulted in a reduction in herd size over recent decades. Similar declines have been noted in the pork, lamb and poultry industry, which prior to the MERCOSUR (the Southern Common Market) trade agreement were complementary activities within the farm. Crop production has also benefited in recent years from an increase in the exchange rate and the high international prices of soybeans and maize.

Methods

Various methods, such as surveys, interviews, and workshops with farmers and other agricultural actors (public officials, leaders of farmer unions, rural infrastructure specialists and academics) were used to obtain information in both regions. The farm level survey was designed to collect data on farm characteristics (for example, type of production system, landholding size, agricultural practices and income sources), farm-level resources hypothesized to be associated with adaptive capacity (for example, education, age, technology use, climate information use, risk perception and finances) and indicators of the farm households' sensitivity to climate impacts (for example, frequency and extent of crop losses) (Table 18.1; see also Gay, 2006). Farmers' interviews and stakeholder workshops (for farmers and public officials) in both regions were used to determine their climate risk perceptions, the adaptive strategies employed and the principal constraints faced in their implementation.

Table 18.1 *Farmers' socioeconomic characteristics*

Capacity Attribute	Variable	Total South of Córdoba	Total González
Number of cases		227	234
Social/human resources	Potential experience (years)	36.5 (SD 14.1)	n/a
	Education (years)	10.1	3.4 (SD 1.77)
	Age (years)	52.6 (SD 12.0)	51.6 (SD 14.3)
Material resources	Landholding size (ha)	649.5 (SD 716,6)	69.9 (SD 285.2)
	Machinery index	1.91 (SD 1.03)	1.62* (SD 1.86)
	Income (Arg$)	213,075 (SD 329,509)	n/a
Management capacity	Rented land (as % of worked area)	38.2 (SD 34.5	n/a
	Rented land (ha/household)	n/a	13.9 (SD 54.9)
Financial resources	Other sources of income (% of cases)	18.43	n/a
	Other source of income (non-farm income as % of total income)	n/a	45.5 (SD 28.3)
Information	Official technical assistance (% beneficiaries)	30.9	26.9
Diversity	Number of crops	2.4 (SD 0.79)	1.7 (SD 1.0)
	% of land dedicated to cash crops	71.5 (SD 42.3)	n/a
	% livestock income	12.8 (SD 21.8)	12.7 (SD 21.6)
	% of land dedicated to soybeans relative to cash crop area	64.8 (SD 25.1)	n/a

Note: *In Mexico, the machinery index is the sum of six binary variables, representing the ownership of six different farm machines. N/a refers to the fact that the particular variable in question was not measured in the case study. SD is standard deviation.

Source: Survey data.

Together these data were integrated in a vulnerability assessment framework (Eakin et al, 2006) to identify the primary resources and characteristics of farms in each region that were considered necessary for adaptation (adaptive capacities) and the degree to which those characteristics were either present or absent in the population. Resources associated with adaptive capacity were assigned weights in consultation with farmers, based on their importance for facilitating adaptation (Eakin et al, 2006). This allowed adaptive capacity to be evaluated as a product of resources and attributes, in which no one resource or attribute is a substitute for another, but rather different combinations of resources can provide households with similar degrees of flexibility in the face of climate risk. The evaluation thus provided the basis for examining potential obstacles to adaptation in each case study.

Vulnerability to Climate Threats

Farmers in both regions face frequent conditions of climate variability and extremes, which continue to exert a toll on production, indicating a lack of success in adapting to climate risk. In González, rainfall is highly variable and climate extremes tend to follow a pattern of decadal oscillation (Conde, 2005). There is also a possibility that the winter rainfall here is associated with the Pacific North American Oscillation and the El Niño-Southern Oscillation (ENSO) (Cavazos, 1999) phenomena. According to experts, excessive rainfall, floods and cyclonic activity were common in the 1970s, while in the 1980s and 1990s drought and high temperatures were more common (Gay, 2006), with both resulting in crop losses. In general, a greater overall climate variability and a decrease in precipitation have been observed since the 1970s, (Sánchez Torres and Vargas Castilleja, 2005). The year 2000 was reported to be the worst year by a majority (64 per cent) of the farmers surveyed, particularly for sorghum and safflower production. Pest outbreaks are also particularly problematic and believed to be associated with the magnitude of the midsummer drought (July–August).

In interviews, farmers reported experiencing recent changes in climate in terms of increasing temperatures and some associated an increase in precipitation in September with greater moisture available for winter planting. Although climate models for the region are inconclusive in terms of future rainfall changes, they do consistently indicate that the area will likely experience higher temperatures and, possibly, a consequent decrease in soil moisture availability (Sánchez Torres and Vargas Castilleja, 2005; Gay, 2006). The city of González already experiences deficits in water availability, which is expected to increase even without climate change according to recent modelling studies (Sánchez Torres and Vargas Castilleja, 2005).

In the South of Córdoba, thermal and water conditions are important variables affecting crop yields, with soil moisture being the primary limiting factor. Winters are mild and short, characterized by frost events coupled with soil moisture deficits. Although there is a water surplus in the average balance of the region, the inter-annual variability of precipitation generates occasional droughts of varying frequency and severity (Rivarola et al, 2004a; Vinocur et al, 2004). According to farmers, such droughts, followed by hail storms, have the biggest impact on production.

The impact of climate events in the region is further complicated by the soil properties and topography in certain areas (depressed areas and flood-prone basins, salty soils, drainage difficulties, soil water capacity, and so on), which cause varying levels of risk of drought or flood and generate varying environmental responses. Drought risk increases from the east of the region to the west and south (Rivarola et al, 2004b), whereas floods are more common in the south. In that area three major flood episodes have occurred during the past 25 years, affecting agricultural production and the economy for several years after each episode (Seiler et al, 2002).

Linkages of the present climate variability with the El Niño Southern Oscillation are a good possibility and strong evidence exists of a La Niña signal associated with significant decreases in rainfall in the west and east of the region over the rainfall periods analysed (Seiler and Vinocur, 2004). On the other hand, there was insufficient evidence linking increases in rainfall during El Niño years with a clear El Niño signal, as compared to neutral years. Climate change scenarios for the region project an overall increase in mean temperature for all seasons and an increase in precipitation during summer, spring and autumn. The greater rainfall projected during the summer and autumn may contribute to an increased flood risk in the flood-prone basin of the south of the region (Gay, 2006).

Other factors that contribute to climate risks

It is important to recognize that climate is only one of several factors to which farmers are making intra-seasonal, inter-annual and longer-term adjustments in their production strategies (Risbey et al, 1999; Smit et al, 1996). In Mexico, the price of grains – principally maize, but also wheat and sorghum – declined during the 1990s and further declines are projected in the future (Claridades Agropecuarias, 2004). The liberalization of Mexico's grain import markets during this period resulted in increased competition for González farmers and necessitated financial support from the government to help farmers commercialize their sorghum harvests. A restructuring of public agricultural institutions has paralleled market liberalization, reducing the availability of publicly subsidized credit, insurance and technical assistance for smallholders (Appendini, 2001; de Janvry et al, 1995). Nevertheless, current agricultural plans for the region focus on facilitating farmers' risk management through the promotion of contract farming (to provide greater price stability) and private insurance schemes (to address climatic risk).

In Argentina, trade liberalization and a retrenchment of state roles in the agriculture sector were instituted to revive national economic growth. As a result, farmers' resources and their production decisions now have greater weight in determining their economic stability. In the 1990s, a fixed exchange rate translated into declining relative prices of traded to non-traded goods and high real interest rates, producing a 60 per cent decline in farmers' purchasing power. Despite devaluation of the Argentinean peso in 2001 and the consequent economic recovery of farmers, rising costs of production and finance and newly incorporated export taxes have prevented smaller agricultural enterprises from maintaining agricultural equipment and acquiring sufficient capital to finance their production. This situation, in a sector characterized by greater competition and economic consolidation, has increased the economic vulnerability of lower-scale producers (Lattuada, 2000) and reduced the demand for locally sourced agricultural inputs and services, thus producing a drop in local economic activity.

These socioeconomic and institutional changes in Mexico and Argentina have overall greatly impacted the sustainability of farmers' livelihoods, especially affecting the small producers, and further compounded their vulnerability to the impacts of climate variability and change.

Current Farm-level Adaptations and Public Sector Support

In both regions, farmers currently practise a variety of production strategies that represent their different capacities to manage risk and to take advantage of new opportunities in their respective agricultural sectors. From the farmers' perspective, production, income and investment decisions are rarely made in response to a single stressor such as drought risk, but are rather the outcome of simultaneously considering a wide variety of stressors, including, but not limited to, climatic factors. The degree to which households are able to and do respond to a specific climatic threat is, in part, determined by their perception of the threat, as well as the relative importance they place on climatic risk compared to other sources of stress and the range of choice and opportunity available from their particular socioeconomic conditions.

To assess farmers' climate adaptation options, we evaluated the factors considered in their production decisions and the specific role of climate and climate information in those decisions. We also recorded their current climate risk management strategies, such as crop diversification and seasonal crop switching, the potential of cattle-raising and pasture as a more sustainable alternative under drier conditions, the use of irrigation, and financial mechanisms such as insurance.

González, Mexico

In González, farmers reported a range of adjustments to the drought conditions of the 1990s, including changing crop-planting dates, switching crops, changing crop varieties or livestock breeds, modifying infrastructure or inputs, or a combination of these strategies. On an inter-annual basis, over one third of the surveyed farmers reported adjusting their crop choice, according to their observations of the timing and quantity of the initial rains of the season. The total range of crops planted in González is, however, relatively small (averaging between 1 and 2 crops per household per year in the survey), and because of this homogeneity of production, local markets are often saturated with the same crop variety, resulting in problems with their commercialization.

The government is therefore now promoting diversification into non-traditional crops and livestock, to address both environmental challenges to production (for example, soil degradation due to sorghum mono-cropping) and the lack of commercial opportunities in grain farming (Secretaría de Desarrollo

Económico y del Empleo, 2001). Some of the alternatives include crops suited to drier and warmer climates such as tequila *ágave* (*Agave tequiliana* Weber azul) and aloe (*Aloe barbadensis* Miller). The planting of buffle grass is also being encouraged through a national programme of crop conversion (PIASRE, Programa Integral de Agricultura Sostenible y Reconversión Productiva en Zonas de Sinestralidad Recurrente) (Yarrington Ruvalcaba, 2004). As a result the area under planted pasture increased by 63 per cent between 1999 and 2002 in the Rural Development District of González (which includes González's neighbouring *municipios* of Altamirano and Mante), although only a handful of farmers reported receiving support through this programme. The smaller-scale farmers were particularly dissatisfied and perceived the government's support for these alternatives to be inadequate and found the investment necessary to be prohibitive. Even much of the land planted with *ágave* and aloe was rented out by *ejidatarios* to outside investors because the *ejidatarios* often found it difficult to obtain credit and commercialize their harvests due to the small scale of their production and the variable quality of their products.

Nevertheless, an increasing number of farmers, mostly smallholder *ejidatarios*, are now investing in livestock in response to repeated crop losses and problems in commercializing their harvests. The government's Program of Incentives for Livestock Productivity (PROGAN) partly supports such livestock rearing activity. However, not all experts interviewed favoured the livestock-pasture strategy as an appropriate response to drought given that livestock farmers have been the most affected by drought in recent years. This is because, in the event of drought, livestock farmers have to resort to either culling their animals or selling them at very low prices due to a scarcity of feed or buying sorghum as feed from their neighbours, which is again expensive. Some local farmers interviewed also agreed with this assessment of the liability of owning cattle in the event of drought. According to our survey farmers who had planted pasture reported some of the highest losses due to drought in 2002, and many sold cattle as a result. Additionally the influx of live cattle from the US due to the liberalization of the cattle market has also further driven down local cattle prices.

In response to drought, some wealthy farmers, usually with larger land-holdings and private tenure, have constructed small earthen dams to capture rainwater for auxiliary irrigation during dry spells. Interviews with some *ejidatarios* who had constructed such dams revealed much scepticism about their effectiveness because in the event of insufficient rain little water is captured, meaning that the investment has been in vain.

In an effort to help farmers address climatic contingencies and price volatility, the state and federal governments are actively promoting crop insurance and contract farming to reduce the financial burden of crop loss compensation programmes (Yarrington Ruvalcaba, 2004). Very few (9 per cent) of the surveyed farmers, however, had crop insurance and the majority of farmers with insurance were *pequeños propietarios* (large landholders), although a handful of *ejidatarios* in the irrigation district also had insurance. Lack of affordability, lack of information and general distrust were commonly cited reasons for not having contracted insurance. Some *ejidatarios* who had purchased insurance in the 1980s under a government scheme reported repeated difficulties in receiving insurance payments. The recent declining value of smallholder harvests also leaves little incentive for purchasing insurance.

Other factors that inhibit farmers from experimenting with new tools such as crop insurance or new commercial crops include the relatively low education levels coupled with the absence of extension services (either private or public). Less than one quarter of farmers reported being members of agricultural organizations through which they could conceivably acquire information on public and private agricultural services and opportunities, as well as lobby for programme changes to meet their common goals.

South of Córdoba, Argentina

The farmers in South of Córdoba differed from those in González not only in their exposure to specific climate stressors but also in terms of production activities, soil conditions and use, material assets, perception of risk, and landholding size (and therefore income). As a result, their specific responses to climate impacts also differ, although direct relationships are difficult to quantify (Eakin et al, 2006).

From the survey data, the most common agronomic adaptations of farmers were adjusting planting dates (36 per cent of surveyed farmers); spatially distributing risk through geographically separated plots (52 per cent); changing crops (12 per cent); accumulating commodities as an economic reserve (85 per cent); and maintaining livestock (70 per cent). Many of these strategies were not always specifically in response to climate conditions but rather as economic responses to general changes in the agricultural sector.

Drought is perceived not only as an event in and of itself but also the result of a combination of climate events, namely increasing temperature, decreasing precipitation and wind, with livelihood impacts that potentially last over a year. Although irrigation is an obvious technical solution for drought risk mitigation, its exorbitant cost (especially for smaller farmers) and the quality of available groundwater are important barriers. Only 1 per cent of farmers in the region, mostly large landholders, have installed irrigation systems.

Cattle-rearing activities, on the other hand, are generally perceived to offer greater security than cash cropping because cattle rearing is considered to be less sensitive to climatic anomalies (for example, hail storms) and cattle also serve as an economic reserve. The provincial government has also introduced programmes, including credit support, to promote livestock rearing.

Commercial hail insurance is typically used to address the impact of hail storms, although only 65 per cent of farmers surveyed reported having contracted hail insurance and, of these, only 53 per cent contracted insurance annually. Another type of insurance, 'climate risk insurance', is even less used, largely because it is costly and also poorly implemented. Public officials interviewed reported that the implementation of subsidized climate risk insurance was often problematic due to oligopolistic practices by insurance companies, which have created a general sense of distrust among farmers. In response, a group of farmers have recently established a new cooperative programme called Seguro Solidario, wherein participating farmers contribute a fixed sum to a collective fund for coping with climatic events. This local insurance mechanism is, however, not widespread in the regions studied and its effectiveness depends largely on the severity of climate impacts. The programme is now being promoted at the provincial level as a pilot programme.

The primary source of government support for dealing with climate impacts is from a highly controversial mechanism under the Agricultural Emergency Law (AEL) of 1983 under which farmers may publicly declare their losses. With the objective of diminishing impacts from climatic, telluric, biological, physical and unforeseeable or inevitable events, the AEL allows farmers access to benefits such as delaying fiscal obligations, acquiring tax extensions or exemptions and accessing credit, among others. However, opinion about this mechanism among farmers is once again generally low, largely because it is a tedious process and in most cases only serves to delay payments.

In contrast, participating in farmers' organizations or associations is generally viewed as highly positive, necessary, useful and powerful because of its ability to promote common interests. However, farmer interviews revealed that the advantages of organization depend largely on personal experience and member attitudes. Moreover, with the improved economic situation due to the devaluation of the peso, organizations are now considered necessary only as a temporary response in periods of difficulty. In our analysis we found that only 50 per cent of farmers participated in such organizations, while the rest considered them either not useful (13 per cent), associated with bad experiences (12 per cent), of little interest (27 per cent) or lacking capacity (39 per cent).

Besides formal mechanisms, adaptation to climate risks in the South of Córdoba is also facilitated by the use of climate information from various sources. However, climate information, especially seasonal forecasts, is usually only accessed through private seed or chemical providers, the internet and special workshops/seminars organized by farmers' organizations. It is largely used to inform short-term decisions such as determining planting and harvesting dates, while the major production decisions are based primarily on market signals, soil conditions and the availability of working capital.

Overall, the common opinion among farmers is that any action necessary to resolve local problems such as repeated negative climate impacts would benefit more from local action rather than interventions from the national government. Recent changes in macroeconomic and sector policy have made farmers distrustful of any support or protection from the national government, and the added burden of export taxes as a result of these market changes represents a fundamental concern. Thus, both factors, state interventions and climate, are commonly perceived as very unpredictable by the majority of the farmers.

Opportunities for Intervention

The determination of the resources and attributes of adaptive capacity specific to each region discussed above enabled the identification of possible priorities for public sector interventions that could enhance farm-level capacities (a systematization of these priorities is presented in Table 18.2).

Table 18.2 *Synthesis of adaptation options*

Measures	Irrigation	Insurance	Infrastructure	Technology	Information
Timing of measure (before/after and for what hazard)	Before – drought	After – hail, drought, floods	Before – flood	Before – general	Before – general
Type of measure	Individual or system development; groundwater or surface water	Commercial, publicly subsidized or cooperative	Drainage containment infrastructure, roads	Inputs (seeds, fertilizers, etc) and management (conservation tillage etc)	Climate trends, variability, forecasts; markets; prices; new technologies
Who would implement?	Farmer	Farmer	Government (national or state)	Local government, public research institutions	All levels of government, farmer associations, extension services
Conditions	Hydrological studies, credit	Guarantees of contracts, market transparency, information, high value crops	Public funds	Time for technology development, institutional coordination	Information networks and intermediaries, extension services, human resources
Capacity to implement	Low to medium*	Medium (Arg) Low (Mex)	Medium (Arg)	High (Arg) Medium-low (for those technologies that require public investment)	Medium**
Potential obstacles	Cost of equipment, cost of maintenance, economies of scale, scepticism (Mex)	Political will (Arg), scepticism, distrust, low value crops (Mex)	Competition for public funds, regional priorities	Cost, decline in public investment in research, lack of explicit demand from social sector	Lack of organizational capacity, lack of funding, lack of interest (Mex), lack of 'culture of information' (Mex)
Benefits	Improved yields, reduced drought impacts, additional subsistence benefits (aquaculture, Mex), reduced risk in new crop investment (Mex)	Enables cost recovery after loss (Mex and Arg), facilitates agricultural diversification (Mex)	Reduced uncertainty over production in flood-prone areas	Reduces productivity gap between farmer groups, increases economic margins	Better risk management and improved decision making, improved dissemination of technology, greater access to public support programmes

Notes: *With financial and technical support;   **Information is available but the network for distribution is not established.

Critical components of adaptive capacity in González include financial resources, such as credit and insurance; material resources, such as land, irrigation and equipment; the degree of economic and agricultural diversification of the farm; and access to resources, such as by means of technical assistance (Eakin et al, 2006). In South of Córdoba, adaptive capacity was observed to be more a function of material and financial resources (such as credit, soil quality, landholding size and type of activity) and to a lesser extent of human/social resources (for example, personal experience, availability of technical assistance and participation in organizations). Other coping strategies such as maintaining crop diversity and alternative non-farm income sources were also important, together with specific climate adaptations, such as the use of insurance and climate information. The prioritization of potential interventions by the public sector to boost this existing farm-level adaptive capacity and address the various issues that contribute to the increasing vulnerability to future climate impacts are discussed below.

Mexico

The majority of the adaptation options described above require finance and, although there are limited credit windows for smallholders, the support is generally not extensive. Most households therefore increasingly depend on alternative income sources to finance their agricultural activities. Public intervention to improve and expand access to credit, especially for the small farmers, could help ease their financial burden and enable more effective adaptive responses to climate risks.

For instance, according to some of the larger-scale producers in the region, diversification into alternative commercial crops is only possible with appropriate tools and capital, for example, the construction of greenhouses, *mayas de sombra* (artificial shade cover) and private auxiliary irrigation networks for supporting new crops under warmer and drier conditions. Obtaining public support for such projects is tedious and farmers still need to invest substantially in terms of financial and labour contributions. As a result, few small-scale farmers risk investment in infrastructure projects given the uncertainty of their harvests. Adaptation in this case can thus be facilitated with targeted public support for specific infrastructure projects at the farm level, combined with a guaranteed living wage for farmers and greater security in marketing their harvests.

Another element critical for adaptation in agriculture is irrigation, but the future availability of irrigation water is uncertain given the increasing urban and industrial demand (Sánchez Torres and Vargas Castilleja, 2005). Currently, *ejidatarios* with irrigation generally belong to the irrigation districts of the Río Guayalejo, depending on infrastructure (often unlined canals) from the 1970s. Investment in auxiliary irrigation or improved irrigation efficiency is only viable if the crop is sufficiently high value (for example, *ágave*, onions and other vegetables, and fruit trees), yet such high-value crops are again associated with new and often high economic risks (Eakin, 2003). Many farmers are thus increasingly renting their land to outside investors and have little personal interest in investing in their water works. Improved management of current irrigation networks and greater efficiency of new infrastructure are thus likely to be critical adaptation interventions.

Furthermore, many farmers also do not possess any insurance against climate risks, once again due to the high costs as well as the overall lack of confidence in insurance mechanisms. However, investing in new crops such as aloe without insurance also represents a high financial risk, despite the fact that the crop may be better adapted to warmer or drier conditions. Public initiatives to make insurance services more affordable and reliable would therefore enable more farmers to be protected against investment risks in the implementation of adaptive strategies to cope with climate impacts.

Finally, farm organizations and producer associations could potentially play a key role in climate adaptation by facilitating access to markets and providing services and information. Public support for the development of their administrative and technical capacity would, however, be important. The trend of using such associations for meeting political ends would also need to be discouraged since it prevents the achievement of any technical and productive goals by participating farmers (Eakin, 2004).

Argentina

In the case of Argentina, under the current policy environment, access to credit is likely to continue to be restricted to the private banking system and input suppliers, despite the fact that most farmers believe that increasing credit availability and diminishing export taxes could resolve their problems. Important investments in fixed capital would possibly be required in the near future for the installation of supplementary irrigation systems to address the impacts of drought caused by climate change (Wehbe et al, 2005; Eakin et al, 2006; Gay, 2006). Public interventions (tax incentives or interest rate subsidies) to overcome the high cost of private banking credit as well as educating farmers to use supplementary irrigation technologies would therefore assume importance. There is also need for an accurate analysis of the capacity of regional surface and groundwater as potential sources of irrigation.

Smallholder farms in Argentina also face significant barriers in obtaining insurance to cover climate-related agricultural losses due to the lack of guarantees by insurance companies and the exorbitant premiums. Government interventions to address these issues and facilitate insurance use could include supervising the completion of contract obligations, providing information and subsidizing insurance for lower-scale farmers. Unfortunately, a lack of political will and the absence of the necessary political infrastructure have so far prevented any control over the industry. To date, the primary government interventions have been restricted to limited subsidies of premiums and the declaration of an agriculture emergency only when an event affects an important geographic area.

Some farmers in the region, mainly in the south of the region, are highly vulnerable to an increased risk of floods. Floods are also a principal source of conflict among neighbouring farmers and between rural and urban areas.

Interventions to support flood risk management such as infrastructure works (additional drainage or containment structures, the diversion of excess water, and road construction) as well as improved rezoning of crops and improvements in land use practices would entail substantial investments that would require support from the national or provincial government. In addition, support from local governments for smaller flood management initiatives and maintenance work would also be essential.

Last but not least, the use of advanced agricultural technology could additionally play a very important role in adapting to climate risks in this region. Unfortunately, most agricultural technology is presently only commercially available and is therefore unaffordable for many small farmers. Public sector support for technology, research and development could help increase the accessibility of technology for such farmers and increase the likelihood of adaptation. However, this sort of public intervention currently faces significant barriers which would first need to be addressed, for example, the high investment costs of technology, the lack of institutional coordination, and a lack of participation by farmers in producer associations to help articulate their technology needs.

Comparing the two regions

Despite significant differences in the scale of production and agricultural histories between the *municipio* of González in Mexico and the South of Córdoba Province in Argentina, there are also important similarities in the opportunities and constraints for adaptation. In both cases, one could argue for improved access to climate, market and technological information as an important means for enabling farmers to respond rapidly to economic and environmental change. Enhancing the accessibility of information entails supporting farmers' social and professional networks, as well as investment from public sector institutions in the synthesis and systematization of available information.

It can also be concluded that targeted interventions from the public sector and from farm associations are necessary to ensure capacity building for adaptation. For smaller-scale farmers to be able to sustain their agricultural livelihoods under a potentially more variable future climate, specific technical support would be necessary to facilitate their access to appropriate technological packages, markets for alternative cash crops and formal insurance mechanisms, and to support improvements in irrigation, drainage and other productive infrastructure. In the absence of such support, it appears probable that the ability to adapt to future climatic and economic changes will be restricted to larger-scale farmers or external agribusinesses with the capital to acquire credit, technology and insurance. Many of these producers will likely be outside investors renting or purchasing land. Small-scale farmers struggling with crop losses and commercialization issues today may choose adaptive options entirely outside the agricultural sector, a move which would stimulate not only a social transformation of the sector but might also entail important landscape and ecological changes in both regions.

Conclusions

Our investigation of the agricultural sector in González, Mexico and the South of Córdoba Province, Argentina, has primarily highlighted the importance of the local context, existing practices and farmers' perceptions in the determination of potential adaptation options to the impacts of climate change in these regions. We have shown that although some sections of the population are currently engaging in a variety of agricultural practices that could be helpful in mitigating climate risk, such as crop and economic diversification, insurance coverage and irrigation development, widespread adoption of these practices and technologies are limited by a lack of access to finance, poor information networks and market failures. Of particular concern is the differential access to specific coping strategies between large- and small-scale farmers and, in the case of Argentina, between smaller family-run farms and large agribusinesses. Current policy trends in both countries indicate little government support for resolving specific agricultural sector problems. Instead, the focus of public policy is on the development of an enabling environment for private investment and economic growth, with less attention to the distributive implications of such policies. Although improved economic conditions will undoubtedly increase the flexibility of some farmers in responding to environmental change, problems with resource access and technology adoption in vulnerable subsectors demand more specific local action. Ideally such local action would be the result of collaboration between farmers, producer associations, the private sector and local government. Given the significance of economic and political obstacles to the implementation of various adaptation options, the possible interventions identified above require rigorous evaluation within a participatory and collaborative local context. This would ensure the selection of adaptive strategies that have the greatest potential to foster the sustainability of the farm sector and thus positively impact the economic, social and environmental conditions of communities.

References

Appendini, K. (2001) *De la Milpa a los Tortibonos: La Restructuracion de la Politica Alimentaria en Mexico* [*From Milpa to Tortibonos: Food Policy Restructuring in Mexico*], El Colegio de Mexico, Mexico City, Mexico and United Nations Research Institute for Social Development (UNRISD), Geneva, Switzerland

Claridades Agropecuarias (2004) 'Perspectivas agrícolas de la OCDE 2003–2008' ['Agriculture Perspectives from the OCDE 2003–2008'], *Claridades Agropecuarias*, vol 133 (September), pp16–30

Cavazos, T. (1999) 'Large-scale circulation anomalies conducive to extreme precipitation events and derivation of daily rainfall in northeastern Mexico and southeastern Texas', *Journal of Climate*, vol 12, pp1506–1522

Conde, C. (2005) 'Climate change and variability in southern Tamaulipas', paper presented at Stakeholder Workshop, Project Closure, AIACC Project LA29, González, Tamaulipas, 6 May

de Janvry, A., M. Chiriboga, H. Colmenares, A. Hintermeister, G. Howe, R. Irigoyen, A. Monares, F. Rello, E. Sadoulet, J. Secco, T. van der Pluijm and S. Varese (1995)

Reformas del Sector Agricola y el Campesinado en Mexico [*Agriculture Sector Reforms and Peasantry in Mexico*], Fondo Internacional de Desarollo Agrícola y Instituto Interamericano de Cooperacíon para la Agricultura, San José, Costa Rica

Eakin, H. (2003) 'The social vulnerability of irrigated vegetable farming households in Central Puebla', *Journal of Environment and Development*, vol 12, pp414–429

Eakin, H. (2004) 'Waiting to recover: Adaptation to the coffee crisis in two Mexican coffee communities', paper presented at the Annual Meetings of the Association of American Geographers, Philadelphia, PA, 14–19 March

Eakin, H., M. Wehbe, C. Avila, G. Sánchez Torres and L. Bojórquez (2006) 'A comparison of the social vulnerability of grain farmers in Mexico and Argentina', AIACC Working Paper No 29, International START Secretariat, Washington, DC, www.aiaccproject.org

Gay, C. (2006) 'Vulnerability and adaptation to climate change: The case of farmers in Mexico and Argentina', final report, Project LA29, Assessments of Impacts and Adaptations to Climate Change, International START Secretariat, Washington, DC, www.aiaccproject.org

Houghton, J. T., Y. Ding, D. J. Griggs, M. Noguer, P. J. van der Linden and D. Xiaosu (eds) (2001) 'Technical summary', in *Climate Change 2001: The Scientific Basis*, contribution of Working Group I to the Third Assessment Report, Intergovernmental Panel on Climate Change, Cambridge University Press, Cambridge, UK

INDEC (Instituto Nacional de Estadística y Censos) (2002) 'Censo nacional agropecuario 2002' ['National Agriculture Census 2002'], preliminary data, available at www.indc.mecon.ar

INEGI (Instituto Nacional de Estadística Geografía e Informática) (2000) 'Censos económicos 1999' ['Economic Census 1999'], Government of México, available at www.inegi.gob.mx/

IUCN (World Conservation Union)/IISD (International Institute of Sustainable Development) (2004) 'Sustainable livelihoods and climate change adaptation. A review of phase one activities for the project "Climate Change, Vulnerable Communities and Adaptation"', available at www.iisd.org/pdf/2004/envsec_sustainable_livelihoods.pdf

Lattuada, M. (2000) 'El crecimiento económico y el desarrollo sustentable en los pequeños productores agropecuarios argentinos de fines del siglo XX' ['Economic growth and sustainable development of small Argentinean agriculture producers at the end of the 20th Century'], Conferencia electrónica sobre Políticas Públicas, Institucionalidad y Desarrollo Rural en América Latina, available at www.rlc.fao.org/foro/institucionalidad

Ravelo, A. C. and R. A. Seiler (1979) 'Agroclima de la provincia de Córdoba: Expectativa de precipitación en el curso del año' ['Agroclimate of the Province of Córdoba: expected amount of annual rainfall'], *Revista de Investigaciones Agropecuarias*, INTA, vol 14, pp15–36

Risbey, J., M. Kandlikar and H. Dowlatabadi (1999) 'Scale, context and decision making in agricultural adaptation to climate variability and change', *Mitigation and Adaptation Strategies for Global Change*, vol 4, pp137–165

Rivarola, A., R. Seiler and M. Vinocur (2004a) 'Vulnerabilidad y adaptación de los productores agropecuarios al cambio y a la variabilidad climática: El uso de la información agrometeorológica' ['Vulnerability and adaptation of farmers to climate variability and change: The use of agroclimatic information'], *Revista Reflexiones Geograficas*, vol 11, pp109–120

Rivarola, A., R. Seiler and M. Vinocur (2004b) 'Vulnerabilidad agroclimática a las sequías en el sur de la provincia de Córdoba' ['Agroclimatic vulnerability to droughts in the south of Córdoba, Argentina'], paper presented at Actas X Reunión Argentina y IV Reunión Latinoamericana de Agrometeorología, Mar del Plata, Argentina, 13–15 October

Sánchez Torres, G. and R. Vargas Castilleja (2005) 'Integrated assessment of social vulnerability and adaptation to climate variability and climate change among farmers in Mexico and Argentina. Case Study: Municipality of González, Tamaulipas, Mexico', final report, Graduate Division, School of Engineering, Universidad Autónoma de Tamaulipas, Tampico, Tamaulipas, Mexico

Secretaria de Desarrollo Económico y del Empleo (2001) *Programa Regional de Conversión de Cultivos* [*Regional Programme for Crop Conversion*], Ciudad Victoria: SAGARPA y Gobierno de Tamaulipas, Tamaulipas, Mexico

Seiler, R., M. Hayes and L. Bressan (2002) 'Using the standardized precipitation index for flood risk monitoring', *International Journal of Climatology*, vol 22, pp1365–1376

Seiler, R. and M. Vinocur (2004) 'ENSO events, rainfall variability and the potential of SOI for the seasonal precipitation predictions in the south of Cordoba, Argentina', 14th Conference on Applied Climatology, Seattle, WA, 11–15 January, available at http://ams.confex.com/ams/84Annual/techprogram /paper_71002.htm

Smit, B., D. McNabb and J. Smithers (1996) 'Agricultural adaptation to climatic variation', *Climatic Change*, vol 33, pp7–29

SMN (Servicio Meteorológico Nacional) (1992) 'Estadísticas climatológicas' ['Climatic statistics'], Servicio Meteorológico Nacional, Fuerza Aérea Argentina, Buenos Aires

Vinocur, M., A. Rivarola and R. Seiler (2004) 'Use of climate information in agriculture decision making: Experience from farmers in central Argentina', Second International Conference on Climate Impacts Assessment, SICCIA, Grainau, Germany, 28 June–2 July, available at www.cses.wash ington.edu/cig/ outreach/ workshopfiles/ SICCIA /program .shtml

Wehbe M. B, R. A. Seiler, M. R. Vinocur, H. Eakin, C. Santos and H. M. Civitaresi (2005a) 'Social methods for assessing agricultural producers' vulnerability to climate variability and change based on the notion of sustainability', AIACC Working Paper No 19, International START Secretariat, Washington, DC, www.aiaccproject.org

Wehbe, M. B., H. Eakin and A. Geymonat (2005b) 'Macroeconomic reforms and agricultural policies in developing countries: Impacts on the social vulnerability of family farmers in Argentina and Mexico', paper presented at the IHDP Workshop on Human Security and Climate Change, Oslo, 21–23 June

Yarrington Ruvalcaba, T. (2004) *V Informe de Gobierno 1999–2004 Tamaulipas* [*Fifth Report of the Government of Tamaulipas 1999–2004*], Government of the State of Tamaulipas, Ciudad Victoria, Mexico

19

Maize and Soybean Cultivation in Southeastern South America: Adapting to Climate Change

Maria I. Travasso, Graciela O. Magrin, Walter E. Baethgen,
José P. Castaño, Gabriel R. Rodriguez, João L. Pires,
Agustin Gimenez, Gilberto Cunha and Mauricio Fernandes

Introduction

Several important changes in climate and crop production trends have been detected in Southeastern South America since the late 20th century. The climatic changes are characterized by increases in precipitation (up to 50 per cent in some areas); decreases in maximum temperature, especially during spring and summer; and increases in minimum temperature during most of the year (Castañeda and Barros, 1994; Barros et al, 2000; Pinto et al, 2002; Bidegain et al, 2005; Magrin et al, 2005). In response to the favourable climate trends, a subsequent significant increase in crop production, especially in the yield of rain-fed crops, has been noted. Magrin et al (2005), in comparing the period from 1950 to 1970 with that from 1971 to 1999, calculated a 38 per cent increase in soybean yields and an 18 per cent increase in maize yields attributable to the changes in climate (isolated by using crop simulation models with the same production technology).

Added to the influence of changing temperature and precipitation, changes in land use over the past few years have further contributed towards the increase in crop yields. Encouraged by favourable climatic and economic conditions, farmers in the region have begun to devote more and more land to agriculture, especially soybean farming. The recent expansion of soybean cultivation is particularly remarkable, with an increase of 133 per cent (from 6 to 14 million hectares, Mha) in the area devoted to this crop between 1995 and 2003 in Argentina alone (SAGPyA, 2005). A similar trend has also been observed in Brazil, where, in the traditional soybean growing areas such as Rio Grande do Sul, lands devoted to soybean have increased by 38 per cent over the past five years (IBGE-LSPA, 2005). More recently, a huge expansion of this crop has also been observed in Uruguay, with an increase in cultivated area from 12,000ha in 2000 to 278,000ha in 2004 (MGAP-DIEA, 2005).

According to future projections, the total area under soybean cultivation in South America is expected to grow from 38Mha in 2003/2004 to 59Mha in 2019/2020. Thus an increase of 85 per cent (172 million tons) in the total production of Argentina, Brazil, Bolivia and Paraguay is expected (Maarten Dros, 2004).

Some previous studies have used crop production models[1] based on climate scenarios from general circulation models (GCMs) to assess future changes in crop yields specifically for the southeastern South America region. According to model projections, when CO_2 enrichment effects on crop growth stimulation are accounted for, soybean production in the Pampas region of Argentina during the 21st century could range between a reduction of 22 per cent and increases of 3 per cent, 18 per cent and 21 per cent under the UKMO (Wilson and Mitchell, 1987), GFDL (Manabe and Wetherald, 1987), GISS (Hansen et al, 1989) and MPI-DS (Magrin et al, 1998) scenarios respectively.[2] Maize production, on the other hand, is expected to decline under most GCM scenarios; results for the overall Pampas region were -8 per cent, -16 per cent, -8 per cent and +2 per cent under the UKMO, GFDL, GISS and MPI-DS respectively (Magrin and Travasso, 2002), although an earlier study for the main maize production zone estimated reductions of between 20 per cent and 25 per cent based on UKMO, GFDL and GISS projections (Paruelo and Sala, 1993). For Brazil (de Siqueira et al, 2000), a 16 per cent reduction in maize yields and a 27 per cent increase in soybean yields were reported under the GISS scenario when CO_2 effects are taken into consideration. Finally, for Uruguay, reductions of 14 per cent and 25 per cent in maize yields have been reported for increases of 2°C and 4°C in mean temperature respectively (Sawchik, 2001), although in this case CO_2 effects were not considered.

Some degree of uncertainty is invariably associated with such crop model projections, largely because the interactions between various climatic and non-climatic elements and their impacts on crop production are not fully understood. However, regardless of the level of uncertainty, it appears that future climate conditions will probably be more favourable for soybeans than for maize in southeastern South America. Therefore, based on the current trend, a further expansion of soybean growing areas in this region can be anticipated.

Such large-scale expansion of soybean farming, with a subsequent decline in maize production, could have significant negative consequences in the medium and long term. Potential ecological implications of this trend in soybean monoculture could include reductions in soil organic matter (García, 2003), soil compaction in superficial layers (Diaz-Zorita et al, 2002) and a noticeable increase in the use of herbicides like glyphosate (for example, in Argentina, use was 28Ml in 1997 and 100Ml in 2002) (Joensen and Ho, 2004). Such impacts could seriously threaten the system's sustainability if current practices are not reconsidered. Effective adaptation strategies to sustain the viability of agricultural production systems in this region have therefore also become a critical necessity for the immediate future.

Keeping these findings in mind, our specific objective in this study was to assess the impacts of future climate scenarios (based on the latest GCM projections – HadCM3 – under different SRES scenarios and time periods) on crop production systems in southeastern South America, specifically the Pampas region in Argentina, Uruguay and the states of Rio Grande do Sul in Brazil (Figure 19.1).[3] Six sites in this region were selected representing areas with contrasting environmental conditions (from the humid subtropics in Brazil to the humid and semiarid Pampas), namely Azul, Pergamino, Santa Rosa and Tres Arroyos in Argentina; La Estanzuela in Uruguay; and Passo Fundo in

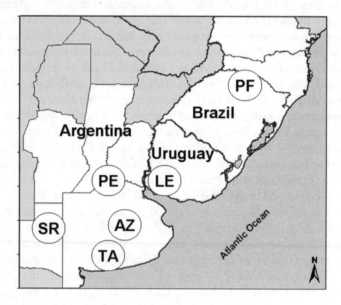

Figure 19.1 *Study area and study sites*

Note: AZ = Azul; PE = Pergamino; SR = Santa Rosa; TA = Tres Arroyos;
LE = La Estanzuela; PF = Passo Fundo.

Brazil (see Figure 19.1).

Crop simulation models were used to examine the influence of increasing temperature and precipitation on future crop yields. The effect of CO_2 enrichment was also assessed since some studies have shown that increased CO_2 concentration in the atmosphere promotes plant growth by boosting photosynthetic activity and increasing water use efficiency (see Kimball et al, 2003). The crop yield results generated were next used in the same crop models to evaluate potential adaptation measures that could help reduce the negative impacts of climate change on maize and soybean production in this region. This study thus played an important role not only in determining the vulnerability of agriculture in this region, but also in identifying strategies that can potentially address this vulnerability.

Future Climate Scenarios in Southeastern South America

Future climatic scenarios for each of the six study sites were determined with the help of the HadCM3 (Johns et al, 2003) GCM model, which was found to reproduce local conditions in this region with greater confidence than other GCMs (Camilloni and Bidegain, 2005). Model runs were conducted for the two socioeconomic scenarios, A2 and B2, from the IPCC Special Report on Emissions Scenarios (SRES). Monthly values for the variables maximum temperature, minimum temperature and precipitation were obtained from the model results for three 30-year time periods (centred on 2020, 2050 and 2080), and the monthly rate of change of each variable was obtained by comparison with the baseline period 1960–1990. These coefficients of change were next applied to the observed data (1971–2000) to obtain the future climatic scenario on a daily basis.

It was observed that larger increases in temperature and precipitation, particularly for 2050 and 2080, could be expected for the SRES A2 scenario (which assumes a higher CO_2 concentration) than for the SRES B2 scenario (Table 19.1 and Figure 19.2). Increases in mean temperature for the warm semester (October–March) ranged from 0.8°C to 4.1°C under SRES A2 and from 0.7°C to 2.9°C under SRES B2, depending on site and time period (Table 19.1).

Table 19.1 *Projected changes in mean temperature (°C) for the warm semester (October–March) according to HadCM3 under SRES A2 and B2 scenarios for 2020, 2050, and 2080*

| | Mean temperature changes (October–March) | | | | | |
| | HadCM3 A2 | | | HadCM3 B2 | | |
	2020	2050	2080	2020	2050	2080
SR	0.9	2.1	3.4	0.7	1.7	2.5
TR	0.8	1.9	3.1	0.7	1.6	2.4
AZ	0.8	1.9	3.1	0.7	1.6	2.4
PE	0.9	2.1	3.4	0.8	1.7	2.7
LE	0.8	2.0	3.2	0.8	1.5	2.6
PF	0.9	2.4	4.1	0.9	1.8	2.9
Mean	**0.9**	**2.1**	**3.4**	**0.8**	**1.7**	**2.6**

Note: AZ = Azul; PE = Pergamino; SR = Santa Rosa; TA = Tres Arroyos; LE = La Estanzuela; PF = Passo Fundo.

With respect to precipitation (Figure 19.2), the general pattern showed an increasing trend during the warm semester (October–March), with up to 253mm and 172mm rainfall increases for SRES A2 and B2 respectively. A decreasing trend was observed for the coldest months (May–August), with up to 46mm and 34mm reductions in rainfall for SRES A2 and B2 respectively.

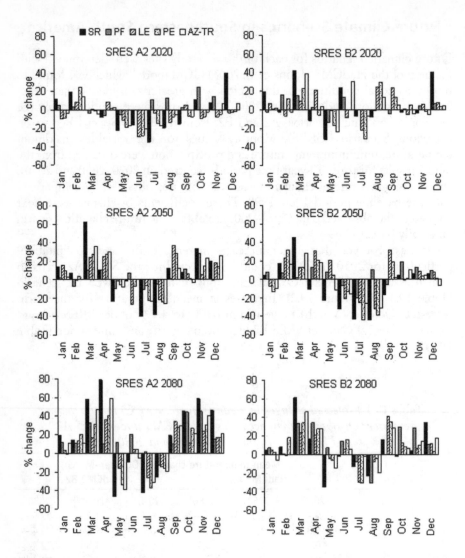

Figure 19.2 *Changes in monthly precipitation (%) projected by HadCM3 under SRES A2 and B2 for 2020, 2050 and 2080*

Impact on Crop Production

The expected impacts of these climate scenarios on crop yields in each of the six study sites were next assessed using crop simulation models included in the Decision Support System for Agrotechnology Transfer (DSSAT) computer program[4] (Tsuji et al, 1994). The crop models in DSSAT (including CERES for maize and CROPGRO for soybean) are detailed biological simulation models of crop growth and development that operate on a daily time step. They simulate dry matter production as a function of climate conditions, soil properties and management practices. These models are also able to simulate crop growth for variable atmospheric CO_2 concentrations. The inputs required to run the models are daily weather variables, crop and soil management information (planting date, fertilizer use, irrigation and so on), cultivar characteristics, and soil profile data. Output from the models includes final grain yield, total biomass and biomass partitioning between the different plant components at harvest, among others.

These models have previously been exhaustively tested in the study region at the plot and field levels with relatively low estimation errors (Guevara and Meira, 1995; Meira and Guevara, 1995; Travasso and Magrin, 2001; Sawchik, 2001; de Siqueira et al, 2000). Subsequently, they have also been used to assess the impacts of interannual climate variability and climate change in the agricultural sector of this region (Magrin et al, 1997 and 1998; Travasso et al, 1999; Magrin and Travasso, 2002; de Siqueira et al, 2000; Sawchik, 2001).

Agronomic model inputs used in our study included initial water and nitrogen content in the soil profile, date of planting, plant density, sowing depth, date and rate of fertilizer application, and cultivars. Climatic inputs for the crop simulation models included observed daily maximum and minimum temperatures, precipitation, and solar radiation corresponding to the period 1971–2000 and the climate change scenarios obtained from the HadCM3 runs described above. Crop models were run under rain-fed and irrigated (water and nutrients non-limiting) conditions for different atmospheric CO_2 concentrations: 330ppm (current) and those corresponding to each SRES scenario (417, 532 and 698ppm for A2, and 408, 478 and 559ppm for B2 in 2020, 2050 and 2080 respectively).

Changes in crop yields

The results show a decrease in irrigated maize yields at almost all sites and under all scenarios in comparison to the baseline (1971–2000) when the direct effects of CO_2 on crop growth were not considered (Figure 19.3). Yield reductions were larger for the later time periods, and were stronger under SRES A2 (up to –23 per cent) than under B2. A significant correlation was observed between changes in maize yield and temperature increases during the crop growing season ($R^2 = 0.74$), resulting in a reduction of 5 per cent in yields per °C temperature increase.

Under the same conditions, irrigated soybean yields were, however, less affected, with yield changes varying between –8 per cent and +5 per cent (Figure 19.3). The correlation between yield changes and temperature increases was weaker ($R^2 = 0.4$) than in the case of maize and therefore the yield reduction was also smaller (a decrease of 1.8 per cent per °C temperature increase).

When the direct CO_2 effects on crop growth were considered, maize yields under irrigated conditions were somewhat higher than those obtained in the absence of CO_2 effects, but the increase was insufficient to offset the negative temperature effects (Figure 19.3). In contrast, huge increases in irrigated soybean yields were detected under both SRES scenarios (of up to 43 per cent and 38 per cent for A2 and B2 respectively).

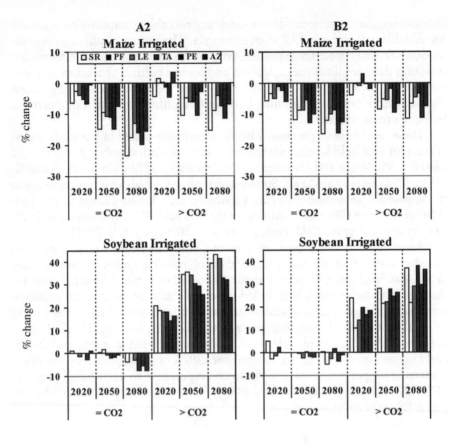

Figure 19.3 *Changes in irrigated maize and soybean yields (%) under different scenarios and CO₂ concentrations*

Under rain-fed conditions and without considering the direct CO_2 effects, maize yield changes ranged between –9 per cent and +9 per cent for SRES A2, and –12 per cent and 3 per cent for SRES B2. Rain-fed soybean yield changes varied between –22 per cent and 10.5 per cent for SRES A2, and between –18 per cent and 0.5 per cent for SRES B2 (Figure 19.4). When the direct effect of CO_2 on crop growth was taken into account, grain yields increased for both crops but once again a greater impact was observed on soybean yields (up to 62.5 per cent increase).

Thus, under the A2 scenario, when the direct effects of CO_2 were considered, irrigated and rain-fed soybean yields and rain-fed maize yields were higher than current climate yields: the direct effects of high CO_2 concentration and the higher spring and summer precipitation more than compensated for the negative effect of increased temperature. As expected, the changes in crop yield under the B2 scenario were in the same direction as those under the A2 scenario but smaller in magnitude.

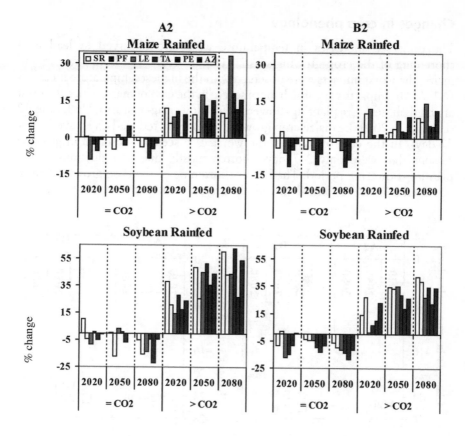

Figure 19.4 *Changes in rain-fed maize and soybean yields (%) under different scenarios and CO₂ concentrations*

The differences in the response between soybean and maize can be attributed to the differences in the influence of temperature and CO_2 concentration on crop growth. In soybean (a C3 plant) CO_2 effects on photosynthesis are greater than in the case of maize (a C4 plant) (Derner et al, 2003).[5] The soybean simulation model used in our research assumes about a 40 per cent increase in photosynthesis efficiency at a CO_2 concentration of 660ppm, while the corresponding value for maize is only about 10 per cent. Consequently, the effect of temperature is more dominant in irrigated maize crops than the effect of CO_2 concentration, which explains the obtained yields. On the other hand, the effect of CO_2 on stomatal resistance is known to be higher in C4 than in C3 plants (Kimball et al, 2003), which contributes to a higher water use efficiency in rain-fed maize and explains the increase in yield, especially when the direct effects of CO_2 are not considered.

Changes in crop phenology

The projected increases in temperature were also observed to lead to a shortening of the crop growing seasons (Figure 19.5). For both soybean and maize, the worst impacts were observed with the highest temperature increases (A2, 2080). Impacts were much more severe in the case of maize, since the most affected phases were planting–flowering and flowering–maturity. Under the A2 scenario, in 2080 the maize crop growing season was reduced on an average by 27 days. In the case of soybean the worst case scenario resulted in growing seasons that were only 2–7 days shorter, mostly due to reductions in the planting–flowering period. The bigger shortening of the crop growing season observed in maize is coincident with the greater reductions in grain yields.

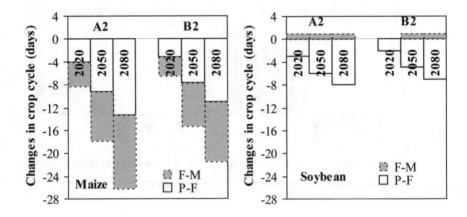

Figure 19.5 *Changes in the duration of planting–flowering (P–F) and flowering–maturity (F–M) periods, expressed as mean values for the six sites, for maize and soybean crops under different SRES scenarios and time periods*

Assessment of Potential Adaptive Measures

The DSSAT crop models used to assess crop yield vulnerability to changes in temperature and precipitation were also capable of evaluating climate adaptation measures. We therefore used the same crop models to examine the impact of several adaptive management practices on crop yields, which include changing planting dates, supplementary irrigation and increasing nitrogen application rates. At each site, alternate planting dates were tested for maize and soybean by advancing/delaying planting from the actual dates. Supplementary irrigation was added to both crops during the reproductive period, beginning 20 days before flowering at a rate of 20mm every 20 days. Incremental nitrogen application rates were tested in all sites for maize only since nitrogen application is not a current practice for soybean in this region. These measures were tested both in the presence and absence of CO_2 enrichment effects.

Considering CO_2 effects

Changing planting dates for maize

On average, earlier planting dates were observed to lead to increased maize yields under both SRES scenarios, especially for 2050 and 2080, although there were differences between individual sites (Figure 19.6). Earlier planting dates contributed to longer planting–flowering periods (Table 19.2) and earlier maturity dates. This measure thus allows maize crops to develop under more favourable thermal conditions, increasing the duration of the vegetative phase, which in turn increases grain yield. An additional possible advantage of earlier planting dates relates to the corresponding earlier crop maturity dates and therefore the earlier harvest period. Under current planting dates, maize crops are usually harvested during March–April or later, depending on the region. Future climate scenarios project important increases in rainfall for these months (see Figure 19.2), which could lead to excess water episodes that could, in turn, affect harvest and cause yield losses. The CERES model is unable to account for this effect and therefore the impact of earlier sowing dates could possibly be even higher under the climate conditions predicted by the HadCM3 GCM.

Table 19.2 *Length (days) of planting–flowering (P–F) and flowering–maturity (F–M) periods for maize at current planting date and 20 and 40 days earlier under SRES A2 scenario for 2020, 2050 and 2080*

	Current				20 days before			40 days before		
	1971–2000	2020	2050	2080	2020	2050	2080	2020	2050	2080
P–F	86	82	77	73	91	85	81	101	94	88
F–M	59	54	50	46	53	49	46	53	50	47

Changing planting dates in soybean

Even though soybean yields were less affected by temperature increases than maize, changing planting dates did result in higher yields. In three of the six sites evaluated (Azul, Santa Rosa and Passo Fundo), earlier planting dates were found to be beneficial, while in the others, delayed planting dates were found to be the best option under future conditions (Figure 19.7).

Effects of nitrogen fertilization (maize only)

Changes in nitrogen application rates along with optimal planting dates resulted in increased maize yields in only two of the six study sites (Passo Fundo and Santa Rosa). At these sites increases in nitrogen application rates of

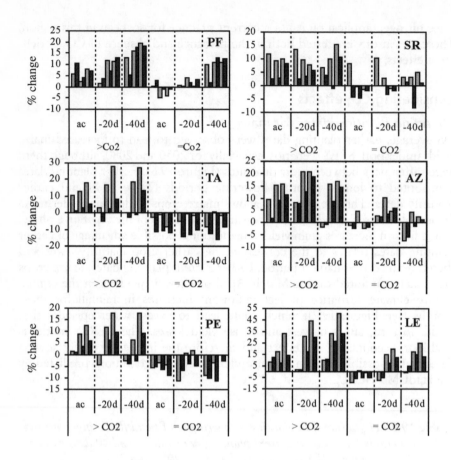

Figure 19.6 *Maize: yield changes (%) for different planting dates*
(ac = current, -20 and -40 days) in the six sites under different scenarios
(A2 in grey, B2 in black for 2020, 2050 and 2080) and CO_2 concentrations

20 and 45kg N/ha respectively were found to be optimum for encouraging increased yields, possibly due to the more favourable future environmental conditions allowing a positive crop response to increases in nitrogen.

Therefore, in the case of maize, given optimal planting dates and nitrogen application rates, mean yield increases of 14 per cent, 23 per cent and 31 per cent for 2020, 2050 and 2080 respectively are possible under the SRES A2 scenario and mean yield increases of 11 per cent, 15 per cent and 21 per cent are possible under the SRES B2 scenario. For soybean, when optimal planting dates are considered, mean yield increases of 35 per cent, 52 per cent and 63 per cent for 2020, 2050 and 2080 respectively are possible under the SRES A2 scenario, while under the SRES B2 scenario mean yield increases of 24 per

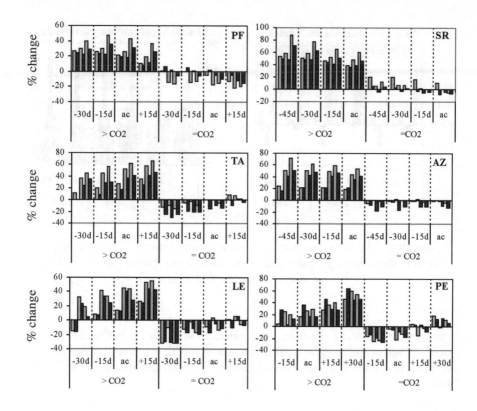

Figure 19.7 *Soybean: yield changes (%) for different planting dates*
(ac = current, ± 15, 30 days) in the six sites under different scenarios
(A2 in grey, B2 in black for 2020, 2050 and 2080) and CO$_2$ concentrations

cent, 38 per cent and 47 per cent are possible for the same time periods.

Without considering CO$_2$ effects

Maize

When CO$_2$ effects were not taken into consideration, yields decreased between 1 per cent and 5 per cent under future scenarios in the absence of adaptation measures. Under optimal planting dates and nitrogen application rates, the overall yield response was higher but differed from the case where CO$_2$ effects were considered, even though the adaptation measures were exactly the same. In this case, changes in maize yields were positive under all scenarios and time periods only in Passo Fundo, Santa Rosa and Azul. Conversely, in Tres Arroyos, a generalized yield decrease was obtained, while in Pergamino and La Estanzuela both positive and negative changes were found depending on the scenario (Figure 19.8). Mean changes for the six sites ranged between 4 per cent (B2 2020) and 12 per cent (A2 2050 and 2080).

These results suggest that without CO$_2$ fertilization simple measures such

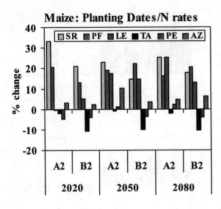
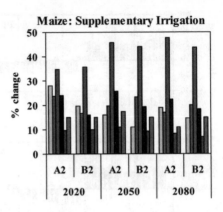

Figure 19.8 *Adaptation measures for maize: Yield change (%) under optimal planting dates/nitrogen rates and supplementary irrigation for the six sites without considering CO$_2$ effects*

as changes in planting dates or nitrogen rates will not be sufficient in some places. When supplementary irrigation was applied, an overall yield increase was observed with changes close to 20 per cent under all scenarios (Figure 19.8).

Soybean

In the absence of adaptation measures, soybean yields decreased under all scenarios (1–12 per cent). Changing planting dates led to a modest increase in yields (2–9 per cent) only for 2020 and 2050 (Figure 19.9), but the addition of supplementary irrigation strongly reverted this situation, increasing yields

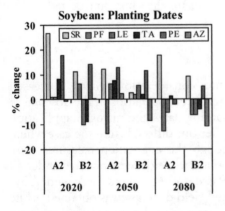
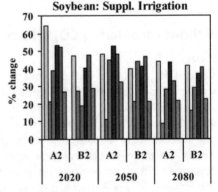

Figure 19.9 *Adaptation measures for soybean: Yield changes (%) under optimal planting dates and supplementary irrigation for the six sites without considering CO$_2$ effects*

between 30 per cent (A2 2080) and 43 per cent (A2 2020) (Figure 19.9).

From these results it is obvious that soybean crops would benefit much more in comparison to maize from the application of simple adaptation measures under the projected future climate conditions. This is true even when CO_2 effects are not taken into consideration. However, the positive crop responses to CO_2 enrichment must be treated with caution since such responses have yet to be fully understood under field conditions. Though the positive effect of increasing atmospheric CO_2 concentration on photosynthesis and water use efficiency has been demonstrated for several crops (Kimball et al, 2003), such simulation studies have also been criticized by some scientists (Long et al, 2005; Morgan et al, 2005) on the grounds that they are carried out under controlled or semi-controlled conditions, which could lead to an overestimation of the effects, in particular, for soybeans, where photosynthetic efficiency was assumed to be higher than 30 per cent in model simulations. Similarly, Leakey et al (2006) have reported that under field conditions, maize crops growing under ample water and nutrient conditions showed a lack of response to increased CO_2. In addition, there is also uncertainty about the response of crops to environments slowly enriched with CO_2 (as would happen in the case of climate change) because of the likely acclimation (Ainsworth and Long, 2005). Past research has suggested that the initial stimulation of photosynthesis observed when plants grow at elevated CO_2 concentrations may be counterbalanced by a long-term decline in the level and activity of photosynthetic enzymes as the plants acclimate to their environment, a phenomenon referred to as 'down-regulation' (Ainsworth and Long, 2005). Acclimation to CO_2 enrichment is not included in the crop models used in our study.

Irrespective of the uncertainties, the climatic changes so far have proven to be favourable to the expansion of soybean cultivation in the study area, as is obvious from current trends. However, the present high growth rate of soybean cropping and its possible continuation in the future, provided continued favourable climatic conditions, is raising important concerns with respect to its impacts on the long-term sustainability of agriculture in the region. The ecological impacts that could potentially affect the growth trends in soybean farming are discussed in greater detail in the following section.

Ecological Impacts of Soybean Cultivation

The significant growth of agricultural production in southeastern South America, especially of soybean, is unfortunately occurring at the expense of the local environment in the region. It has been observed that the large scale expansion of agriculture in Argentina has created negative soil nutrient balances (nutrient removal exceeding nutrient application) and although crop fertilization is a common practice, only 37 per cent of nutrients extracted are actually replenished (García et al, 2005).

Soybean is a highly nutrient extractive crop with a low level of crop residues, and the current trend of extensive soybean monoculture not only creates a deficit in soil nitrogen content (nitrogen fertilization is not a current practice for this crop) but also leads to negative soil carbon balances because of low carbon (C) inputs to the soil. Experiences in Argentina have shown that for a crop yielding 4000kg/ha, some 120kg N/ha/year and 950kg C/ha/year are lost from the system (García, 2003).

Modern agricultural technologies such as direct seeding and herbicide resistant soybeans (Pengue, 2001) have also contributed to the ecological degradation in the region by further encouraging intensive soybean monoculture. The direct seeding technology was initially introduced to address issues of serious soil erosion and the subsequent loss of soil fertility since it requires no tillage and allows residues from the previous crop to remain on the ground. Although this technique has successfully reduced the rate of erosion, other problems have cropped up from the further intensification of agriculture it encourages, namely the emergence of new diseases and pests, a marked reduction in the levels of nitrogen and phosphates in the soil, and, most recently, the emergence of herbicide-resistant weeds.[6] In the Pampas, there are already several types of weeds that are suspected of being tolerant to the recommended doses of glyphosate herbicide. Some of these now require a doubling of the application, leading to increased herbicide use. This further endangers the environment since herbicide run-off from the farms is already affecting adjacent ecosystems, for example, aquatic ecosystems.

In order to address these issues several alternatives have been suggested, one option being the use of fertilizers to restore soil carbon and nitrogen balances. Other potential measures to improve soil nutrient levels include the use of grasses as cover crops and crop rotation using a higher proportion of corn and wheat in the rotation. The traditional method of crop–pasture rotations is also recommended for correcting soil organic matter balances and thus restoring soil carbon and nitrogen levels (García, 2004). Such crop–pasture rotations were previously the primary crop rotation method in the Pampas region of Argentina (García, 2004). Similarly in Uruguay, traditional rotations included 3–4 years of crops alternating with 3–4 years of pasture. The recent expansion of soybean cultivation has unfortunately led to a decrease in the pasture component of this system. For southern Brazil, Costamilan and Bertagnolli (2004) recommend a three-year crop rotation, including the sequence oats/soybean, wheat/soybean and spring vetch/maize. Besides restoring soil nutrient levels, rotating crop varieties also helps to reduce the incidence of diseases, pests and weeds by breaking their cycle.

A somewhat innovative option for maintaining the viability of agriculture in the region is the 'transformation in origin' approach, promoted by Oliverio and Lopez (2005). The transformation in origin approach suggests the cultivation of a mix of oilseeds (for example, soybean) and cereals (for example, maize) with a part of the production (for example, of maize) remaining at the place where it was cultivated. This is either used as animal feed or in local industry. This practice adds value to the primary product, as opposed to the traditional approach in which it is sold as a commodity, which, in some cases, implies important costs of transportation to ports and fiscal retentions, among other things. Additionally, the cultivation of a mix of oilseeds and cereals under this method also ensures that the nutrient content of the soil is better maintained. In order to test this approach, Oliverio and Lopez (2005) analysed two possible scenarios to estimate Argentina's crop production in 2015, assuming that the trend of increasing agricultural production will continue, regardless of climate change. In the first case, they extrapolated the actual trend in planted areas (with increasing importance of oilseeds, especially soybean), and in the second, they proposed a maximum ratio of 2.5:1 between oilseeds and cereals to test the transformation in origin method. They found that even if half the cereal portion of the crop yield (for example, maize) were to be transformed in origin, economic benefits could be more than doubled.

The future outlook for agricultural expansion in southeastern South America will therefore depend not only on future climatic conditions in the region but also on the viability of crop production systems, especially the maintenance of soil nutrient content. Although important remedial strategies have been suggested in studies, further research in this area could help to identify options that effectively address the combined implications of crop responses to future climatic conditions and the ecological impacts of intensive agriculture, specifically soybean cultivation, for the long-term sustainability of farming in the region.

Conclusion

Future climate scenarios based on the runs of HadCM3 suggest that mean temperatures for the entire study region would increase by an average of 0.9, 2.1 and 3.4°C by 2020, 2050 and 2080 respectively, for the SRES A2 scenario. Corresponding figures for SRES B2 are 0.8, 1.7 and 2.6°C. Precipitation projections show an increasing trend during the warm semester (October–March), which encompasses the growing seasons of both maize and soybeans, and a decreasing trend during the coldest months (May–August). Changes in precipitation were stronger for the 2050 and 2080 time periods.

These climatic changes are likely to have important implications for agriculture in the region according to crop model results. In the case of maize, the increased temperatures would result in shorter growing seasons and consequently in lower grain yields. However, if the effect of CO_2 enrichment is accounted for, this negative impact could be greatly diminished by adjusting the crop sowing time to earlier dates. When the effect of CO_2 enrichment was not considered, changes in planting dates or nitrogen application rates were not enough to improve yield, but moderate supplementary irrigation during the reproductive growth phase did lead to significant yield increases (of up to 20 per cent).

The crop that would benefit the most under future climate conditions was found to be soybean. Assuming the effect of CO_2 enrichment, yield increases of greater than 60 per cent could be obtained simply by modifying planting dates. In the absence of CO_2 effects, supplementary irrigation was found to be necessary to ensure yield increases and increases of about 30–40 per cent were still possible.

The positive impact of the future climate on soybean production, nonetheless, has important implications for the local environments. A rapid expansion of soybean farming is already underway in this region, a trend that is only expected to strengthen in the future. However, intensive soybean cultivation tends to drain the soil of critical nutrients such as nitrogen and carbon. Additionally, the use of modern technological tools such as direct seeding and herbicide resistant seeds have generated further problems of new diseases and pests, a reduction in soil nutrient levels and the emergence of herbicide resistant weeds. Contamination of aquatic ecosystems with herbicide run-off from farms is also an issue.

Given the inevitability of continued large-scale expansion in soybean cultivation, the timely implementation of appropriate adaptive measures to tackle its harmful impacts becomes critical. Adaptation measures should thus promote adequate nutrient supply, crop and soil management practices, as well as weed, pest and disease control to sustain the environment and the welfare of farmers in the region. Suggested strategies include the traditional methods of crop rotation, crop–pasture rotation, mixed cropping, the use of grasses as cover crops and transformation in origin, in addition to the more common method of fertilizer application. A failure to initiate timely adaptive responses in the commercial farming sector in southeastern South America could translate into significant economic and ecological losses that could negate the benefits arising out of the new climatic conditions.

Finally, the further improvement of crop models is also important in order to better understand the impact of CO_2 enrichment on crop production, the interaction with weeds, pests and diseases, and the impact of excess water/flooding. This would ensure better accuracy in determining yield estimations under future conditions and help to identify more effective adaptation strategies.

Notes

1   Such models estimate agricultural production as a function of weather and soil conditions, as well as crop management by integrating current knowledge from various disciplines to predict growth, development and yield (Hoogenboom, 2000).
2   United Kingdom Meteorological Office (UKMO); Goddard Institute of Space Studies (GISS); Geophysical Fluid Dynamics Laboratory (GFDL); and Max Plank Institute – Downscaling model (MPI-DS).

   It is important to note that reductions in soybean production are only expected under UKMO, which projects for the main production area a huge increase in mean temperature (7°C) without changes in precipitation during the most sensitive period for the crop (December–February).
3   In the Pampas region of Argentina, arable lands cover approximately 34Mha and mean annual precipitation varies from 600mm in the southwest to more than 1200mm in the northeast. Mean annual temperature displays a north–south gradient from 13.5 to 18.5°C. In Uruguay, the total area under agriculture is approximately 1.5Mha, precipitation is more or less evenly distributed, and the mean annual rainfall ranges from 1050mm in the southwest and west to 1450mm

in the north. Mean annual temperature varies from 18.2°C in the north to 16.8°C in the south and southeast. In Rio Grande do Sul, cultivated area covers about 7Mha. This region is characterized by subtropical humid conditions with rainfall evenly distributed throughout the year. Mean annual rainfall in Passo Fundo is 1788mm and mean annual temperature is 17.5°C.

4   DSSAT is a microcomputer software program combining crop soil and weather data bases and programs with crop models and application programs to simulate multi-year outcomes of crop management strategies. It allows users to ask 'what if' questions and is capable of simulating results in a mater of minutes (www.icasa.net/dssat/).

5   C3 plants produce a three-carbon compound in the photosynthetic process and include most trees and common crops like rice, wheat, barley, soybeans, potatoes and vegetables. C4 plants produce a four-carbon compound in the photosynthetic process and include grasses and crops like maize, sugar cane, sorghum and millet. Under increased atmospheric concentrations of CO_2, C3 plants have been shown to be more responsive than C4 plants (see IPCC, 2001)

6   Intensive monoculture is associated with several known issues: due to the absence of rotation with other crop varieties or pasture, the cycle of diseases, pests and weeds cannot be broken, which then necessitates the intensive application of chemical pesticides, herbicides and so forth, which, in turn, can eventually result in the development of resistance to these chemicals over successive crop cycles. Monocropping also depletes soil nutrient levels that otherwise would have been restored under the traditional methods of crop rotation or mixed cropping.

References

Ainsworth, E. A. and S.P. Long (2005) 'What have we learned from 15 years of free-air CO_2 enrichment (FACE)? A meta-analytic review of the responses of photosynthesis, canopy properties and plant production to rising CO_2', *New Phytologist*, vol 165, pp351–371

Barros, V., M. E. Castañeda and M. Doyle (2000) 'Recent precipitation trends in southern South America to the east of the Andes: An indication of a mode of climatic variability', in *Southern Hemisphere Paleo- and Neo-climates. Concepts, Methods, Problems*, Springer Verlag, New York

Bidegain, M., R. M. Caffera, F. Blixen, V. Pshennikov, J. J. Lagomarsino, E. A. Forbes and G. J. Nagy (2005) 'Tendencias climáticas, hidrológicas y oceanográficas en el Río de la Plata y costa Uruguaya' ['Climatic, hydrological and oceanographical trends in the la Plata River and Uruguayan coast'], Chapter 14 in V. Barros, A. Menéndez and G. Nagy (eds) *The Climate Change in the Plata River*, Centro de Investigaciones del Mar y la Atmósfera (CIMA), Buenos Aires, pp137–143 (in Spanish)

Castañeda, M. E. and V. Barros (1994) 'Las tendencias de la precipitación en el Cono sur de America al este de los Andes' ['Precipitation trends in the Southern cone of America East of the Andes'], *Meteorológica*, vol 19, no 1–2, pp23–32

Camilloni, I. and M. Bidegain (2005) 'Climate scenarios for the 21st century', Chapter 4 in V. Barros, A. Menéndez and G. Nagy (eds) *The Climate Change in the Plata River*, Centro de Investigaciones del Mar y la Atmósfera (CIMA), Buenos Aires, pp33–39 (in Spanish)

Costamilan, L. M. and P. F. Bertagnolli (2004) *Indicações Técnicas para a Cultura de Soja no Rio Grande do Sul e em Santa Catarina – 2004/2005* [*Technical Recommendations for Soybean Crops in Rio Grande do Sul and Santa Catarina – 2004/2005*], Embrapa Trigo, Passo Fundo, Rio Grande do Sul, Brazil

Derner, J. D., H. B. Johnson, B. A. Kimball, P. J. Pinter Jr, H. W. Polley, C. R. Tischler, T. W. Boutton, R. L. Lamorte, G. W. Wall, N. R. Adam, S. W. Leavitt, M. J. Ottman, A. D. Matthias and T. J. Brooks (2003) 'Above- and below-ground responses of C3 C4 species mixtures to elevated CO_2 and soil water availability', *Global Change Biology*, vol 9, no 3, pp452–460

de Siquiera, O., S. Steinmetz, M. Ferreira, A. Castro Costa and M. A. Wozniak (2000) 'Climate changes projected by GISS models and implications on Brazilian agriculture', *Revista Brasileira de Agrometeorologia*, vol 8, pp311–320

Díaz-Zorita, M., G. A. Duarte and J. H. Grove (2002) 'A review of no-till systems and soil management for sustainable crop production in the subhumid and semiarid Pampas of Argentina', *Soil Tillage Research*, vol 65, pp1–18

García, F. (2003) 'Agricultura sustentable y materia orgánica del suelo: Siembra directa, rotaciones y fertilidad' ['Sustainable agricultura and soil organic matter. Direct seeding, rotations and ferility'], www.INPOFOS.org

García, F. (2004) 'Fertilizers to sustain the production of 100 million tonnes of grain in Argentina', www.INPOFOS.org

García, F. O., G. Oliverio, F. Segovia and G. López (2005) 'Fertilizers to sustain production of 100 million metric tons of grain', *Better Crops*, vol 89, no 1

Guevara, E. and S. Meira (1995) 'Application of CERES-Maize model in Argentina', Second International Symposium on Systems Approaches for Agricultural Development (SAAD2), International Rice Research Institute, Los Baños, Philippines

Hansen, J. I., I. Fung, A. Lascis, D. Rind, S. Lebedeff, R. Ruedy and G. Russell (1989) 'Global climate changes as forecasted by Goddard Institute for Space Studies three-dimensional model', *Journal of Geophysical Research*, vol 93, pp9341–9364

Hoogenboom, G. (2000) 'Contribution of agrometeorology to the simulation of crop production and its applications', *Agricultural and Forest Meteorology*, vol 103, pp137–157

IBGE-LSPA (2005) *Levantamento Sistemático da Produção Agrícola [Systematic Survey of Agricultural Production]*, Instituto Brasileiro de Geografia e Estadistica, www.ibge.gov.br

IPCC (Intergovernmental Panel on Climate Change) (2001) 'Glossary of terms', in J. McCarthy, O. Canziani, N. Leary, D. Dokken and K. White (eds) *Climate Change 2001: Impacts, Adaptation, and Vulnerability*, Contribution of Working Group II to the Third Assessment Report of the Intergovernmental Panel on Climate Change, Cambridge University Press, Cambridge, UK and New York, US

Joensen, L. and M. W. Ho (2004) 'La paradoja de los transgénicos en Argentina' ['The paradox of transgenics in Argentina'], *Revista del Sur*, no 147–148, available: at www.redtercermundo.org.uy

Johns, T. C., J. M. Gregory, W. J. Ingram, C. E. Johnson, A. Jones, J. A. Lowe, J. F. B. Mitchell, D. L. Roberts, D. M. H. Sexton, D. S. Stevenson, S. F. B. Tett and M. J. Woodage (2003) 'Anthropogenic climate change for 1860–2100 simulated with the HadCM3 model under updated emissions scenarios', *Climate Dynamics*, vol 20, pp583–612

Kimball, B. A., K. Kobayashi and M. Bindi (2003) 'Responses of agricultural crops to free-air CO_2 enrichment', *Advances in Agronomy*, vol 77, pp293–368

Leakey, A. D. B., M. Uribelarrea, E. A. Ainsworth, S. L. Naidu, A. Rogers, D. R. Ort and S. P. Long (2006) 'Photosynthesis, productivity and yield of maize are not affected by open-air elevation of CO_2 concentration in the absence of drought', *Plant Physiology*, vol 140, pp779–790

Long, S. P., E. A. Ainsworth, A. D. B. Leakey and P. B. Morgan (2005) 'Global food insecurity: Treatment of major food crops with elevated carbon dioxide or ozone under large-scale fully open-air conditions suggests recent models may have overestimated future yields', *Philosophical Transactions of the Royal Society*, vol 360, pp2011–2020

Maarten Dros, J. (2004) 'Managing the soy boom: Two scenarios of soy production expansion in South America', commissioned by WWF Forest Conversion Initiative, AID Environment, Amsterdam, The Netherlands

Magrin, G., M. I. Travasso, R. Díaz and R. Rodríguez (1997) 'Vulnerability of the agricultural systems of Argentina to climate change', *Climate Research.*, vol 9, pp31–36

Magrin, G. O., R. A. Díaz, M. I. Travasso, G. Rodriguez, D. Boullón, M. Nuñez and S. Solman (1998) 'Vulnerabilidad y mitigación relacionada con el impacto del cambio global sobre la producción agrícola' ['Vulnerability and mitigation related to global change impacts on agricultural production'], in V. Barros, J. A. Hoffmann and W. M. Vargas (eds) *Proyecto de Estudio Sobre el Cambio Climático en Argentina* [*Country Study Project on Climate Change in Argentina*], Project ARG/95/G/31-PNUD-SECYT, Secretaría de Ciencia y Tecnología (SECYT), Buenos Aires, Argentina

Magrin, G. O. and M. I. Travasso (2002) 'An integrated climate change assessment from Argentina', Chapter 10 in Otto Doering III, J. C. Randolph, J. Southworth and R. A. Pfeifer (eds) *Effects of Climate Change and Variability on Agricultural Production Systems*, Kluwer Academic Publishers, Boston, MA

Magrin, G. O., M. I. Travasso and G. R. Rodríguez (2005) 'Changes in climate and crop production during the 20th century in Argentina', *Climatic Change*, vol 72, pp229–249

Manabe, S. and R. T. Wetherald (1987) 'Large-scale changes in soil wetness induced by an increase in carbon dioxide', *Journal of the Atmospheric Sciences*, vol 44, no 8, pp1211–1235

Meira, S. and E. Guevara (1995) 'Application of SOYGRO model in Argentina', Second International Symposium on Systems Approaches for Agricultural Development (SAAD2), International Rice Research Institute, Los Baños, Philippines

MGAP-DIEA (2005) Ministerio de Ganadería, Agricultura y Pesca, Dirección de Estadísticas Agropecuarias, Montevideo, Uruguay, www.mgap.gub.uy

Morgan, P. B., G. A. Bollero, R. L. Nelson, F. G. Dohleman and S.P. Long (2005) 'Smaller than predicted increase in above-ground net primary production and yield of field-grown soybean was found when CO_2 is elevated in fully open air', *Global Change Biology*, vol 11, pp1856–1865

Oliverio, G. and G. Lopez (2005) 'El desafío productivo del complejo granario argentino en la próxima década: Potencial y limitantes' ['The productive challenge of the Argentine's grain complex for the next decade: Potential and limitations'], Fundación Producir Conservando, available at www.producirconservando.org.ar

Paruelo, J. M. and O. E. Sala (1993) 'Effect of global change on maize production in the Argentinean Pampas', *Climate Research*, vol 3, pp161-167

Pengue, W. (2001) 'The impact of soya expansion in Argentina', *Seedling*, vol 18, no 3, June, GRAIN Publications

Pinto Hilton, S., E. D. Assad, J. Zullo Jr and O. Brunini (2002) 'Mudanzas climáticas: O aquecimento global e a agricultura' ['Climate change: Global warming and agriculture'], www.comciencia.br/

SAGPyA (2005) website of Secretaría de Agricultura, Ganadería, Pesca y Alimentos [Secretariat of Agriculture, Cattle, Fisheries and Food], Buenos Aires, Argentina, www.sagpya.mecon.gov.ar

Sawchik, J. (2001) 'Vulnerabilidad y adaptación del maíz al cambio climático en el Uruguay' ['Vulnerability and adaptation of maize to climate change in Uruguay'], available at www.inia.org.uy/disciplinas/agroclima/publicaciones/ambiente/

Travasso, M. I., G. O. Magrin and M. O. Grondona (1999) 'Relations between climatic variability related to ENSO and maize production in Argentina', *Proceedings of 10th Symposium on Global Change Studies*, American Meteorological Society, Boston, MA, pp.67–68

Travasso, M. I. and G. O. Magrin (2001) 'Testing crop models at the field level in Argentina', in *Proceedings of the 2nd International Symposium Modelling Cropping Systems, 16–18 July*, Florence, Italy, pp89–90

Tsuji, G. Y., G. Uehara and S. Balas (1994) *Decision Support System for Agrotechnology Transfer (DSSAT v3.0)*, International Benchmark Sites Network for Agrotechnology Transfer (IBSNAT), Department of Agronomy and Soils, University of Hawaii, Honolulu, HI

Wilson, C. A. and J. F. B. Mitchell (1987) 'A doubled CO_2 climate sensitivity experiment with a global climate model including a simple ocean', *Journal of Geophysical Research*, vol 92, pp13,315–13,343

20

Fishing Strategies for Managing Climate Variability and Change in the Estuarine Front of the Río de la Plata

Gustavo J. Nagy, Mario Bidegain, Rubén M. Caffera,
Walter Norbis, Alvaro Ponce, Valentina Pshennikov
and Dimitri N. Severov

Introduction

The Río de la Plata estuary supports important artisanal and coastal fisheries in Uruguay and Argentina. The estuary and fisheries have been substantially influenced by human activities in recent decades and are vulnerable to climate extremes and changing precipitation patterns caused by climate change and variability (Camilloni and Barros, 2000; Nagy et al, 2002a; Nagy et al, 2006a). A self-sufficient artisanal fleet based on the Uruguayan shore at Pajas Blancas exploits fisheries within the estuarine frontal system. The fishermen of this artisanal fleet are impacted by changes in the fisheries that are driven by shifts in the location of the front, which are related to climate variations, and by symptoms of eutrophication, which are associated with human activities in the watershed but can be triggered by climatic events. Though subject to highly variable fish catch and incomes, until recently, the fishermen showed resilience over a wide range of conditions. In more recent years, however, they have been less resilient to the stresses to which they have been exposed.

We examine in this chapter strategies used by the artisanal fishing fleet to cope with past and current variability, called Type I adaptation. In a previous paper (Nagy et al, 2006a) we analysed the vulnerability of the fishery and the fishing fleet to the effects of the El Niño Southern Oscillation (ENSO). Here, we focus on strategies for coping with and adapting to the effects of ENSO. Our motivation for examining Type I adaptation is a belief that observed responses to past socioeconomic and environmental changes can serve as analogues for social adaptation to future climate change (Stockholm Environment Institute, 2001; Easterling et al, 2004). Based on our assessment of experiences with Type I adaptation, we suggest strategies for adapting to future climate change, or Type II adaptation.

The main questions addressed in the chapter are as follows:

- Is Type I adaptation adequate to cope with current climate variability in the coastal fishery?
- Do temperature increases of 1–2°C and precipitation and river flow changes of ~5–20 per cent, which are plausible scenarios by 2050, represent threats to the sustainable livelihoods of coastal fishermen?
- Are information and knowledge generation, access and communication adequate to enable Types I and II adaptation?

Adaptation, Sustainable Livelihood and Local Knowledge

Adaptation is the process by which stakeholders reduce the adverse effects of climate on their livelihoods. This process involves passive, reactive and anticipatory adjustments of behaviour and economic structure in order to increase sustainability and reduce vulnerability to climate change, variability and weather/climate extremes (modified from Burton, 1996 and 1997; Smit, 1993; Smit et al, 2000). An action is effective when it avoids a potential impact (Ionescu et al, 2005).

In the context of our study of an artisanal fishery of the Río de la Plata estuary, adaptation occurs in response to two physical processes: changes in river flow and winds, and associated displacement of the frontal zone. We examine two categories of adaptation called Type I and Type II by Burton (2004). The former refers to strategies for managing past and current climate-related stresses without considering future climate change. Most of the adaptation that is presently done is Type I. Type II adaptation refers to strategies that explicitly take into account potential future changes in climate. Because climate change risks, future scenarios and uncertainty have still not been factored into many development decisions, not much Type II adaptation has taken place.

A sustainable livelihood is a dynamic set of capabilities, assets and activities required for a means of living (DFID, 2000). The assets are human, physical, social, financial and natural capitals. Livelihoods are considered to be sustainable when they are resilient in the face of external shocks and stresses, are not dependent on external supports, and maintain the long-term productivity of natural resources.

The sustainable livelihood of small-scale and subsistence fisheries is strongly associated with local and traditional knowledge, the way they organize themselves to manage natural resources, and the improvement of participatory processes and governance (FAO, 2000). Knowledge is at the heart of economic growth and sustainable development. Understanding how people and societies acquire and use it – and why they sometimes fail to do so – is essential to improving people's lives, especially the lives of the poor.

The fisheries sector is particularly rich in custom, tradition and local knowledge, reflecting these in its communities and their established beliefs and practices. The location and seasonality of fishing grounds and fishing are all facets of this knowledge fund. The proximity to the natural resource base has a dominating influence on the culture and thinking of the fishing community (FAO, 2000). Conservation and management decisions for fisheries should be based on the best scientific evidence available, also taking into account local knowledge of the resources and their habitat (FAO, 2000). Some of the characteristics of local knowledge (Studley, 1998) which are attributes of the Pajas Blancas community of fishermen are 1) it is linked to a specific place, 2) it is dynamic in nature, and 3) it belongs to a group of people who live in close contact with natural systems.

Framework and Methods

Our approach is based on a coupled human–environment system framework developed at Clark University and the Stockholm Environment Institute for assessment of vulnerability and adaptation (Kasperson and Kasperson, 2001; Kasperson et al, 2002). The approach, shown schematically in Figure 20.1, has been called a second-generation method for climate change assessment (Leary and Beresford, 2002). It is based on six concepts: 1) to determine what responses reduce risks, 2) to investigate causes of vulnerability, 3) to take into account social causes of vulnerability, 4) to consider multiple stresses, 5) to use recent experiences as analogues and 6) to treat adaptation measures centrally.

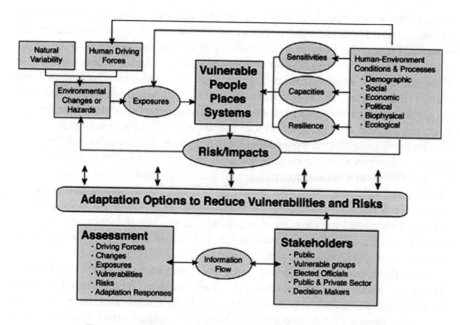

Figure 20.1 *Vulnerability and adaptation framework*

Source: Leary and Beresford (2002), adapted from Kasperson et al (2002).

We also took into account the FAO Report No 639 on fisheries (FAO, 2000), Adaptation Policy Frameworks for Climate Change (Lim et al, 2005), our previous assessment of vulnerabilities (Nagy et al, 2006a) within the context of the multilevel indicator of vulnerability to climate variability and change developed by Moss (1999), and a classical driver-Pressure-State-Response (d-PSIR) framework (Nagy et al, 2006a and b). Likewise, we assume that reducing vulnerability requires enhancement of adaptive capacity.

We used a combination of primary and secondary information described in Nagy et al (2006a), including:

1   in situ and remote observations to follow the location of the estuarine fronts;
2   climate scenarios from global circulation models;
3   expert judgement of the authors based on primary and secondary information, indicators such as ENSO SST 3,4, salinity, fish yields and wind speed; and
4   cost–benefit valuation of direct use of artisanal fisheries/fishing activity.

Social vulnerability was assessed from secondary information (Hernández and Rossi, 2003; Spinetti et al, 2003; Norbis et al, 2005; Nagy et al, 2006a).

Variability of the Estuarine Front and Fishery Resources

The estuary of the Río de la Plata is characterized by a circulation and stratification pattern formed by the penetration of saltier and denser marine waters riverward along the bottom while freshwater inflow moves seaward along the surface. The salt intrusion limit is located within two front lines, shown in Figure 20.2: the main turbidity front (MTF) and the secondary main front (SMF) (Lappo et al, 2005; Severov et al, 2003). This feature sustains ecological processes (for example, nutrient assimilation and denitrification), services (for example, carbon dioxide fixation and fish reproduction) and fisheries.

The estuarine frontal system displaces anywhere between 10 and 200 kilometres either towards the river or towards the sea, as a function of variations in river inflow at seasonal to interannual timescales and variations in onshore (S–SE) and offshore (W–NW) winds at synoptic (1–10 days) and seasonal timescales (Framiñán and Brown, 1996; Severov et al, 2003, 2004; Nagy et al, 2006a). Frequency patterns of these winds have changed over the last few decades, with an increase in onshore (E–SE) winds (Pshennikov et al, 2003; Escobar et al, 2004; Bischoff, 2005).

Variations in river inflow are associated in part with ENSO variability. Strong El Niño (in other words >2.0 Sea Surface Temperature-SST 3.4) sustained for three or more months induces high rainfall and increased river flow. This increase in freshwater displaces the frontal system seaward away from the fishermen's land site. During strong La Niña (<2.0 SST 3.4), rainfall and river flows are below average and the front displaces riverward. Freshwater inflow to the estuary, from the River Uruguay and total are shown

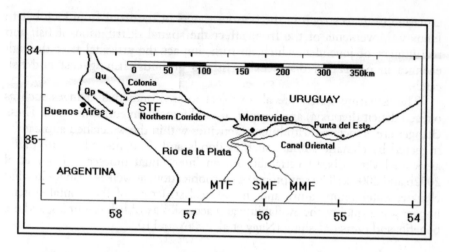

Figure 20.2 *River flow corridors and fronts of the Río de la Plata*

Note: The estuarine frontal system is demarked by MTF and SMF, the shown locations of which are based on a composite image from Sea-WIFS satellite observations and in situ measurements for 17 November 2003. The Pajas Blancas and San Luis fishing sites are located 140 and 240km downriver from Colonia respectively.

for selected years in Table 20.1. Usually, the percentage of inflow from the River Uruguay increases with total freshwater inflow. Flooding of the River Uruguay occurs mainly in the northern corridor in October–November, followed by low water from December to February. Flood events were triggered by El Niño in 1997 and 2002, and low flow events by La Niña in 1988 and 1999, which respectively pushed the frontal zone far downriver and upriver from the land base.

Table 20.1 *Freshwater inflow to the Río de la Plata from the River Uruguay and total*

Period	ENSO state	River Uruguay flow ($10^3 m^3/s$)	Total Inflow ($10^3 m^3/s$)	Percentage from River Uruguay
1987–1988		2.9	19.2	15
1988–1989	La Niña	1.5	14.0	11
1990–1999*		5.6	26.0	21
2000	La Niña	4.7	20.1	23
2002	El Niño	9.2	26.8	35

Note: *Average annual flows for 1990–1999.

Fish populations track movements of the front. For example, whitemouth croakers (*Micropogonias furnieri*), the main fish resource, migrate to the estuarine front to spawn in the bottom waters of the front from October to January. Movements of the front affect the spatial distribution of fish and recruitment of juveniles, which, in turn, impact the artisanal fleet through changes in navigation distances to fishing grounds, fishing cost and fish catch.

The estuarine waters are also subject to environmental changes such as oxygen deficit (hypoxia) and harmful algal blooms from eutrophication. These changes are associated with human activities within the watershed and can be triggered by climatic stimuli, such as floods and droughts, which are partly associated with ENSO variability on an interannual timescale (Nagy et al 2002b and 2006a). The symptoms of eutrophication, as well as changes in wind climate, water temperature and the vertical structure of the frontal system, impact water quality, the availability and accessibility of fish resources, and the benefits and costs of fishing (Nagy et al, 2006a and b).

Impacts of Variability on Fishing Activity and Incomes

The artisanal fleet exploits the croaker and other fisheries within 3–4km of the Uruguayan coast at the Northern corridor close to the MTF. The main fishing community is based at Pajas Blancas, about 140km downriver from Colonia. Fishing activity is carried out year-round, the peak being associated with the croakers' migration to the estuarine front to spawn during spring and early summer. This period accounts for more than 80 per cent of annual catch and incomes (Figure 20.3), after which, from January to September, many fishermen migrate to seaward fishing sites at San Luis, 240km downriver from Colonia, or look for alternative income sources.

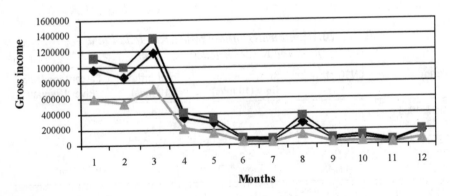

Figure 20.3 *Long-term gross income of fishermen (local currency, 1999)*

Note: Average: black; strong ENSO years: light grey; maximum: dark grey. 1 = October, 12 = September (from Nagy et al, 2003). The primary peak, October through December, is associated with croakers; the secondary peak in May is associated with other species.

The first studies of the Pajas Blancas fishery were carried out from 1987 to 1989. During the 1988–1989 fishing period, both the length of the reproductive period and the number of captured fishes were less than in 1987–1988 and total catch was 60 per cent down (Acuña et al, 1992). Norbis (1995) analysed captures from the 1987–1988 fishing period and concluded that the catch was greater after southeast winds (postfrontal period) and when north/northeast winds blow, but diminished when south/southeast winds greater than 8m/s or west/southwest winds, which oppose river discharge, prevailed. These studied fishing periods were characterized by moderate (1987–1988) and very low river discharge (1988–1989), especially of River Uruguay flow, the latter associated with the strong La Niña in 1988 (see Table 20.1).

As described previously, movements of the front impact the fisheries and the economic returns on fishing. This is most evident and dramatic in strong ENSO years. During these years, net incomes of the fishermen are reduced to about 40–70 per cent, or 60 per cent on average, relative to the long-term average for the 1988–2001 period (Norbis et al, 2005; Nagy et al, 2006a and b). Decreases in net income are mainly the result of a shortened peak fishing period, due both to the inaccessibility and small size of fish. Strong La Niña events such as 1988–1989 and 1999–2000 are severe shocks that pose a big threat to fishermen's adaptive capacity and sustainable livelihood (Norbis et al, 2005).

Until 2001–2002, fishermen had shown that they were resilient to stresses within a wide coping range and maintained the long-term productivity of resources. However, they were partly dependent on external support (they received some money in advance for fuel in exchange for future catch). According to the criteria of Pittaluga et al (2005) for poverty and vulnerability of livelihood systems the Pajas Blancas community of fishermen can be placed between moderately poor and self-sufficient groups.

Coping with Current Variability: Type I Adaptation

Fishermen have acquired local knowledge of the interactions between the environment and resources, as well as adjusted their own mode of behaviour in response to such environmental and resource interactions. This accumulated human capital is the basis of their adaptive capacity. Fishermen have developed adaptation strategies to cope with the increase in ENSO-related river flow variability and related locations of front lines, evolving from autonomous and reactive actions to planned private ones, without any participation of public managers or local authorities. These strategies include migration to follow movements of fish stocks and cautious fishing behaviour to avoid fishing effort on days when conditions are unfavourable. But, in spite of their human capital and good availability of resources, the fishermen lack sufficient social and financial capital to cope with climate extremes and remain vulnerable to changes in the spatial distribution and quantities of fish stocks (Nagy et al, 2006a). Typically, fishermen do not notice fluctuations of resources until they perceive changes of availability as a consequence of river flow and/or winds.

Thus, extreme events drive adaptive prevention of loss (Norbis, 2003). It is probable that the very low catch attributable to La Niña in 1988 induced autonomous behavioural changes.

Migration can be divided into seasonal (temporary, once a year), relocation of fishing site (longer-term) and spontaneous. The first two are planned actions and the third one is a reactive response that is forced once each 3–5 years by the lack of resources usually associated with ENSO-related river flow variability. The main causes of migration are changes in the availability of resources and migration decisions are influenced by the structure of the fishing unit. Because of the spatial and temporal changes of the estuarine front and resource availability, as well as the increasing trend of river flows, many fishermen have migrated seasonally or permanently along the coast following resources in order to reduce their long-term vulnerability to hydroclimatic fluctuations and avoid bad catch years (Hernández and Rossi, 2003; Norbis et al, 2005; Nagy et al, 2006a).

A typical fishing unit is a family business composed of 4 or 5 persons with the head of family acting as the crew skipper and sometimes the owner. The boat, radio, engine and fishing gear belong to the owner, who receives 40 per cent of gross income, while the skipper and sailors receive 60 and 40 per cent of net income respectively (Spinetti et al, 2003). Fuel is supplied in advance by middlemen and the owner is in charge of maintenance costs, about 2.3 per cent of mean gross value. Cash flow is managed domestically, which leads in certain times to the fishermen not having enough working capital to perform fishing activity and needing funds in advance from the middleman. During bad years, the fishermen exchange labour and raw materials without having any contact with money, a condition called pre-economy by Spinetti et al (2003).

About 50 per cent of fishermen migrate by February or March of each year as an anticipatory response to the expected seasonal decrease in croakers (Spinetti et al, 2003). Most fishermen migrate 60–100km seaward of the front following resources along the coast, going mainly to San Luis, 80km to the east of Pajas Blancas, to start the sea trout season (Hernández and Rossi, 2003; Spinetti et al, 2003). This behaviour of seasonal migration began, we believe, from fishermen looking for new resources following the very bad season of 1988/89, a strong La Niña year with extremely low river flow. During bad fishing years more than 50 per cent of fishermen migrate spontaneously riverward or seaward of the front, sometimes as early as December, as a reaction to an unexpected decrease in fish capture often associated with ENSO extremes or at other times due to very low river flows that are not associated with ENSO.

Some fishermen that used to migrate seasonally to San Luis relocated permanently there during the 1980s and 1990s, as well as more recently. The causes of this planned, long-term relocation were the search for resources the whole year, lower dependence on hydro-climate variability, and increased experience and skills among sailors and skippers. The cost of this relocation was not quantified but seems to be small as there are no big obstacles to migration and many migrants have relatives and friends who host them. They do not need port facilities, only a small bay with a large beach to place their fishing boats and gear. Spinetti et al (2003) report that net monthly income at Pajas

Blancas is 66 per cent higher than at San Luis during the peak of the fishing season (three months) but several times lower the rest of the year. Only 9 per cent of San Luis fishermen have alternative jobs, whereas this figure reaches 45 per cent at Pajas Blancas (Spinetti et al, 2003).

Generally, those that migrate are young, unmarried and have less economic assets. Many of them are fishing boat sailors who receive only 20 per cent of net boat income and some of them are skippers; they are seldom owners. Their opportunity is to have resources available the whole year and a lower exposure to the effects of hydro-climatic variability.

Aside from these seasonal and climate-induced migration strategies, fishermen have learned that cautious behaviour to avoid weather-related risks also reduces their vulnerability (Nagy et al, 2006a). Most skippers began their fishing careers very young; they have an average experience of 21 years and all of them took courses on navigation and safety (Spinetti et al, 2003). They have acquired the knowledge that the availability of croakers decreases when there are fresh southern to eastern winds, even before navigation is made difficult or impeded by these winds. Skippers developed and widely adopted practices of not fishing the day after fresh breezes come from the south or southeast. This behaviour, which sacrifices income but also reduces fishing costs and risks, is interpreted by Norbis (1995) and Norbis et al (2003 and 2005) as a 'cautious behaviour' regarding weather.

The frequency of unfavourable winds has increased over the last few decades. For instance, strong E/SE winds at Buenos Aires increased in frequency by 80 per cent from 1961–70 to 1991–2000 (Bischoff, 2005). But the impact of this increase was not clearly perceived by fishermen since they regard changes in fish catch as a consequence of factors other than weather (Wells and Daborn, 1997; Spinetti et al, 2003; Nagy et al, 2006a).

Because of cautious fishing behaviour, the average time spent fishing during the fishing season is 15 days per month (Spinetti et al, 2003). However, analyses of wind conditions and fish availability indicate that fishermen could increase their net incomes by increasing the number of fishing days. Norbis (1995) found that 8m/s is the lower threshold at which fish availability is decreased by southern to eastern winds. Applying this threshold to the period 1977–1986, the frequency of days with unfavourable winds during the fishing season is 24 per cent, with a range of 6–9 days and 4–6 events per month (see Figure 20.4). Based on this, we believe that the cautious practice of the fishermen could be modified by increasing the average fishing days from the current 15 (Spinetti et al, 2003) to about 18 (Norbis; 1995; Norbis et al, 2003). An early scouting of sea conditions and fish availability based on real time fishing-oriented forecasting and flow of information should allow keeping safe navigation and increasing net income by at least 15 per cent.

Adaptation is a risk-management strategy that is not free of cost (Easterling et al, 2004). However, there is evidence that, up to the year 2002, fishermen could afford the residual loss that occurred even after the use of reactive and anticipatory adaptation measures (Norbis et al, 2005; Nagy et al, 2006a). The latter suggested that the balance between climate and socioeconomic drivers of impacts and autonomous adaptation lay between the coping ranges. However,

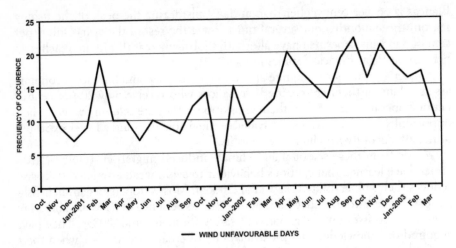

Figure 20.4 *Unfavourable days for fishing activity on a monthly basis from October 2000 to March 2003, based on a lower threshold of 8m/s wind speed*

adaptation failed and cautious behaviour disappeared during the bad fishing season of 2002/03 due to three different pressures. First, increased river flow due to a moderate to strong El Niño events displaced the estuarine front seaward by up to 150–200km from Colonia. Second, days with unfavourable winds increased by 57 per cent from 11 days per month in 2001/02 to 17.6 days per month in 2002/03 (Norbis et al, 2005; Nagy et al, 2006a). Compared to the 1970s, which averaged 7.2 unfavourable days per month, this increase is 150 per cent. The third factor was a deterioration of socioeconomic conditions due to a regional economic crisis during 2001–2003 that reduced economic welfare, increased fuel price and limited alternative sources of income (Nagy et al, 2003).

Thus, many fishermen took the risk of fishing under unfavourable conditions (southeast and eastern winds), which ultimately resulted in maladaptive practice (Norbis et al, 2005; Nagy et al, 2006a). Fish yields were very bad because of several factors – displacement of the front; increased cost of navigation (due to both fuel price and distance), shortened fishing period, lower number of good fishing days and reduced availability of resources – all of which led to a high number of 'non-fishing trips' at the beginning of the season and forced many fishermen to migrate to eastern sites such as San Luis before January (non-fishing trips are defined as those where the cost of navigation equals or exceed the benefits of catch).

Was this failure of human capital an anticipation of future difficulties to adapt (Type II adaptation) to increasing changes? It seems that fishermen will need to adopt further adaptation strategies in the event of climate variability and change, as well as increases in non-climate stresses. A key question for future research is thus whether adaptation benefits from avoiding damage losses are equal or greater than adaptation costs and loss of benefits. The potential adaptation options available to fishermen are discussed in the forthcoming sections using the reference period 2002–2003.

Future Climate and Environmental Scenarios

To examine whether the increases in temperature, precipitation and river flow pose a threat to the sustainable livelihood of coastal fishermen, we examined both recent experiences with unfavourable climatic conditions and projections of the future from the literature. In order to construct scenarios of future climate for the southeastern South America region, we use outputs from climate experiments of general circulation models that are available from the Intergovernmental Panel on Climate Change (IPCC) Data Distribution Center. The selected models are HadCM3 (UK) and ECHAM4/OPYC3 (Germany), which have acceptable agreement with the observed sea level pressure field and are able to represent the position and intensity of the pressure systems of the region, both on monthly and annual bases (Bidegain and Camilloni, 2004; Nagy et al, 2006a). Although observed climate fields indicate that both models underestimate precipitation within the Río de la Plata basin, monthly and annual temperature fields show that, in general, both models have acceptable agreement with the observed fields.

The selected climate experiments of the HadCM3 and ECHAM4/OPYC3 models are forced with the A2 and B2 scenarios of greenhouse gas emissions, which are characterized by a globalized world with high emissions and a regionalized world with low emissions respectively. They correspond to middle–high and middle–low views of future emissions respectively.

Future regional climate scenarios for precipitation and temperature were constructed for 2050 and 2080 based on the range of values generated by the aforementioned models and socioeconomic scenarios. Current climate and future scenarios for the Río de la Plata basin and estuary for 2050 suggest a change in precipitation within the range +5 to +20 per cent and in temperature from +1 to +2°C. In comparison, during the last few decades precipitation increased by 20 to 25 per cent, river flows increased by 25 to 40 per cent and average temperatures rose by 0.5 to 0.8°C.

Trends for future river flows are very difficult to estimate because of both the uncertainty of regional human drivers and because of the varied regional scenarios from different GCMs. Under a future scenario in which stream flow remains similar to or slightly lower than at present (no change to -10 per cent), we do not expect a significant increase in current environmental stresses on the estuarine system from the already moderately high level, with the exception of nutrient inputs (a pressure indicator for eutrophication).

Of greater concern is a future scenario where river flow increases by 10 to 25 per cent, together with projected temperature increases and economic and population growth, for which significant impacts are expected in the estuarine system. Considering the fact that seasonal temperature, precipitation and stream flow cycles are not superposed, any changes should modify seasonal circulation, stratification and estuarine front location, inducing further environmental shifts with a probable increase in the degree and occurrence of symptoms of eutrophication such as hypoxia (oxygen deficit) in bottom waters and harmful algal blooms (Nagy et al, 2002a and b and 2006a).

Scenarios of Sustainability and Future Adaptation

A model of fishing activity was developed by Nagy et al (2006b) based on direct use costs estimated from secondary information for 1998–1999, a good and long fishing season. Usually, the fishing period lasts 3–5 months, average favourable fishing days are 15–16/month and total number of boats is about 31 (Norbis et al, 2005; Nagy et al, 2006a). We assume, on the basis of climatic conditions and seasonal yield compared with long-term yields and income figures that the studied case – the 1998/99 fishing season – allowed the fishermen's livelihood to be sustainable. These figures are 923 fishing navigations in 64 days with an average catch of 22 boxes, that is to say about 20 per cent greater than the long-term average capture.

Modelled scenarios presented in Table 20.2 are based on observed seasons and show the long-term catch, observations for the 1998/99 fishing season, and model simulations for the minimum activity of a typical season and a bad season (as would be typical in a strong ENSO year) and for improved performance with changes in fishing behaviour. This latter case corresponds to a typical season with normal flow and wind conditions but with an adjustment in fishing behaviour to increase the number of fishing days to 18 per month over a 4 to 5 month fishing season. The model is based on the hypothesis that the actual number of fishing days is about 80 per cent of the maximum on a monthly basis within a 5 months season and estimates the number of boxes that fishermen would have under increasing number of fishing days on both monthly and seasonal bases. Both the availability of resources and current demand for fresh fish by local and regional markets allow increasing fish catch during typical years.

The increase in economic return from changes in fishing behaviour is an easy low-cost adaptation. The increase in capture estimated as the maximum possible with low-cost measures should allow fishermen to recover both investment capital and losses during bad years. An information system to support such changes in behaviour is described in the next section.

It must be noted that average boxes and navigation days are not independent because usually a few leading fishermen decide to navigate and are followed by others, and this number should increase if adaptation measures were taken. The model, which takes into account conservative numbers of catch per day (20 boxes), suggests that an increase in the number of fishing navigations is a key factor (provided the number of non-fishing trips remains stable). Thus, because neither weather nor climate conditions can be managed, both real-time forecasting and communication with fishermen, which should allow an increase in the number of navigations under favourable fishing conditions, seem to be wise and low-cost adaptation practices. For this, an increase

Table 20.2 *Fishing activity, capture and income: Comparison between a good year (1988–89), long-term average and model results for a low-typical year (1), a bad year (2) and results with change in fishing behaviour (3)*

	Observed (1998/99)	Long-term average	Model 1	Model 2	Model 3
Number of boxes per boat per day	22	21	20	16	20
Fishing days	64	57	50	40	70–75
Fishing navigations	963	850	640	500–600	1150–1300
Boats per day	15	15	10	10	17–18
Income (% relative to 1999)	100	85	70	45	115–125

Source: Nagy et al (2006b).

in security of navigation and good relations with the Coast Guard and the Directorate of Aquatic Resources are needed.

Regarding the question of whether the coastal fisheries' Type I adaptation is adequate in the face of climate variability, we can say that it is adequate for coping with current variability, but it is not adequate for coping with anticipated changes in climate, climate variability and other factors that affect the fishery and livelihoods of fishermen. For this, Type II adaptations are needed.

Adapting to Future Climate Change: Type II Adaptation

Past experience shows that under severe climatic pressures, fishermen are strongly impacted and reactive measures have poor results independent of socioeconomic and institutional responses. Under projected climatic, environmental and economic changes, the artisanal fishery is likely unsustainable without policy and behavioural changes to adapt to climate change. Yet nothing has been done to mobilize or enable Type II adaptations. The lack of a social network is identified as a main structural vulnerability and an obstacle to climate change adaptation (Nagy et al, 2006a). Lack of information, knowledge, capacity and financial resources are also obstacles to adaptation.

Measures to enable Type II adaptation are needed at national, regional and local levels to increase resilience to climatic and environmental threats, decrease economic vulnerability, increase adaptive capacity, and decrease physical vulnerability and exposure. Potential measures, identified from a field survey by Spinetti et al (2003) and the expert judgment of the authors, are presented in Table 20.3.

Table 20.3 *Type II adaptation measures by scale of implementation and objectives*

	Measures and Policies to			
Level	Increase resilience to natural threats	Decrease economic vulnerability	Increase adaptive capacity	Decrease physical vulnerability and exposure
National	Weather and climate forecasting	Increase economic growth and employment	Integrate adaptation into planning; coastal zone management	Integrate adaptation into planning; coastal zone management
Regional	Fishing-oriented monitoring and early warning	Subsidize fuel price; increase access to fresh fish markets; refrigeration; insurance	Promote awareness, association, education and flow of information	Facilitate migration
Local	Scouting	Increase access to credit and fishing activity	Stakeholder participation	Cautious fishing behaviour; improve safety, communications, boats and engines

Source: Based on priorities of fishermen (Spinetti et al, 2003), authors' expert knowledge, vulnerability analysis (Nagy et al, 2006a and b) and UNDP-GEF criteria (Ebi et al, 2005).

Regarding the question of whether access to information and knowledge (generation, demand, information and outreach) is adequate to enable Types I and II adaptation, fishermen using wise fishing practices have had an adaptive potential that has proven to be sufficient to cope with past climate and non-climate scenarios, and they have shown that they neither depend on the flow of information from managers and scientists nor demand it. However, empirical evidence suggests that if business-as-usual management scenarios continue, it is likely that current adaptation will not be sufficient under increased climate pressures. Management strategies will need to be periodically revised and adapted to the dynamic conditions of the climate, fish stocks, the environment and resource users, as well as to changes in the intertemporal preferences of the fishing sector. In this dynamic and uncertain environment, knowledge and information will become increasingly important.

Fishermen and policymakers will need to adopt an adaptive management strategy that is supported by a system to integrate, communicate and apply multiple sources of information. Such a system, which we call an adaptation control information system (ACIS), should prioritize generation and access to knowledge and information on weather and climate forecasts, fishery resources, frontal dynamics (satellite data-based), and water quality; education, learning processes and participatory processes; early warning systems; and real-time flow of information to fishermen in an appropriated language.

Local and scientific knowledge are shared between managers, institutions in charge of observations, scientists and fishermen, but there is a lack of interaction and trust between them. Furthermore, some coastal observations have been discontinued in Uruguay (daily salinity in several sites). Knowledge is accumulated through observation, monitoring and analysis, which degrades if the data collection system collapses, if literacy and education levels diminish, or if basic societal infrastructure diminishes (Easterling et al, 2004). States should assign priority to undertake research and data collection in order to improve scientific knowledge of fisheries (FAO, 2000).

Adaptation measures will only be effective if education, the generation of and access to information, and communications among stakeholders (fishermen, managers, local authorities and scientists) are improved. Neither the acquired scientific and local knowledge nor the improvement of early warning systems will be enough until fishermen are able to make effective use of them. An important constraint is the failure of fit between time and space scales between institutions responsible for management and actors (Norbis, 2003).

Translation of climate scenarios and forecasting to advise appropriate action are not simple matters. Most of the success will depend on the adaptive potential of stakeholders, that is to say the ability to innovate and create new strategies and actions outside the actor's customary network (Downing et al, 2004). It is imperative that fishermen participate in this process from the very beginning. These measures should be taken with the agreement of all stakeholders, including fishermen, the coast guard, the directorate of aquatic resources, the directorate of meteorology and the EcoPlata Program of Coastal Management.

Conclusions

The last two decades have been characterized by increasing trends in means and variability of river flow and temperature, changes in front location and salinity, and resource availability. These environmental factors have led to an increase in interannual fluctuations of fish yields and fishermen's income. All of these facts have imposed new challenges and threats to subsistence fishermen.

Past experiences indicate that subsistence fishermen have successfully adopted autonomous adaptation options summarized as follows:

1   seasonal, definitive and reactive migrations along the coast following the resources associated with frontal displacement; and
2   cautious fishing behaviour under non-favourable wind conditions and acceptance of income loss.

However, in spite of these wise Type I adaptation practices, fishermen remain highly vulnerable to severe weather and climatic conditions and their livelihood is likely to be unsustainable under increasing climate variability, environmental changes and economic pressures in the absence of proactive adaptations. Projected scenarios for 2050 will increase vulnerabilities of the artisanal fisheries, which would be heavily impacted.

The current level of uncertainty about near-future climatic change and socioeconomic trends does not seem to be the main constraint to adaptation, rather it is the lack of both access to information and knowledge and public awareness about the impacts of current climate variability and extremes. This statement could be extended for the coastal zone as a whole. Education and training, participatory processes, dialogue and communication between stakeholders are needed to implement effective measures to take advantage of the generation of knowledge and information, forecasting and early warning systems. ENSO events are recurrent, and once the first indicators are known (SST 3.4), anticipatory adaptation measures should start (in other words real-time communication and early warning).

Adaptive management should emphasize the integration of local and scientific knowledge, training, enhancement of data collection systems, weather and climate forecasting, and real-time communication to users (fishermen and the coast guard). The implementation of the suggested easy adaptation measures will improve livelihood quality and should augment the financial and social capitals fishermen need to access credit and markets at the national level. As a consequence of research on recent severe ENSO events and the National Communications to the UNFCCC, public awareness has been increased and new regulations on practices and plans have been planned in several sectors in order to adapt to the new climate variability conditions.

References

Acuña, A., J. Verocai and S. Márquez (1992) 'Aspectos biológicos de Micropogonias furnieri (Desmarest 1823) durante dos zafras en una pesquería artesanal al Oeste de Montevideo' ['Biological aspects of Micropogonias furnieri (Desmarest 1823) during two fishing years of an artisanal fisheries at Western Montevideo'], *Revista de Biología Marina, University of Valparaíso*, vol 27, pp113–132

Bidegain, M. and I. Camillloni (2004) 'Performance of GCMs and climate baselines scenarios for southeastern South America', paper presented at the Latin America and Caribbean AIACC Workshop, Buenos Aires

Bischoff, S. (2005) 'Sudestadas', in V. Barros, A. Menéndez and G. J. Nagy (eds) *El Cambio Climático en el Río de la Plata [Climate Change in the Río de la Plata]*, Project Assessments of Impacts and Adaptation to Climate Change, CIMA-CONICET-UBA, Buenos Aires, pp53–67

Burton, I. (2004) 'Climate change and the adaptation deficit', Occasional Paper 1, Adaptation and Impacts Research Group, Environment Canada, Ottawa, Canada

Burton, I. (1996) 'The growth of adaptation capacity: Practice and policy', in J. Smith, N. Bhatti, G. Menzhulin, R. Benioff, M. Campos, B. Jallow, F. Rijsberman, M. I. Budyko and R. Dixon (eds) *Adapting to Climate Change: An International Perspective*, Springer, New York, pp55–67

Burton, I. (1997) 'Vulnerability and adaptive response in the context of climate change', *Climate Change*, vol 36, pp185–196

Camilloni, I. and V. Barros (2000) 'The Paraná river response to El Niño 1982–1883 and 1997–1998 events', *Journal of Hydrometeorology*, vol 1, pp412–430

DFID (Department for International Development) (2000) 'The sustainable livelihoods approach (SLA) and the code of conduct for responsible fisheries (CCRF)', Sustainable Fisheries Livelihoods Programme, www.sflp.org/eng/007/pub1/bul1a_art1.htm

Downing, T., S. Bharwani, S. Franklin, C. Warwick and G. Ziervogel (2004) 'Climate adaptation: Actions, strategies and capacity from an actor-oriented perspective', unpublished manuscript, Stockholm Environment Institute, Oxford, UK

Easterling, W. A., B. H. Hurd and J. B. Smith (2004) *Coping with Global Climate Change. The Role of Adaptation in the United States*, Pew Center on Global Climate Change, Arlington, VA

Ebi, K., B. Lim and Y. Aguilar (2005) 'Scoping and designing an adaptation project', in B. Lim, E. Spanger-Sigfried, I. Burton, E. Malone and S. Huq (eds) *Adaptation Policy Frameworks for Climate Change, Developing Strategies, Policies and Measures*, Cambridge University Press, Cambridge, UK, and New York

Escobar, G., W. Vargas and S. Bischoff (2004) 'Wind tides in the Río de la Plata estuary: Meteorological conditions', *International Journal of Climatology*, vol 24, pp1159–1169

FAO (2000) 'Fisheries Report No 639 FIPL/R639', report of the Third Commission of the Advisory Committee on Fisheries Research, 5–8 December, Food and Agricultural Organisation, Rome

Lappo, S. S., E. Morozov, D. N. Severov, A. V. Sokov, A. A. Kluivitkin and G. Nagy (2005) 'Frontal mixing of river and sea waters in Río de la Plata', *Transactions of the Russian Academy of Sciences* (Earth Science Section), vol 401, pp267–269

Leary, N. and S. Beresford (2002) *Vulnerability of People, Places and Systems to Environmental Change*, International START Secretariat, Washington, DC

Kasperson, J. X. and R. E. Kasperson (2001) 'International Workshop on Vulnerability and Global Environmental Change: A workshop summary', Stockholm Environment Institute, Stockholm, Sweden

Kasperson, J. X., R. E. Kasperson, B. L. Turner II, W. Hsieh and A. Schiller (2002) 'Vulnerability to global environmental change', in A. Diekmann, T. Dietz, C. Jaeger and E. Rosa (eds) *The Human Dimensions of Global Environmental Change*, MIT Press, Cambridge, MA

Lim, B., E. Spanger-Siegfried, I. Burton, E. Malone and S. Huq (eds) (2005) *Adaptation Policy Frameworks for Climate Change: Development Strategies, Policies and Measures*, Cambridge University Press, Cambridge, UK, and New York

Moss, R. (1999) 'Vulnerability to climate variability and change: Framework for synthesis and modeling: Project description', Battelle Pacific Northwest National Laboratory, Richland, WA

Nagy, G. J., M. Gómez-Erache and A. C. Perdomo (2002a) 'Río de la Plata', in T. Munn (ed) *The Encyclopedia of Global Environmental Change. Volume 3: Water Resources*, John Wiley & Sons, New York

Nagy, G. J., M. Gómez-Erache, C. H. López and A. C. Perdomo (2002b) 'Distribution patterns of nutrients and symptoms of eutrophication in the Río de la Plata estuarine system', *Hydrobiology*, vol 475–476, pp125–139

Nagy, G. J., G. Sención, W. Norbis, A. Ponce, G. Saona, R. Silva, M. Bidegain and V. Pshennikov (2003) 'Overall vulnerability of the Uruguayan coastal fishery system to global change in the estuarine front of the Río de la Plata', The Open Meeting Human Dimensions of Global Environmental Change, Montreal, Canada

Nagy, G. J., M. Bidegain, R. M. Caffera, F. Blixen, G. Ferrari, J. J. Lagomarsino, C. H. López, W. Norbis, A. Ponce, M. C. Presentado, V. Pshennikov, K. Sans and G. Sención (2006a) 'Assessing climate variability and change vulnerability for estuarine waters and coastal fisheries of the Río de la Plata', AIACC Working Series Paper No 22, International START Secretariat, Washington, DC

Nagy, G. M. M., Bidegain, R. M. Caffera, J. J. Lagomarsino, W. Norbis, A. Ponce and G. Sención (2006b) 'Adaptive capacity for responding to climate variability and change in estuarine fisheries of the Río de la Plata', AIACC Working Paper No 36, International START Secretariat, Washington, DC

Norbis, W., J. Verocai and V. Pshennikov (2003) 'Activity of the artisanal fishing fleet in relation to meteorological conditions during the October 1998 to March 1999 fishing season', in Vizziano et al (eds) *Research for Environmental Management, Fisheries Resources and Fisheries in the Saline Front*, vol 14, EcoPlata, Montevideo, Uruguay, pp191–197

Norbis, W., A. Ponce, D. N. Severov, G. Saona, J. Verocai, V. Pshennikov, R. Silva, G.Sención and G. J. Nagy (2005) 'Vulnerabilidad y capacidad de adaptación de la pesca artesanal del Río de la Plata a la variabilidad climática' ['Vulnerability and adaptive capacity of the artisanal fisheries of the Río de la Plata to climate variability'], Chapter 18 in V. Barros, A. Menéndez and G. J. Nagy (eds) *El Cambio Climático en el Río de la Plata [Climate Change in the Río de la Plata]*, Project Assessments of Impacts and Adaptation to Climate Change, CIMA-CONICET-UBA, Buenos Aires, Argentina, pp173–187

Norbis, W. (1995) 'Influence of wind, behaviour and characteristics of the coraker (*Micropogonias fournieri*) artisanal fishery in the Río de la Plata', *Fisheries Research*, vol 22, pp43–58

Norbis, W. (2003) 'Adaptive management of fisheries in the Río de la Plata', paper presented at the AIACC Latin America and Caribbean Workshop, San José de Costa Rica, May, available at www.aiaccproject.org/meetings/San Jose_03

Pittaluga, F., N. Salvati and C Seghieri (2005) 'Livelihood systems profiling: Mixed methods for the analysis of poverty and vulnerability', LSP working series paper, FAO. Rome

Pshennikov, V., M. Bidegain, F. Blixen, E. A. Forbes, J. J. Lagomarsino and G. J. Nagy (2003) 'Climate extremes and changes in precipitation and wind-patterns in the vicinities of Montevideo, Uruguay', paper presented at the AIACC Latin America and Caribbean Workshop, San José de Costa Rica, May, available at www.aiaccproject.org/meetings/San Jose_03/Session 6/

Severov, D. N., G. J. Nagy and V. Pshennikov (2003) 'SeaWiFS fronts of the Río de la Plata estuarine system', *Geophysical Research Letters*, vol 5, p1914

Severov, D. N., G. J. Nagy, V. Pshennikov and E. Morozov (2004) 'Río de la Plata estuarine system: Relationship between river flow and frontal variability', paper presented at 35th COSPAR Scientific Assembly, Paris, France, July

Smit, B. (1993) *Adaptation to Climatic Variability and Change*, Environment Canada, Guelph, Ontario, Canada

Smit, B., I. Burton, R. J. T. Klein and J Wandel (2000) 'An anatomy of adaptation to climate change and variability', *Climatic Change*, vol 45, no 1, pp233–251

Spinetti, M., G. Riestra, R.Foti and A. Fernández (2003) 'Activity of the artisanal fishery in the Río de la Plata: Socio-economic structure and situation', in Vizziano et al (eds) *Research for Environmental Management, Fisheries Resources and Fisheries in the Saline Front*, vol 17, EcoPlata, Montevideo, Uruguay, pp231–259

Studley, J. (1998) 'Dominant knowledge systems and local knowledge', Mountain Forum Online Library document, www.mtnforum.org/resources/library/stud98a2.htm

Wells, P. G. and G. Daborn (1997) 'The Río de la Plata: An environmental overview', EcoPlata Project background report, Dalhosie University, Halifax, Nova Scotia, Canada

Index

acceptability, social 288, 289
adapt now 2–3
administrative level 75, 83, 101–104, 105, 176–177
Africa
  cereal production 19, 131–146, 181–195
  conservation strategies 28–52
  disease 109–130
  drought 9, 90–108, 147–162
  obstacles to adaptation 11, 12
  water management 53–70
  weather forecasting 163–180
agriculture
  cereal production 19, 131–146, 181–195
  drought 78, 79–80, 101
  frontier expansion 297–302
  rice production 228–246
  variable weather 164–166
  vulnerability 153
  water resources 59, 181, 221, 253, 254–257, 262
agropastoralism 96
AHP see analytic hierarchical process
AIACC see Assessments of Impacts and Adaptations to Climate Change
alternative strategies 159, 234, 238, 240
analytic hierarchical process (AHP) 19–20, 214, 217, 222, 223–225
anticipatory strategies 286, 291, 293, 360, 368
  see also planned adaptation
Arba'at, Red Sea State, Sudan 96–100
Argentina 8, 11, 15, 296–314, 315–331, 332, 333, 345

artisanal fishing 11, 353–370
Assessments of Impacts and Adaptations to Climate Change (AIACC) 2, 7, 10, 13, 16, 19–21
assets, livelihood 92, 93, 94–95, 97
autonomous adaptation 100–101, 154, 271, 273, 297–302, 312, 359, 361–362, 367–368
autonomous dispersers 29, 35, 39
awareness
  cereal farmers 185
  disease 121–122, 123, 124
  estuarine fishing 368
  increasing 4
  lack of 11, 22
  long-term trends 302, 304, 307, 312–313
  Pacific Islands 270–271, 274, 276
  risk 25
  see also knowledge

banks 304
Bara Province, Sudan 92–96
bed nets 11, 117
behavioural change strategies 290, 293
Beja pastoralists 96–100
benefits and costs see cost–benefit analysis
Berg river basin, South Africa 10, 19, 53–70
biocapacity 203, 206, 209
biodiversity conservation 9, 28–52
biotic adaptation 39
Bolivia 333
Botswana 12, 71–89, 72–86
bottom–up strategies 17–18, 106, 149–151, 177, 264–278
Brazil 332, 333
breeding and selection 133, 138, 143

Buenos Aires, Argentina 15, 298, 310, 311
business-as-usual strategies 136, 137, 138, 143

capability poverty 71
capacity
    biocapacity 203, 206, 209
    building 84–85, 106, 328
    disease 115–118, 121–122
    drought 148
    estuarine fishing 359
    floods 271
    increasing 7, 17, 24
    lack of 80, 244–245, 276
    Pacific Islands 265
    poverty 81
    reservoir 19, 55, 56, 57, 60, 61, 62–63, 64, 65, 66
    water resources 12, 211
    weather forecasting 168–170, 173–174
Cape Town, South Africa 53
capital 11, 12, 93, 95, 97, 99, 103, 155–157
    see also human capital; social capital
carbon dioxide
    fuel wood trade 82
    predictions 30
    yields 333, 334, 335, 337, 338, 339, 341–342, 343–344, 345, 347
The Caribbean 21, 279–295
cautious behaviour 361, 367
CBNRM *see* Community Based Natural Resources Management
CCD *see* climate change damages
cereal production 19, 131–146, 181–195
CEREBAL 131, 136
Chile 307
China 15, 19–20, 211–227
cholera 109–110, 111, 113, 119–122, 122, 123
civil society organizations 125
climate change damages (CCD) 19,

54–58, 64–65, 67, 136–137, 139–142, 140, 143
cocoa 166
collective strategies 15, 17, 234, 237, 241, 245, 311, 313
commercial development 81–82, 236, 238, 240, 242–243, 321, 322
common property resources 12, 44–45, 45–46, 197–198
communication 169, 174, 175–177, 276
Community Based Natural Resources Management (CBNRM) 44–45, 80, 84
community level
    development 17, 90–108
    indigenous knowledge 71–86
    Pacific Islands 273
    rice production 234, 237, 238–240, 240, 241, 245
    watersheds 256, 258, 259, 262
    see also local level
complementary indicators 216
conflict 100, 105, 218, 219
conservation 4–5, 32–33, 45, 189–190, 191, 203, 205–206
    see also biodiversity conservation
consultations 201–202, 204–205, 215, 225, 251–252
consumption 62, 63, 67–68, 141, 142
contract farming 320, 322–323
contractual reserves 37, 40, 42, 45
Córdoba Province, Argentina 315, 317, 319, 323–324, 326, 327–328, 329
costs
    caution and precaution 58, 65, 67
    conservation strategies 36–38
    dengue fever 287, 288, 291
    extreme weather 264, 266–267
    irrigation 141–142
    reforestation 260–261
cost–benefit analysis 19, 53–70, 139–142, 143
credit systems 11, 24, 326, 327
crop–pasture rotations 346, 348
cross-sectoral effects 247–263

dams 53, 58, 60, 62–63, 104,
248–250, 322
decision making
AHP method 217
community engagement 84
consultations 204–205
evaluation options 19
participation 275
tools 212
traditional 12, 75, 76, 83
water resources 221, 223, 224–225
weather forecasting 167–168
decision trees 35
deficit, adaptation 2–3, 10, 159
deforestation 301
delayed actions 18
dengue fever 10, 21, 279–295
dependency, export 85–86
determination to adapt 21–23, 25
deterministic scenarios 68
developing countries 2, 4–5
development
capacity 84–85
community 17, 90–108
economic 77–81
human 11, 13–14
integrating adaptation 3, 15–18,
23, 25, 105–106
research 134, 160, 186, 209
urban water demand 60, 63, 64,
65, 66, 67
direct seeding 346, 348
disaster risk 14
disease
dengue fever 21, 279–295
malaria and cholera 11, 109–130
dispersers 9, 29, 35–38, 39
diversification 159, 185, 236, 242,
321–322, 326
domain-based frameworks 216
do nothing approaches 34, 36
dredging 270, 272, 277
drinking water 120, 121, 124
droughts
Africa 147, 148
coping strategies 90–108, 154–158
crop production 132

El Niño 268
government intervention 77–81
local actions 322
pastoralists 6, 196–210
vulnerability 152–153
dryland cereal farming 181–195
dynamic spatial equilibrium models
54–68

early maturing crop varieties 155,
156, 159
early warning systems
disease 118, 122, 124, 291–292,
293
weather forecasting 12, 167–168,
169, 177, 234
East Africa 11, 109–130
ECHAM4 model 133, 134, 137, 138,
139, 363
economic aspects
agricultural expansion 300–301
analysis tools 214
cereal production 131–146
conservation 39–40, 43–44
development 77–81
disease 114–115
estuarine fishing 364
export dependency 85–86
natural disasters 13
policy-planning models 54–68
reform 223–224
transition 197–198
education
biodiversity conservation 43
cereal production 183, 185
disease 118, 285–286, 288, 290
estuarine fishing 367, 368
rangelands 11, 202–204
rural households 160
vulnerability 153
Egypt 181–195, 187–192
El Fashir Rural Council, North
Darfur, Sudan 100–101, 102–103,
105
El Niño Southern Oscillation
(ENSO)
disease 113–114, 119, 282–283

drought 268
estuarine fishing 353, 356, 357, 359, 360, 362, 368
floods 302, 303
rainfall 320
emissions scenarios *see* scenarios; Special Report on Emissions Scenarios, IPCC
employment 78–79
enabling conditions 3, 4, 16, 21–25, 84
engineering options 220, 221
ENSO *see* El Niño Southern Oscillation
environmental aspects
autonomous adaptation 301, 312
disease 114, 119, 288, 289
estuarine fishing 358, 363–364
soybean cultivation 333, 345–347, 347–348
epidemics 112–113, 114, 119, 279, 283
equity assessment 92
estuarine fishing 353–370
Expert Choice 2000 223
export dependency 85–86
ex situ conservation 9, 34, 38, 39, 40, 46
extended-range weather forecasting 168–170, 177
extension services 158, 159, 178, 185, 186, 192
extreme weather 8, 10, 264, 274, 308–310
see also droughts; El Niño Southern Oscillation; floods; La Niña

facilitated dispersers 18, 29, 36–38, 39, 40, 45, 46
family institutions 75, 76
farming *see* agriculture
feasibility analysis 136
fertilization 19, 133, 138, 139–141, 143, 158, 341–342, 343, 344, 346
Fiji 15, 17–18, 265–266
financial level

assets 93, 94, 97, 98, 101, 102, 104, 155–156
conservation strategies 36–38
constraints 12–13, 207–208
credit access 326, 327
dengue fever 286–287
international 5, 24–25, 33, 208, 304
rice production 238, 239, 240, 242, 245
fishing 11, 153, 159, 353–370
floods
Argentina 8, 301, 309
Lower Mekong river basin 6, 229–230
Pacific Islands 266–270, 271, 276, 277
planned adaptation 302–304
vulnerability 327–328
water resources 254
food
security 41, 100–101, 134–135, 140, 143, 144, 155, 203, 209, 242
shortages 147–148
storage 154, 155, 156, 159
veld products 81, 82
weather forecasting 163–180
forage production 198, 199–200
forecasts *see* weather forecasting
forests 20, 253, 254–257, 258, 259, 260–261, 262, 269–270, 272
fuel wood trade 76–77, 82
fynbos biome 29, 41

The Gambia 19, 131–146
game-ranches 31, 42, 45, 78
gated communities 311
GEF *see* Global Environment Facility
general circulation models 59, 133, 188, 199, 333, 363
Gireighikh Rural Council, Bara Province, Sudan 92–96
global circulation models 29, 59, 133, 134, 356
Global Environment Facility (GEF) 208
goal-based indicators 216

González, Tamaulipas, Mexico 315, 316, 319, 321–323, 326–327, 329
governance 12–13, 75, 76, 83, 84, 85, 96, 275–276
government level
  cereal production 133, 143–144
  cooperation leadership 23
  disease 125
  floods 271
  intervention 77–81
  obstacles to adaptation 12–13
  planning 208
  priorities 274–275
  support 160, 242, 322, 324, 327, 328, 329
  top–down approaches 272
  see also national level; policy level
grasslands 196
growing seasons 340

HadCM3 models 133–134, 137, 138, 139, 143, 334, 335, 347, 363
hail storms 323
hardveld, Botswana 72–86
hazards, adapting to 1–2
health care 6, 114–115, 117–118, 285–287
Heihe river basin, China 15, 19–20, 211–227
herbicide resistance 346, 348
heterogeneity 41
horticultural production 144
household level
  Nigeria 11, 147–162
  rice production 231, 232–233, 235, 236, 240, 241
housing 10, 270, 311
human capital
  drought 93, 95, 97, 99, 101, 103, 104, 106
  estuarine fishing 359, 362–363
  vulnerability 153, 157
human development 11, 13–14
human–environment system frameworks 355

incentives 43–44, 272, 291, 322

income 71, 78–79, 84–85, 358–359, 360, 365
indicators
  adaptation options 216–217, 219, 223
  drought resilience 92
  economic 140, 141, 142
  vulnerability 151, 152, 284
indigenous knowledge 71–89, 149–150, 231–234
information
  accessibility of 328
  cereal production 186
  disease 122, 124
  estuarine fishing 366, 368
  interventions 325
  lack of 11
  long-term trends 313
  weather forecasting 158, 169, 174, 175–177, 324
infrastructure 11, 124, 161, 309, 310, 325
initial survey tools 213–214
insecticides 286, 288
insecticide treated nets 11, 117
institutions
  change 160, 161, 207
  governance 275–276
  strengthening 4, 104, 105
  traditional 12, 15, 71–89
  watersheds 252, 255, 256, 257–261, 262
insurance 24, 267, 322–323, 323–324, 325, 327
integrated approaches 3, 15–18, 23, 25, 105–106, 214–225
international level
  finance 24–25, 33, 208, 304
  weather forecasting 168, 169
International Union for the Conservation of Nature (IUCN) 30, 31, 32
Intertropical Convergence Zone (ITCZ) 170, 172, 173
intervention, government 77–81
irrigation
  cereal production 19, 133,

138–139, 141–142, 143, 144, 181–195
flooding 270
maize and soybean yields 337, 338, 340, 344
water management 12, 59–60
ITCZ *see* Intertropical Convergence Zone
IUCN *see* International Union for the Conservation of Nature

Jamaica 10, 283–284, 284, 285–287, 288

Kairouan, Tunisia 182–186
KARP *see* Khor Arba'at Rehabilitation Project
Kenya 114, 117
Kgotla system 12, 75, 76, 83, 84, 85
Khor Arba'at Rehabilitation Project (KARP) 97–99
kinship networks 11
knowledge
  increasing 4, 25
  indigenous 71–89, 149–150, 231–234
  lack of 11–12, 22
  local 354–355, 366, 367, 368
  scientific 297, 367, 368
  traditional 12, 354–355
  water resources 220
  see also awareness
Kula Field, Thailand 236–240

Lake Victoria, East Africa 109–130
land
  availability 134
  ownership 197, 202, 207, 316, 317
  use 253, 297–302, 332
language 158, 175–176
La Niña 255, 268, 283, 320, 356, 357, 359, 360
Lao People's Democratic Republic 17, 231–236, 232, 233, 241–245
La Plata river basin 11, 297, 302–304, 353–370
leadership 23, 101–104, 105

leaf simulations 135, 136
Limpopo river basin, Botswana 12, 71–89
livelihood assets 92, 93, 94–95, 97
livelihood systems approaches 71–89, 151–161
livestock
  changing practices 159, 160
  diversification 185, 322
  herders vulnerability 153
  options evaluation 20, 196–210
  ownership 12, 75, 76, 79, 179
  see also pastoralism
local level 177, 221, 241, 275, 315–331, 354–355, 366, 367, 368
  see also community level
logging 269–270
long-term climate trends 296–307, 312–313
Lower Mekong river basin 6, 228–246

mafisa system 12, 75, 76
maize cultivation 332–352
malaria 11, 109, 110, 111–118, 122, 123–124
markets
  biodiversity conservation 43, 46
  crop value 142
  efficient water 58, 60, 62, 65, 66, 67
  liberalization 320
marriage 75, 76
masked climate trends 304–310
matrix management 34, 37, 41–45
means to adapt 24, 25
medical treatment 117–118, 121
medicinal plants 77, 81, 82
Mersa Matrouh, Egypt 192
meteorological organizations 168–170, 176
Mexico 11, 315, 316, 319, 320
midseason dry spells 229
migration
  corridors 40–41, 45
  estuarine fishing 359, 360–361, 367

obligatory dispersers 35–36
obstacles to adaptation 105
seasonal 202, 236, 360, 367
Millennium Development Goals
13–14
millet production 137, 139–141
modelling, crop 135–136
Mongolia 11, 12, 20, 196–210
monitoring 291–292
mortality 109, 115, 116, 121, 199
mosquitos 116–117, 286
multi-criteria evaluation approaches
214, 215, 217, 225
multiple cropping 166–167
multiple measures approaches 21,
105
multi-stakeholder approaches
251–252

national level 236, 238, 239,
240–241, 241–242, 366
see also government level; policy
level
natural resources
community-based management
44–45, 80, 84, 93, 94, 97, 98,
102
conservation 4–5, 203, 205–206
degradation 12, 92–93
dependent communities 71–72,
85–86
lack of 104
Navua township, Fiji 15, 265–266
NBA *see* net benefits of adaptation
neighbour effects 288, 289
net benefits of adaptation (NBA) 58,
136–137
networks 11, 22, 206, 311, 365
NGOs *see* non-governmental
organizations
Nigeria 11, 12, 147–162, 163–180
Nile Delta region 187–192
nitrogen fertilization 138, 341–342,
343, 344
no-hopers 9, 29, 39
non-cereal production 144
non-engineering options 220–221

non-governmental organizations
(NGOs) 85, 273
non-reserve land 42, 43
North Africa 181–195
North Darfur, Sudan 11, 100–101,
102–103, 105
Northern Nigeria 147–162

obligatory dispersers 9, 29, 35–38
obstacles to adaptation 10–13, 21,
101–105
off-farm measures 235
on-farm measures 232–233,
236–238, 240
opportunities for adaptation
101–105, 158–161
ownership
conservation areas 32–33, 45
land 197, 202, 207, 316, 317
livestock 12, 75, 76, 79, 197

Pacific Island 17–18, 264–278
palm fruits 165
Pantabangan–Carranglan watershed,
The Philippines 10, 20, 247–263
partial dispersers 9, 29
participation
bottom–up approaches 273
decision-making 275
determination to adapt 22–23
estuarine fishing 367
farmers organizations 324
integrated assessment 215
natural resource management 93
people at risk 5
watersheds 251
pastoralism 6, 96–100, 196–210
see also livestock
peaceparks 33–34
perceptions 118, 120, 121, 228–229,
274–275, 306, 323
persisters 9, 29, 34–35, 39
Phane caterpillars 77, 82
phenology of crops 340
The Philippines 10, 20, 247–263
photosynthetic activity 135–136,
345

physical assets 93, 94, 97, 98, 101, 102, 104, 153, 157
place-based approaches 5–6, 228–246
planned adaptation 297, 302–304
    see also anticipatory strategies
planning 40–45, 46, 176, 208
planting dates 340, 341, 342, 343, 344
policy level
    analysis 212, 213–225
    consultations 204–205
    drought 77–79
    estuarine fishing 366–367
    flooding 302–303
    lack of support 329
    local needs 315
    matrix management 43–44
    obstacles 104
    planning models 54–68
    rice production 238
    rural poverty 157–158
    scenarios 134–135
    support 160
    watersheds 261–262
    see also government level; national level
political level 75, 160, 196, 197–198
population 114, 134
poverty
    development 14, 15
    disease 114–115, 119–120, 124, 285
    obstacles to adaptation 11
    relief measures 81
    rural policies 157–158
    sustainable livelihoods 91–92
    vulnerability 5, 14, 71, 152–153
precipitation
    Argentina 8, 305, 308, 319
    China 219
    crop production 184, 188, 192, 193, 338
    disease 119, 280, 281, 282, 283
    global changes 7–8
    Lower Mekong 244
    Mexico 319, 320
    Mongolia 198, 199, 200

Nigeria 148
Pacific Islands 264, 267, 268–269
Pantabangan–Carranglan watershed 250
Río de la Plata 363, 364
South America 297–298, 332, 335, 336, 347
southern Africa 29, 74
weather forecasts 173–174, 174–175
yield 165, 166
pre-economy 360
private land 43–44, 45
productivity 92, 101, 189, 199–200
Proteaceae 9, 29–30, 41
protected areas 30–32, 33, 257
    see also reserves
public information campaigns 124
public sector support 321–324

quint forecast categories 170–171

rainfall *see* precipitation
range expanders 9, 29
rangelands 11, 12, 20, 196–210
rapid rural appraisal 149
reactive strategies 286, 293, 360
reconfiguration of reserves 34, 36, 39–40, 41
Red Sea State, Sudan 11, 96–100
reforestation 250, 260–261
regional level 173, 174, 366
regrets 55–56, 57, 65, 68
regulations 44, 60, 221, 269–270, 272
relief 77–79, 80–81, 96–97, 274
relocation 360
research and development 134, 160, 186, 209
reserves 34–40, 41, 45
    see also protected areas
reservoir storage capacity 19, 55, 56, 57, 60, 61, 62–63, 64, 65, 66
resettlement 250
residential development 10, 270, 311
rice production 6, 17, 229–230, 231–245, 266, 270
Río de la Plata 11, 297, 302–304, 353–370

risk
adaptation and development 16
awareness of 25
capacity to cope 7
development 13
disaster 14
drought 101–105
estuarine fishing 361, 362
floods 6
involving those at 5
malaria and cholera 109–130
management 12
place-based approaches 228–246
rivers
basins 6, 10, 11, 12, 19, 53–70,
71–89, 228–246, 297, 302–304,
353–370
dredging 270, 272
flow 305–306, 356–357, 362,
363–364, 367
River Uruguay 8, 302, 303, 356–357,
359
road density 301
runoff 59, 60–61, 62, 63, 68

Sahel 147
St James, Western Jamaica 283–284
sanitary systems 120, 121, 124
Savannakhet, PDR Lao 231–236
scenarios
cereal production 133–135
driven approaches 213
estuarine fishing 363–365
floods 244
maize and soybean cultivation
333–334, 335–336
water management 19, 58–67
scientific knowledge 204, 297, 367,
368
screening options 201
sea level rise 264, 310
seasonal aspects
disease 281–282, 293
migration 202, 236, 360, 367
weather forecasts 12, 163–180,
234, 324
sea surface temperature anomalies

(SSTA) 168, 172
sedimentation 270, 277
selective breeding 133, 138, 143
sensitivity 148
siltation 269
simulations 54–68, 135–139,
188–192, 252–254, 334
skill assessments 170–173
slow trends 304–307, 312
social level
acceptability 288, 289
capital 12, 93, 95, 97, 99, 101, 103,
104, 106, 157
change 197–198
networks 206, 365
protection 79
vulnerability 153
socioeconomic level
cereal production 134–135
change 320–321
dengue fever 289
estuarine fishing 362
farmer characteristics 318
flooding 276
rice production 243
watersheds 250–251
Western Jamaica 284, 288
soil quality 345–346, 347, 348
soil water atmosphere plant model
(SWAP) 131, 135, 136
South Africa 10, 19, 30, 53–70
South America 296–297, 297–298,
332–352
southern Africa 9, 28–52, 74
southwesterly winds 173
sowing dates 164, 165, 190, 191,
193
soybean production 8, 299–300,
332–352
Special Report on Emissions
Scenarios (SRES), IPCC 59, 133,
188, 334, 335–336, 337–338, 340,
341, 342, 347
species level 9, 28–29, 29–30, 34–39
spillover effects 20, 245, 247–263
SRES *see* Special Report on
Emissions Scenarios, IPCC

SSTA *see* sea surface temperature anomalies
stock utilization ratios 139, 140–141, 141
streamflow 302, 303, 305–307, 363
subsidies 44
substitution of cultivars 138, 143
Sudan 11, 17, 90–108
supplementary irrigation 184, 340, 344
surveillance 286, 288, 292
sustainability 92, 216, 217, 364–365
sustainable livelihood approaches 17, 91–92, 101–102, 155, 354–355
SWAP *see* soil water atmosphere plant model

Tamaulipas, Mexico 11, 315, 316, 319, 321–323, 326–327, 329
Tanzania 114, 115, 116, 117, 120
temperature
    cereal production 134, 188, 189, 190
    China 219
    disease 113–114, 119, 280, 281–282, 283
    global change 7
    maize and soybean production 337, 339, 340
    Mexico 319
    Mongolia 198, 199, 200
    Nigeria 148
    Pacific Islands 264
    Pantabangan–Carranglan watershed 250
    Río de la Plata 363, 364
    South America 332, 335, 347
    southern Africa 29
tercile forecast categories 171
Thailand 17, 232, 233, 236–240, 241–245
top–down approaches 17, 151, 272, 275–276, 277
tourism 82–83
tradable development rights 44
trade 15, 76–77, 320
trade-offs 20, 216, 247–263, 252–261

traditional level
    disease treatment 117, 118, 120
    farming practices 166–167
    knowledge 12, 71–89, 354–355, 359
    pastoral systems 197, 205, 206, 209
trans-boundary effects 245
transformation in origin approaches 346–347, 348
translocation 34, 38, 40
transnational megaparks 33–34
treatment of disease 117–118, 121
tree food crops 167
Trinidad and Tobago 282, 285–287
tropical rainstorms 165–166
Tunisia 11, 12, 181–195
two-tiered screening processes 20

Ubonratchathani Province, Thailand 236–240
Uganda 114
uncertainty 8–9, 18, 107
UNEP *see* United Nations Environment Programme
UNFCCC *see* United Nations Framework Convention on Climate Change
United Nations Environment Programme (UNEP) 213
United Nations Framework Convention on Climate Change (UNFCCC) 212, 213–214
upland cereal production 131–146
urban water demand 53, 60–63, 64, 65, 66, 67
Uruguay 332, 333

variability
    Argentina 319, 320
    cereal production 137, 138, 139, 183
    disease 113–114, 280, 282
    estuarine fisheries 356–363, 365
    food production 163–180
    long-term trends 305–307, 312–313

Lower Mekong 243–244
Mexico 319, 320
Mongolia 200
North Africa 182
Pacific Islands 274
traditional strategies 74–76
vulnerability 300
variety changes 232, 234, 240
veld products 12, 76–77, 81
Vientiane Plain, Lao PDR 231–236
Vietnam 17, 232, 233, 240–241,
    241–245
visual presentations 262
vulnerability
    agricultural expansion 300
    cereal production 132
    changing practices 159–160
    definitions 148
    development 14–15
    disease 110, 114–115, 122,
        123–124, 277, 281, 284–285
    drought 90, 91
    estuarine fishing 355–356, 359, 368
    farm level 318–321
    floods 270, 271, 327–328
    livelihood systems approaches 151
    livestock herding 198–200, 207
    Lower Mekong 244–245
    natural resources 12, 71–72, 85–86
    poverty 5, 84
    rural households 151–153,
        244–245
    small islands 264, 265
    water resources 219, 225, 248–251

water
    conservation 189–190, 191
    cross-sectoral impacts 258, 259,
        260, 262
    demand 15
    management 19–20, 53–70, 97
    scarcity 152–153

spillovers and trade-offs 254, 257
storage 166, 285, 291, 293
transport 135
use practices 159, 211, 218,
    219–225
    see also irrigation
watersheds 10, 20, 247–263, 269,
    316
weather
    cautious behaviour 361, 367
    extreme 8, 10, 264, 274, 308–310
    forecasting 12, 158, 163–180, 234
    predictions 7–8
    stations 158
    see also droughts; El Niño
        Southern Oscillation; floods; La
        Niña
welfare 63, 64, 65, 67
West Africa 12, 168–170
wheat 184, 187–192, 193
will to adopt 21–23, 159–160
winds 173, 361, 362, 367
winter severity 199, 200
withdrawal ratios 219
WOFOST *see* world food studies
    model
world food studies model
    (WOFOST) 131, 135, 136

yam production 165
yield
    cereal production 137–139, 183,
        184
    high yield varieties 155, 156, 159
    fertilization 139–140
    irrigation 141
    maize and soybean 336–345, 347
    weather forecasts 165, 166, 176
    wheat cultivation 190, 192, 193

zud 199, 200